New Trends in Asymmetric Catalysis

New Trends in Asymmetric Catalysis

Editors

Giorgio Della Sala
Rosaria Schettini

MDPI • Basel • Beijing • Wuhan • Barcelona • Belgrade • Manchester • Tokyo • Cluj • Tianjin

Editors

Giorgio Della Sala
Dipartimento di Chimica e
Biologia "Adolfo
Zambelli"/DCB,
Università di Salerno
Italy

Rosaria Schettini
Dipartimento di Chimica e
Biologia "A. Zambelli",
Università di Salerno,
Via Giovanni Paolo II
Italy

Editorial Office
MDPI
St. Alban-Anlage 66
4052 Basel, Switzerland

This is a reprint of articles from the Special Issue published online in the open access journal *Catalysts* (ISSN 2073-4344) (available at: https://www.mdpi.com/journal/catalysts/special_issues/Asymmet_Catal).

For citation purposes, cite each article independently as indicated on the article page online and as indicated below:

LastName, A.A.; LastName, B.B.; LastName, C.C. Article Title. *Journal Name* **Year**, *Volume Number*, Page Range.

ISBN 978-3-0365-0972-3 (Hbk)
ISBN 978-3-0365-0973-0 (PDF)

Contents

About the Editors . . . . **vii**

Rosaria Schettini and Giorgio Della Sala
New Trends in Asymmetric Catalysis
Reprinted from: *Catalysts* **2021**, *11*, 306, doi:10.3390/catal11030306 . . . . **1**

Marco Giuseppe Emma, Alice Tamburrini, Ada Martinelli, Marco Lombardo, Arianna Quintavalla and Claudio Trombini
A Simple and Efficient Protocol for Proline-Catalysed Asymmetric Aldol Reaction
Reprinted from: *Catalysts* **2020**, *10*, 649, doi:10.3390/catal10060649 . . . . **5**

Cristiana Margarita and Helena Lundberg
Recent Advances in Asymmetric Catalytic Electrosynthesis
Reprinted from: *Catalysts* **2020**, *10*, 982, doi:10.3390/catal10090982 . . . . **27**

Anna Lidskog, Yutang Li and Kenneth Wärnmark
Asymmetric Ring-Opening of Epoxides Catalyzed by Metal–Salen Complexes
Reprinted from: *Catalysts* **2020**, *10*, 705, doi:10.3390/catal10060705 . . . . **51**

Alberto F. Garrido-Castro, M. Carmen Maestro and José Alemán
α-Functionalization of Imines via Visible Light Photoredox Catalysis
Reprinted from: *Catalysts* **2020**, *10*, 562, doi:10.3390/catal10050562 . . . . **117**

Qianfa Jia, Yaqiong Li, Yinhe Lin and Qiao Ren
The Combination of Lewis Acid with *N*-Heterocyclic Carbene (NHC) Catalysis
Reprinted from: *Catalysts* **2019**, *9*, 863, doi:10.3390/catal9100863 . . . . **139**

About the Editors

Giorgio Della Sala is Associate Professor at the University of Salerno. He received a Magna Cum Laude degree in Chemistry in 1997 and his Ph.D. in Chemistry from the University of Salerno in 2002. After a post-doc fellowship in 2002–2003, he held the position of Assistant Professor at the Department of Chemistry of the University of Salerno in 2004. In 2020, he was appointed Associate Professor at the Department of Chemistry and Biology of the University of Salerno. Since 2007, his independent research has mainly focused on the development of stereoselective metal- and organo-catalyzed methodologies, and their applications in asymmetric organic synthesis.

Rosaria Schettini received a Magna Cum Laude degree in Chemistry in 2013 and her Ph.D. in Chemistry in 2017 at the University of Salerno under the guidance of Prof. Irene Izzo and Dr. Giorgio Della Sala. In the period February–April 2016, she joined the research group of Prof. José Alemán Lara at the Universidad Autónoma de Madrid. Between 2017 and 2019, she was a postdoctoral researcher at the Department of Chemistry of University of Salerno in collaboration with Laboratoire de Synthèse Organique et Molécules Bioactives Université de Strasbourg. Her research has been focused on the synthesis of cyclic peptoids, their biological applications and their use as phase-transfer catalysts.

catalysts

MDPI

Editorial

New Trends in Asymmetric Catalysis

Rosaria Schettini and Giorgio Della Sala *

Dipartimento di Chimica e Biologia "Adolfo Zambelli"/DCB, Università di Salerno, 84084 Salerno, Italy; schettinirosaria89@gmail.com
* Correspondence: gdsala@unisa.it

As far back as the mid-nineteenth century, the studies of Louis Pasteur brought to light the essential role of molecular chirality in biology. Since then, organic chemists, biologists and physicians have become more and more aware of the close correlation between the biological activity and chirality of organic compounds, resulting from mediation by chiral biological receptors. This assumption, corroborated by the drawbacks associated with the use of racemic materials (the case of thalidomide is exemplary) led, through the second half of the twentieth century, to a rapid development of methods to selectively obtain the eutomer of a natural product or a synthetic analog for many applications, including pharmaceuticals, food chemicals, health care products and materials. The early efforts in the asymmetric synthesis field were based on the racemic resolution methods and later on the use of enantiopure stoichiometric reagents and chiral auxiliaries. However, the real breakthrough took place in the 1970s and 1980s with the emergence of asymmetric catalysis, marked out by the outstanding achievements by Knowles and Noyori in enantioselective hydrogenation, by Sharpless in enantioselective epoxidation and the subsequent implementation of several industrial processes. Since in principle a single molecule of chiral catalyst can generate tens, hundreds or even thousands of molecules of enantioenriched product, which Noyori referred to as *multiplication of chirality*, asymmetric catalysis can rightly be considered the highest evolution of asymmetric synthesis. The end of the twentieth century has seen an explosion of methods based on the combination of newly designed chiral organic ligands and metal centers, as well as a rapid growth of asymmetric biocatalysis. The beginning of this new century marked a revival of the use of small organic chiral molecules as catalysts. The advantages of organocatalysis in terms of environmental impact, low costs and simplicity of reaction set-up have indeed led to an impressive development of this strategy in the last two decades. To this day, asymmetric catalysis, founded on three main pillars—metal catalysis, organocatalysis and biocatalysis—is undoubtedly the most powerful tool for preparing enantioenriched products, and the ever-increasing research effort made in this field is therefore not surprising.

Although a large volume of work is still devoted to designing novel and more efficient chiral ligands and organocatalysts, and to expanding the scope of accessible enantioselective transformations, more and more attention is being paid to developing more sustainable processes. Accordingly, techniques requiring lower environmental impact, waste reduction, lower energy cost and the use of cheap, readily available and recyclable reagents and catalysts are being pursued. In this respect, the articles collected in the present Special Issue focus on these new trends.

Proline is one of the smallest and most inexpensive and readily available chiral organocatalysts. Unfortunately, a major drawback is its poor solubility in most organic solvents, enabling good enantioselectivities and catalytic activities only in polar aprotic solvents, such as dimethylsulfoxide, dimethylformamide or acetonitrile, which are high-boiling, toxic and expensive. The article by Quintavalla et al. provides a sustainable protocol for the proline-catalyzed asymmetric aldol reaction in methanol and water [1]. Unsatisfactory results were obtained by using pure solvents, since methanol was affected

Citation: Schettini, R.; Della Sala, G. New Trends in Asymmetric Catalysis. *Catalysts* **2021**, *11*, 306. https://doi.org/10.3390/catal11030306

Received: 22 February 2021
Accepted: 24 February 2021
Published: 26 February 2021

by poor stereocontrol, and water led to low reaction rates. The best compromise was achieved by using methanol–water mixtures.

Four review articles offer interesting overviews of some hot topics in asymmetric catalysis. The application of electrochemistry to enantioselective organic synthesis has recently received a renewed interest, being associated with mild reaction conditions and reduced energy consumption. The review of Margarita and Lundberg describes the recent advances in asymmetric metal-catalyzed and organocatalyzed electrosynthesis, reporting examples of C–H and alkene functionalization, carboxylation and cross-coupling reactions, and the related mechanistic insight [2]. The review is divided into oxidative and reductive transformations, highlighting that asymmetric oxidative processes are more explored compared to reductive ones.

The asymmetric ring-opening of epoxides is an important method for the preparation of enantioenriched vicinal difunctionalized organic compounds. The review by Warnmark et al. [3] provides a comprehensive overview of the desymmetrizations of *meso*-epoxides and the kinetic resolution of racemic chiral epoxides, catalyzed by chiral metal–salen complexes, a privileged class of effective and easily tunable chiral catalysts. Throughout their discussion, the authors have stressed the recent progresses made, reducing the catalyst loading and preparing easily recyclable supported catalysts. Particularly relevant is the application of multi-metallic catalysts, which has enabled improved performances and further mechanistic insight. At the end of their review, the authors outline the possible future developments, with particular reference to the substrate scope expansion.

Sunlight is one of the most available and renewable energy sources, and as such, its application in green organic synthesis is extremely attractive. From this perspective, asymmetric visible-light photocatalysis has experienced a remarkable growth in the last decade. Moreover, the mechanism of photocatalyzed redox reactions involves the formation of unusual radical intermediates, enabling synthetic transformations that are difficult to accomplish under conventional conditions. The review by Alemán et al. [4] summarizes the achievements of the last few years in the α-functionalization of imine building blocks, both racemic and stereoselective, by means of visible-light photoredox catalysis. As outlined in the introduction, imines are versatile substrates under photoredox catalysis conditions, which can serve either as nucleophiles or electrophiles. The photocatalytic addition reactions of radicals to imines and reactions involving α-aminoradicals, generated by the single-electron reduction of imines, are covered in different sections. However, as pointed out by the authors, despite the impressive advances made in this field, the asymmetric functionalization of imine bonds remains a formidable challenge.

N-heterocyclic carbenes (NHCs) are valuable Lewis base species largely employed in asymmetric catalysis. Their mode of action involves the generation of intermediates such as acyl anions, homoenolates, enolates and α,β-unsaturated acylazolium equivalents, and their subsequent reaction with electrophiles. Recently, as illustrated in the review by Ren et al. [5], Lewis acids, in combination with chiral NHC catalysts, were proved to increase the reaction yields and enantioselectivity of many reactions, as well as decisively affect (often reversing) regio- and diastereoselectivities. The review examines several examples categorized according to the Lewis acid type involved.

In conclusion, this collection of publications well represents the progresses and latest trends in the constantly evolving area of asymmetric catalysis. We wish to thank the authors for their valuable contributions, and we hope that this Special Issue could be inspiring for many scholars active in this field.

Funding: This research received no external funding.

Conflicts of Interest: The authors declare no conflict of interest.

References

1. Emma, M.G.; Tamburrini, A.; Martinelli, A.; Lombardo, M.; Quintavalla, A.; Trombini, C. A Simple and Efficient Protocol for Proline-Catalysed Asymmetric Aldol Reaction. *Catalysts* **2020**, *10*, 649. [CrossRef]
2. Margarita, C.; Lundberg, H. Recent Advances in Asymmetric Catalytic Electrosynthesis. *Catalysts* **2020**, *10*, 982. [CrossRef]
3. Lidskog, A.; Li, Y.; Wärnmark, K. Asymmetric Ring-Opening of Epoxides Catalyzed by Metal–Salen Complexes. *Catalysts* **2020**, *10*, 705. [CrossRef]
4. Garrido-Castro, A.F.; Maestro, M.C.; Alemán, J. α-Functionalization of Imines via Visible Light Photoredox Catalysis. *Catalysts* **2020**, *10*, 562. [CrossRef]
5. Jia, Q.; Li, Y.; Lin, Y.; Ren, Q. The Combination of Lewis Acid with *N*-Heterocyclic Carbene (NHC) Catalysis. *Catalysts* **2019**, *9*, 863. [CrossRef]

catalysts

MDPI

Article

A Simple and Efficient Protocol for Proline-Catalysed Asymmetric Aldol Reaction

Marco Giuseppe Emma, Alice Tamburrini, Ada Martinelli, Marco Lombardo *, Arianna Quintavalla * and Claudio Trombini

Department of Chemistry "G. Ciamician", Alma Mater Studiorum, University of Bologna, Via Selmi 2, 40126 Bologna, Italy; marcogiuseppeemma@hotmail.it (M.G.E.); alice.tamburrini@hotmail.it (A.T.); ada.martinelli2@unibo.it (A.M.); claudio.trombini@unibo.it (C.T.)

* Correspondence: marco.lombardo@unibo.it (M.L.); arianna.quintavalla@unibo.it (A.Q.); Tel.: +39-0512-099-544 (M.L.); +39-0512-099-462 (A.Q.)

Received: 1 May 2020; Accepted: 7 June 2020; Published: 10 June 2020

Abstract: The proline-catalysed asymmetric aldol reaction is usually carried out in highly dipolar aprotic solvents (dimethylsulfoxide, dimethylformamide, acetonitrile) where proline presents an acceptable solubility. Protic solvents are generally characterized by poor stereocontrol (e.g., methanol) or poor reactivity (e.g., water). Here, we report that water/methanol mixtures are exceptionally simple and effective reaction media for the intermolecular organocatalytic aldol reaction using the simple proline as the catalyst.

Keywords: proline; organocatalysis; asymmetric aldol reaction; methanol/water mixtures; sustainability

1. Introduction

The asymmetric intermolecular aldol addition catalysed by (*S*)-proline, proposed by List and co-workers in 2000 [1], is the prototype of enamine-based organocatalysis [2,3]. Proline is the smallest air- and water-stable bifunctional catalyst; it is inexpensive, non-toxic, and available in both enantiomeric forms. It has been proven to catalyse enantioselective α-functionalizations of carbonyl compounds (aldol and Mannich reactions, Michael additions, α-halogenations, oxygenations, aminations, and so on), adopting reaction protocols that do not require inert atmosphere and anhydrous conditions and are carried out at room temperature [4,5].

The proline-catalysed aldol reaction [4,6–19] has been object of in-depth analyses after the first mechanistic rationale proposed by List and Houck [20], in particular, fundamental contributions have been given by Seebach [21], Armstrong and Blackmond [22,23], Sharma and Sunoj [24], Benaglia [25], and Gschwind [26,27]. However, proline scarce solubility in most organic solvents has limited its use in dimethylsulfoxide (DMSO), acetonitrile, or dimethylformamide (DMF). Moreover, proline often displays poor activity, requiring the use of high catalyst loadings and high reaction times, sometimes with unsatisfactory stereocontrol [1,28–36]. Because, in several cases, proline-catalysed aldol reactions are still underdeveloped, the last 15 years have witnessed an intense effort aimed at modifying the proline scaffold, following two directions: (1) the carboxylic group is replaced by a new hydrogen-bonding donor, such as a tetrazole, or by a sterically demanding group, such as the Hayashi–Jørgensen diarylmethanol and related compounds, as exhaustively reviewed by Trost [37]; and (2) the carboxylic group is retained and a supplementary substituent is bound to the proline scaffold. The new substituent, generally installed on the 4-OH group of 4-hydroxyproline, may play different roles: (*i*) it modifies the solubility profile of the parent amino acid, expanding the solvent choice to further classes [38]; and/or (*ii*) it enhances the catalyst activity and stereoselectivity, allowing a reduction of catalyst loading and reaction time; and/or (*iii*) it allows the catalyst immobilization on a solid support [39–56], adopting a biphasic condition reaction protocol. Despite that high levels of reactivity and selectivity have been achieved with these modified prolines, most of the aforementioned proline derivatives require multi-step syntheses, dramatically increasing the catalyst cost, a severe limitation particularly when the catalyst cannot be efficiently recycled.

After having contributed to the synthesis of prolines, mostly modified with the incorporation of ionic tags on 4-position, and obtaining excellent results in terms of activity and stereoselectivity as well as catalyst recyclability [57–66], we decided to go back to the parent unmodified (*S*)-proline. We envisaged to improve the performance of this small, stable, inexpensive, non-toxic, and easily available organocatalyst exploring new experimental conditions.

The role of the solvent in determining the aldol reaction efficiency was also addressed by other authors. Invariably, the use of unmodified proline (without additives) forced to choose polar aprotic solvents to obtain acceptable yields and selectivities [1,11,13–15,28–36,44,45]. A peculiar case was represented by ionic liquids (ILs) [67–72], which allowed in a few cases to decrease the catalyst loading (up to 1 mol%) and, during the work-up, to confine proline in a separate phase, enabling a simple product isolation and the reuse of the catalytic system. In recent literature, attempts are reported where proline is used in acetone/$CHCl_3$ mixtures [73], in DMF at 4 °C (a condition that often requires several days) [74], in *tert*-butyl methyl ether (MTBE) [75], in deep eutectic solvents [76,77], or under solvent-free conditions, with [78–80] or without [81] the ball milling approach. However, many issues associated with the use of proline remain unsolved and polar aprotic solvents are characterized by several undesirable features (toxicity, high production cost, high environmental impact, difficult product recovery) [82–84].

The use of unmodified proline can be also combined with additives [85,86], such as water [85–87], acids [85,86,88], diols [89,90], amines [85,86], or thioureas [91–93]. In these cases, the additive can tune the solubility, the reactivity, and/or the stereoselectivity of native proline, making the asymmetric aldol process more efficient [18]. In one case, the addition of achiral guanidinium salts as additives allows to switch the diastereoselectivity as a function of the counterion, for example, tetrafluoroborate versus tetraphenylborate [94]. Nevertheless, the achieved performance is not optimal yet (long reaction times, stereocontrol strongly depending on the substrate) and some drawbacks are still present, such as high proline loadings and the cost of the not recovered chiral additive. Significant advances were accomplished employing metal salts as additives [95–106]. In particular, Reiser and co-workers developed a strategy based on cobalt(II)-proline complexes, which ensured excellent results in direct aldol reactions involving aromatic aldehydes [106]. However, several disadvantages lead the avoidance of the use of metals, especially at an industrial level (costs, toxicity, environmental impact, limited sources) [107–109].

In the present work, we aim to avoid the use of both polar aprotic solvents and additives (being sometimes expensive, mostly non-recoverable, and contaminants, used in non-generalizable procedures), in order to develop an efficient and sustainable organocatalyzed aldol condensation protocol, which can be interesting from a scale-up and an industrial point of view. In particular, our goals are as follows: (1) a reduction of the process costs, related to employed solvents and reagents, but also purification and waste disposal; and (2) an improved reactivity, especially for poorly reactive substrates. We planned to achieve these objectives by using the following: (*i*) the native proline, a small, stable, inexpensive, and non-toxic organocatalyst; and (*ii*) the minimum amount of a low-cost, non-toxic reaction solvent, enabling a good process efficiency and a simple and inexpensive final reaction work-up.

A number of research groups noticed that protic media were not suitable for aldol condensations promoted exclusively by native proline [15,29–33,87]. However, despite a plethora of studies focused on the use and the role of water (as solvent, co-solvent, or additive) [31,32,34,87,100–105,110,111], very few authors extended their investigations to alcohols [15,29,31,42,102,106], discouraged by the generally observed poor diastereo- and enantioselectivity. Only when proline was used in combination with metal salts as additives, the use of methanol as solvent [106] or co-solvent [102] afforded acceptable results.

Intrigued by the few data available on the proline-mediated aldol condensation employing methanol, a prototypical green solvent [84,112] also in terms of LCA (life-cycle assessment) [113], we decided to explore in depth the impact of methanol on the asymmetric intermolecular aldol condensation promoted by unmodified (S)-proline. It should be stressed, however, that efficient organocatalyzed aldol condensations invariably require a large excess of a liquid donor ketone (5–10 equivalents) that must thus be considered as a part of the reaction solvent-system.

2. Results and Discussion

2.1. Optimization of the Reaction Protocol

As model reaction, we selected the (S)-proline–catalysed aldol condensation between cyclohexanone **1a** and aromatic aldehydes **2** (Scheme 1). At the outset, we confirmed the low performance of proline in terms of stereocontrol in pure methanol, but soon we realized that the simple use of a hydroalcoholic solution as the reaction medium was highly profitable. Here, we report a comparison among (S)-proline–catalysed reactions between cyclohexanone **1a** and four different aromatic aldehydes **2a–d**, carried out in methanol/water (2/1 V/V), pure water, and pure methanol, respectively, all other parameters being kept identical (Table 1). The 2/1 V/V methanol/water mixture composition ensures that the aldol reaction takes place under homogeneous conditions.

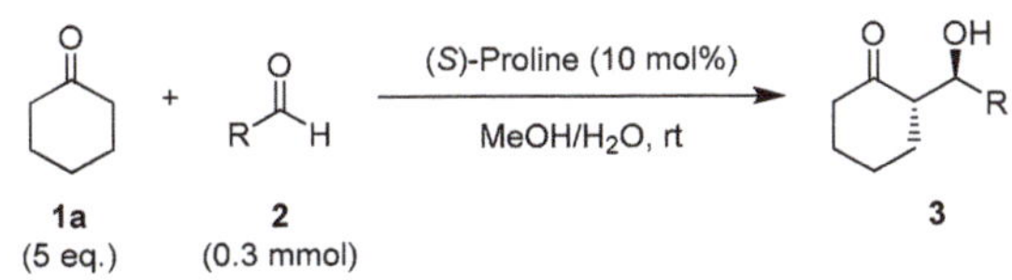

Scheme 1. The benchmark aldol reaction.

Table 1. Comparison of different protic reaction media [1].

R (2)	Solvent	t [h]	3 Conversion [%] [2]	*ee* [%] [3]	*anti/syn* [2]
4-NO_2Ph (**2a**)	MeOH/H_2O	19	>99	98	93:7
	H_2O	19	25	99	95:5
	MeOH	19	>99	76	59:41
4-CNPh (**2b**)	MeOH/H_2O	19	97	98	95:5
	H_2O	19	80	99	95:5
	MeOH	19	>99	87	82:18
4-ClPh (**2c**)	MeOH/H_2O	19	43	99	97:3
	H_2O	19	5	>99	nd
	MeOH	19	64	98	85:15
Ph (**2d**) [4]	MeOH/H_2O	30	58	97	88:12
	H_2O	30	20	>99	>99:1
	MeOH	30	64	83	72:28

[1] Reaction conditions: **1a** (5 equiv.), **2** (0.3 mmol), (S)-proline (10 mol%), rt, MeOH/H_2O (20 μL/10 μL, 2/1 V/V) or H_2O (10 μL), or MeOH (20 μL). [2] Determined by ^1H NMR on the crude mixture. [3] Determined by chiral stationary phase (CSP)-HPLC on the crude mixture. [4] Here, 20 mol% of (S)-proline was used. rt = room temperature, h = hours, nd = not determined.

The data collected in Table 1 demonstrate the crucial role of water; if in pure water conversions are the lowest, enantioselectivity reaches the highest values. On the other hand, pure methanol displays the highest reactivity and the poorest stereocontrol. The 2/1 V/V methanol/water solution is able to combine the pros of the two pure solvents, providing the same conversions of pure methanol and almost the same *ee*s and good *dr*s observed in pure water.

In Table 2, the results are reported when the 2/1 V/V methanol/water solution was applied to aldol reactions between cyclohexanone **1a** and other aromatic aldehydes **2e–i** (Table 2).

Table 2. MeOH/H_2O-based protocol applied to different aromatic aldehydes **2** [1].

R (2)	t [h]	3 Conversion [%] [2]	*ee* [%] [3]	*anti/syn* [2]
C_6F_5 (**2e**)	19	>99	97	>99:1
2-NO_2Ph (**2f**)	19	93	95	95:5
4-BrPh (**2g**)	19	41	99	98:2
2-naphthyl (**2h**)	24	37	93	91:9
4-MeOPh (**2i**) [4]	70	18	90	86:14

[1] Reaction conditions: **1a** (5 equiv.), **2** (0.3 mmol), (*S*)-proline (10 mol%), MeOH/H_2O (20 μL/10 μL), rt. [2] Determined by ^{1}H NMR on the crude mixture. [3] Determined by CSP-HPLC on the crude mixture. [4] Here, 20 mol% of (*S*)-proline was used.

With the most reactive electron-poor aldehydes (**2a**, **2b**, **2e**, and **2f**), high conversion and high stereocontrol were achieved in only 19 h. Moreover, these results are excellent if compared with those reported in the literature for analogous transformations promoted by unmodified (*S*)-proline and exploiting more complex protocols [114–117]. Unfortunately, the limitations of the proline-catalysed aldol reactions were not completely overcome. In fact, electron-rich aromatic aldehydes were confirmed to be less reactive, requiring longer reaction times. More in detail, for substrates **2g** and **2h**, the conversions reached after 19 and 24 h, respectively, were modest; nevertheless, it is worth mentioning that the enantio- and the diastereoselectivities were both higher than those reported so far by proline-based protocols [118,119]. As far as the electron-rich *p*-methoxy benzaldehyde **2i** is concerned, the only example with proline (20 mol%) in DMSO reported a low conversion (15%) and absence of diastereoselection [105]. The effect on product conversion was even poorer when a Lewis acid and water were added. Exploiting our MeOH/H_2O-based protocol, the product conversion remained poor (18%), but the reaction proceeded with good enantio- (90% *ee*) and diastereoselectivity (*anti/syn* = 86:14).

Once the performance of native proline in 2/1 V/V methanol/water solution had been examined, we explored the effect of a more methanol-rich aqueous mixture. In Table 3, aldol reactions of cyclohexanone **1a** and different aldehydes **2** in 2/1 V/V and 4/1 V/V solutions are compared.

Doubling the methanol volume (40 μL), the conversions significantly improved with all the tested aldehydes, while maintaining an excellent to remarkable level of enantiocontrol (Table 3). The most reactive substrates (**2a**, **2b**, **2e**, and **2f**) provided excellent conversions in only 4 h, demonstrating an unprecedented reactivity of proline. Moreover, interesting amounts of product were obtained exploiting these reaction conditions for less electrophilic aldehydes as well (**2c**,**2d**, **2g**,**2i**; Table 3). Concerning the diastereocontrol, a slight drop of the *anti/syn* ratio was observed with some aldehydes when the volume of methanol was increased. Conversely, for benzaldehyde **2d** and 2-naphthyl aldehyde **2h**, the diastereoselection lightly improved. In the case of benzaldehyde **2d**, the better performance could be the result of the reduced amount of catalyst (10 mol%), exploitable thanks to the higher proline reactivity reached with larger amounts of methanol. In the case of the most reactive 4-nitrobenzaldehyde **2a**, we solved the problem of diastereoselectivity drop by adding the methanol amount in two portions (one half after 2 h), completely restoring the diastereocontrol, while maintaining a high reaction rate. In general, a good diastereoselectivity level is retained with this protocol (4/1 V/V methanol/water) compared with the literature data [114–119]. At the same time, reaction rates are significantly enhanced. Therefore, these reaction conditions represent the best trade-off between reactivity and stereoselectivity. In Table 3, this optimized protocol was extended to some other aldehydes (**2j**–**m**), with good results compared with the literature data [120–123]. In particular, aliphatic aldehyde **2j**, known as poorly responsive in this kind of organocatalysed reaction, reached an interesting conversion (63%) and remarkably high stereochemical results (>99% *ee* and *anti/syn* > 99:1), superior to those reported by other authors for unmodified proline [120].

Table 3. Optimization of the MeOH/H_2O-based protocol [1].

R (2)	MeOH/H_2O [μL/μL]	t [h]	3 Conversion [%] [2]	*ee* [%] [3]	*anti/syn* [2]
4-NO_2Ph (**2a**)	20/10	4	**3aa**, 47	97	94:6
	40/10	4	**3aa**, 82	98	92:8
	20 + 20 [4]/10	4	**3aa**, 84	97	94:6
4-CNPh (**2b**)	20/10	4	**3ab**, 65	97	95:5
	40/10	4	**3ab**, 77	95	94:6
4-ClPh (**2c**)	20/10	19	**3ac**, 43	99	97:3
	40/10	19	**3ac**, 64	98	95:5
Ph (**2d**)	20/10 [5]	30	**3ad**, 58	97	88:12
	40/10	30	**3ad**, 75	96	90:10
C_6F_5 (**2e**)	20/10	4	**3ae**, 63	97	>99:1
	40/10	4	**3ae**, 67	96	>99:1
2-NO_2Ph (**2f**)	20/10	4	**3af**, 34	97	98:2
	40/10	4	**3af**, 59	97	97:3
4-BrPh (**2g**)	20/10	19	**3ag**, 41	99	98:2
	40/10	19	**3ag**, 88	96	94:6
2-naphthyl (**2h**)	20/10	24	**3ah**, 37	93	91:9
	40/10	24	**3ah**, 72	93	92:8
4-MeOPh (**2i**) [5]	20/10	68	**3ai**, 18	90	86:14
	40/10	68	**3ai**, 43	89	80:20
i-Pr (**2j**) [5]	40/10	72	**3aj**, 63	>99	>99:1
4-CF_3Ph (**2k**)	40/10	4	**3ak**, 78	97	97:3
2-thiophenyl (**2l**) [5]	40/10	48	**3al**, 65	88	84:16
4-CH_3Ph (**2m**)	40/10	40	**3am**, 64	94	87:13

[1] Reaction conditions: **1a** (5 equiv.), **2** (0.3 mmol), (*S*)-proline (10 mol%), rt. [2] Determined by ^{1}H NMR on the crude mixture. [3] Determined by CSP-HPLC on the crude mixture. [4] Here, 20 μL of MeOH was added after 2 h. [5] Here, 20 mol% of (*S*)-proline was used.

The next step of our investigation was directed to the ketone partner **1** of the asymmetric aldol condensation. In proline-catalysed aldol reactions, a typical drawback is represented by the excess of ketone over the limiting aldehyde generally required to achieve good yields. To increase the sustainability of the process, we planned to lower the ketone excess (Table 4).

Some aldehydes characterized by high or medium reactivity were selected for this study, in which the ketone amount was reduced to 2 equivalents. With almost all the tested substrates, high conversions and excellent *ee* values were again obtained. Although longer reaction times were required, the reaction rates remained worthy of note, especially when compared with the performance of other protocols in similar conditions. The main drawback of this procedure was a slight decrease of diastereoselectivity, an effect that is not immediately obvious. Benaglia, using the reaction progress kinetic analysis (RPKA) approach [25], a technique that allowed Blackmond et al. to define the kinetic rate law of proline-catalysed aldol reactions [23], proved the reversibility of the aldol reaction. Lowering the ketone excess leads to the following: (*i*) longer reaction times to preserve the same level of product conversion; and (*ii*) a less efficient opposition to the retroaldol reaction, which is a slow process within the time scale of our reactions. Both factors promote equilibrium on a little extent, likely accounting for the slightly decreased diastereocontrol observed when reduced amounts of cyclohexanone **1a** were used (Table 4). In conclusion, the high efficiency achieved by MeOH/H_2O/(*S*)-proline-based protocol allows to reduce the ketone excess, involving (*i*) slight adverse effects on aldol reaction performance; and (*ii*) benefits, such as lower costs and easier work up and product purification.

Table 4. Effects of cyclohexanone **1a** amount in the MeOH/H_2O/(*S*)-proline-based protocol [1].

R (2)	1a [eq.]	t [h]	3 Conversion [%] [2]	*ee* [%] [3]	*anti/syn* [2]
4-NO_2Ph (2a)	5	4	82	98	92:8
	5	4	84 [4]	97	94:6
	2	19	99	95	92:8
	1.05	19	92	97	90:10
4-CNPh (2b)	5	4	77	95	94:6
	2	24	98	91	90:10
C_6F_5 (2e)	5	4	67	97	>99:1
	2	24	>99	92	>99:1
2-NO_2Ph (2f)	5	4	59	97	97:3
	2	24	92	93	94:6
4-BrPh (2g)	5	19	88	96	94:6
	2	24	97	91	90:10
4-CF_3Ph (2k)	5	4	78	97	97:3
	3	20	96	96	95:5
	2	24	93	94	93:7

[1] Reaction conditions: **2** (0.3 mmol), (*S*)-proline (10 mol%), MeOH/H_2O (40 µL/10 µL), rt. [2] Determined by ^{1}H NMR on the crude mixture. [3] Determined by CSP-HPLC on the crude mixture. [4] Here, 20 µL of MeOH was added after 2 h.

2.2. Application of the Protocol to Other Ketones

Afterwards, we focused on the application of the developed catalytic protocol to two different donor ketones **1b,1c** (Table 5). Considering the excellent performance (stereoselectivity and reaction rate) achieved employing the MeOH/H_2O/(*S*)-proline protocol in the presence of 5 equivalents of **1a**, we decided for convenience to apply these conditions to the ketones investigation.

Table 5. MeOH/H_2O/(*S*)-proline-based protocol applied to ketones **1b,c** [1].

(1)	R (2)	t [h]	3, Conversion [%] [2]	*ee* [%] [3]	*anti/syn* [2]
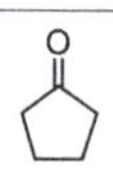	4-NO_2Ph (**2a**)	4	**3ba**, >99	94	61:39
	4-NO_2Ph (**2a**)	19 [4]	**3ba**, 91	94	78:22
	4-BrPh (**2g**)	19	**3bg**, 64	93	75:25
(**1b**)	Ph (**2d**)	30	**3bd**, 70	93	73:27
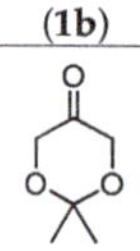 (**1c**)	4-BhR(2d)2g)	28	**3cg**, 80	86	78:23

[1] Reaction conditions: **2** (0.3 mmol), **1** (5 eq.), (*S*)-proline (10 mol%), MeOH/H_2O (40 µL/10 µL), rt. [2] Determined by ^{1}H NMR on the crude mixture. [3] Determined by CSP-HPLC on the crude mixture. [4] Reaction carried out at 0 °C.

At first, we tested cyclopentanone **1b** with highly reactive 4-nitrobenzaldehyde **2a**, observing a particularly high reaction rate, with the transformation being complete in only 4 hours. This result is unprecedented in the presence of unmodified proline or most of its derivatives [124–127], confirming once again the high reactivity achievable employing the MeOH/H_2O protocol. The corresponding product **3ba** was obtained with excellent *ee*, but low diastereoselectivity. This behaviour was expected because poorly diastereoselective aldol reactions catalyzed by proline or its derivatives were regularly reported for cyclopentanone **1b** [87,124–127]. To improve the diastereoselctivity, we lowered the reaction temperature to 0 °C and we obtained a good 78:22 *anti/syn* ratio, maintaining a high conversion in a reasonable time.

Considering the excellent performance achievable with MeOH/H_2O/(*S*)-proline-based protocol, we were particularly interested in the results obtainable with less reactive aldehydes. In fact, 4-Br benzaldehyde **2g** and benzaldehyde **2d** provided the corresponding products (**3bg** and **3bd**, respectively) with good conversions and diastereoselectivities, and, noteworthy, with the best enantioselectivities ever achieved employing unmodified proline as catalyst [128].

As further confirmation, 2,2-dimethyl-1,3-dioxan-5-one **1c** (Table 5) also displayed good reactivity and stereoselectivity when reacted with less reactive aldehydes **2g** and **2d**. In particular, our results represent the first examples of organocatalyzed aldol condensation between 2,2-dimethyl-1, 3-dioxan-5-one **1c** and 4-Br benzaldehyde **2g** or benzaldehyde **2d**, promoted by only 10 mol% of proline [129–133].

At last, we applied our protocol to acetone **1d** as simple aliphatic ketone (Table S3, Section 2, Supplementary Materials). Although with 4-NO_2 benzaldehyde **2a**, we obtained an unprecedented high reaction rate if compared with the published corresponding transformations, the enantioselectivity was poor, as commonly reported for the proline-catalysed aldol additions involving these substrates.

2.3. Large-Scale Application of the Protocol

Our aim is the development of an efficient and sustainable organocatalyzed aldol condensation protocol, which can be interesting from a scale-up perspective. Therefore, as a first step, we confirmed the excellent performance of the MeOH/H_2O/(*S*)-proline-based protocol by carrying out the aldol condensation between moderately reactive benzaldehyde **2d** and cyclohexanone **1a** on a 10 mmol scale (gram scale). The desired product **3ad** was isolated in 78% yield and with high stereocontrol (90:10 *dr*, 95% *ee*), fully confirming the data obtained on small scale (Table 3).

The next step was the accomplishment of the same reaction on a 100 mmol scale of the limiting reagent benzaldehyde **2d** (Scheme 2) in order to study some aspects in more detail.

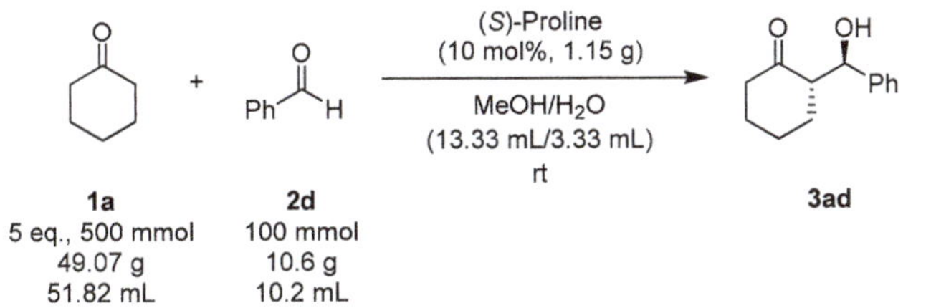

Scheme 2. Process scale-up on 100 mmol of limiting aldehyde **2d**.

At first, we investigated the impact of the aldehyde addition rate on the reaction outcome (Table 6). With benzaldehyde **2d** not being very reactive, good conversions were recorded only after 23 h and we did not observe a significant difference depending on the addition rate of benzaldehyde (Table 6).

Then, we monitored product conversion and stereoselectivity for a longer reaction time (Table 6), to establish if a high conversion could be achieved without a significant loss in stereocontrol exploiting our reaction conditions. Indeed, as previously mentioned, aldol reaction is reversible and longer reaction times could make the retroaldol process competitive, providing a decreased diastereomeric ratio. Actually, we observed a slow increase of the product conversion, achieving 85% after 2 days, without a significant erosion of *anti/syn* ratio (in comparison with small scale-reactions, a slightly lower *dr* was recorded, which remained constant for the first 48 h). We confirmed that even the enantiomeric excess of the product remained at high levels (94% *ee* after 47 h).

Table 6. Process scale-up study [1].

Aldehyde Addition Rate	t [h]	Conversion [%] [2]	*anti/syn* [2]	*ee* [%] [3]
45 min	23	72	86:14	96
	28	78	87:13	-
	47	83	85:15	94
6 h	23	71	87:13	-
	28	80	85:15	-
	47	85	84:16	-

[1] Reaction conditions: **2d** (100 mmol), **1a** (500 mmol), (*S*)-proline (10 mol%), $MeOH/H_2O$ (13.33 mL/3.33 mL), rt. Total volume = 79 mL. [2] Determined by ^1H NMR on the crude mixture. [3] Determined by CSP-HPLC on the crude mixture.

The results reported in Table 6 clearly show that the best reaction outcome is obtained at a reaction time representing the best balance between product conversion and stereocontrol. To further explore this effect, we compared the data obtained with different moderately or poorly reactive aldehydes (Table 7).

Table 7. Study of the reaction outcome as a function of the reaction time [1].

R (2)	t [h]	Conversion [%] [2]	*anti/syn* [2]
Ph (**2d**)	24	75	86:14
	46	81	84:16
	71	85	79:21
	94	85	71:29
4-CH_3Ph (**2m**)	23	53	90:10
	45	65	87:13
	71	74	84:16
	138	75	79:21
4-MeOPh (**2i**) [3]	68	43	80:20
	115	49	72:28
	164	52	66:34

[1] Reaction conditions: **2** (50 mmol), **1a** (5 eq.), (*S*)-proline (10 mol%), $MeOH/H_2O$ (6.67 mL/1.67 mL), rt. [2] Determined by ^1H NMR on the crude mixture. [3] Here, 20 mol% of (*S*)-proline was used.

Concerning the stereoselectivity, in this study, we focused our attention on diastereoselectivity variation, which is much more impaired by retroaldol reaction (see Supplementary Materials for a study on enantioselectivity variation). The data collected in Table 7 demonstrate that the aldol transformation reaches a position, after which the product conversion no longer grows, while the *dr* continues to drop. The time required to achieve this situation depends on the aldehyde reactivity. On the other hand, the rate of retroaldol process is less affected by the aldehyde nature; therefore, the retroaldol effects are less troublesome for reactive aldehydes (high conversion in short time with high *dr*) and more marked for poorly reactive aldehydes (long time required to reach acceptable conversion with low *dr*). This study proves that, in the asymmetric aldol process promoted by proline, the reaction time providing the best performance (balance between conversion and stereoselectivity) strongly depends on the substrate; therefore, a careful investigation should be done before tackling a large-scale application.

A further point that we evaluated to increase the sustainability of our large-scale protocol was the reduction of the ketone excess. For this purpose, we applied the $MeOH/H_2O$/(*S*)-proline-based protocol to moderately reactive benzaldehyde **2d** (50 mmol) in the presence of only 2 equivalents of cyclohexanone **1a**, monitoring the results over the time. After 71 h, we achieved the highest product conversion (83%) with an excellent 89:11 *dr*. Prolonging the reaction time (98 h) only led to a drop in *dr* (84:16). These findings suggest that, exploiting our protocol, a large excess of ketone (5 equivalents) only enhances the initial reaction rate, but it is not necessary for the achievement of an excellent final performance.

2.4. Work-Up Investigations

As the last point, we investigated some different work-up approaches, in order to (*i*) compare the results (also in terms of sustainability), and (*ii*) establish if part of the organocatalyst could be easily recovered. The first 100 mmol-scale aldol condensation (Table 6, 45 min long aldehyde addition) was stopped after 49 h and the reaction mixture (total volume = 79 mL) was divided in six portions, treated as described in Table 8.

Table 8. Process work-up study [1].

Method	Conditions	Final Volume (mL)	Crude Analysis [2]
A	Reaction mixture = 18 mL (22.8 mmol) filtered on a silica-pad, mobile phase = EtOAc	242	Conv. = 87% *dr* = 84:16
B	Reaction mixture = 18 mL (22.8 mmol) diluted with EtOAc, quenched with aqueous NH_4Cl, extracted with EtOAc, dried with Na_2SO_4 (washed with EtOAc)	90	Conv. = 86% *dr* = 83:17
C	Reaction mixture = 10 mL (12.6 mmol) diluted with EtOAc and placed at −15 °C for 36 h. 1° vacuum filtration. At −15 °C for 36 h. 2° vacuum filtration. Solution dried with Na_2SO_4 (washed with EtOAc)	46	Conv. = 89% *dr* = 84:16 Proline recovery: 113.3 mg (78%)
D	Reaction mixture = 10 mL (12.6 mmol) diluted with Et_2O and placed at −15 °C for 36 h. 1° vacuum filtration. At −15 °C for 36 h. 2° vacuum filtration. Solution dried with Na_2SO_4 (washed with Et_2O)	52	Conv. = 90% *dr* = 85:15 Proline recovery: 124.8 mg (86%)
E	Reaction mixture = 10 mL (12.6 mmol) diluted with DCM and placed at −15 °C for 36 h. Two liquid phases obtained, dried with Na_2SO_4 (washed with DCM)	36	Conv. = 88% *dr* = 85:15
F	Reaction mixture = 10 mL (12.6 mmol) diluted with *n*-hexane and placed at −15 °C for 36 h. Two liquid phases obtained, dried with Na_2SO_4 (washed with *n*-hexane)	43	Conv. = 89% *dr* = 86:14

[1] Aldol condensation carried out on 100 mmol of **2d**, reaction conditions described in Scheme 2, reaction stopped after 49 h, reaction mixture (total volume = 79 mL) divided in six portions and treated with six different work-up methods.
[2] Determined by ^1H NMR on the crude mixture. EtOAc = ethyl acetate, Et_2O = diethyl ether, DCM = dichloromethane.

The first portion of reaction mixture (18 mL, 22.8 mmol) was filtered through a short pad of silica to remove water and proline (method **A**, Table 8). EtOAc was used as mobile phase to elute product **3ad** (along with residual reagents **1a** and **2d**). Despite the significant polarity of EtOAc, a large amount of solvent was required to recover all the product and an undesirable high volume of organic solvent (242 mL) had to be evaporated under reduced pressure.

The second portion of reaction mixture (18 mL, 22.8 mmol) was subjected to a typical aqueous work-up to remove water and proline (method **B**, Table 8). NH_4Cl (2 equivalents with respect to proline, solved in 20 mL of H_2O) was employed to quench proline, the two phases were separated, and the aqueous phase was extracted two additional times with EtOAc, until complete recovery of the product (checked by thin-layer chromatography). The solution was dried with Na_2SO_4, which restrained a significant amount of aldol product **3ad**, so that it was necessary to wash it three times with EtOAc. A considerable volume of organic solvent (90 mL) had to be evaporated.

At this point, we tried to develop a work-up method that allowed us in a simple way not only to remove the catalyst, but also to recover it, at least partially. Exploiting the very low amount of protic polar solvents used in our MeOH/H_2O/(*S*)-proline-based protocol, we envisaged that the addition of a small portion of organic solvent could make the reaction environment sufficiently lipophilic to trigger the catalyst precipitation. Four different organic solvents were tested: EtOAc (method **C**, Table 8),

Et_2O (method **D**), dichloromethane (method **E**), and *n*-hexane (method **F**). The minimum amount of solvent able to provide an opalescent solution was added to each portion (10 mL, 12.6 mmol) and the mixtures were stored at −15 °C for 36 h. In the portions treated with EtOAc and Et_2O (methods **C** and **D**, respectively), a white precipitate, corresponding to proline, was clearly observed; therefore, it was filtered under vacuum and washed with a small amount of cold solvent. The filtered solutions were stored at −15 °C for further 36 h and a second portion of catalyst was recovered in both cases. Afterwards, the mixtures were dried with Na_2SO_4, which was required to be washed three times with organic solvent. In these two cases (methods **C** and **D**), a considerable amount of organic solvent also had to be evaporated (46 and 52 mL, respectively). In the portions treated with DCM and *n*-hexane (methods **E** and **F**, respectively), two liquid phases were observed and we decided to directly use Na_2SO_4 to remove the small water-based phase. Na_2SO_4 was washed with solvent until complete recovery of the product (three times for DCM, four times for *n*-hexane). In these two cases (methods **E** and **F**), the lowest amounts of organic solvent were employed (36 and 43 mL, respectively).

By comparing the results obtained with the tested work-up methods, we can infer the following: (1) no work-up approach adversely affects reaction conversion and diastereoselectivity, in fact all the crude mixtures showed comparable good results (Table 8). (2) The least suitable and sustainable method to remove water and proline seems to be the silica-pad (method **A**), owing to the large amount of solvent required to recover all the desired product; (3) when two organic and aqueous phases are formed (methods **B**, **E** and **F**), the simplest and cheapest work-up appears the dilution with a very small amount of DCM, cooling, and directly drying with Na_2SO_4 (method **E**). This approach is practicable only thanks to the very low amount of protic polar solvents used in our MeOH/H_2O/(*S*)-proline-based protocol. (4) Among the tested work-up approaches, the most convenient are those allowing an easy recovery of a large part of the organocatalyst (methods **C** and **D**). In particular, method **D** employing Et_2O reached 86% of proline recovery using an acceptable volume of organic solvent. Moreover, this result is obtainable on the basis of very low amount of protic polar solvents used in our protocol.

Although the tested work-up methods are not optimized and, therefore, can be further improved, they give a clear indication of the advantages that our protocol can offer.

3. Materials and Methods

3.1. General Information

^1H and ^{13}C NMR spectra were recorded on Inova 400 NMR instrument (Agilent, Santa Clara, CA, United States) with a 5 mm probe. Chemical shifts (δ) are reported in ppm, relative to the residual peaks of deuterated solvent signals.

HPLC-MS analyses were performed on an Agilent Technologies HP1100 instrument (Agilent, Santa Clara, CA, United States) coupled with an Agilent Technologies MSD1100 single-quadrupole mass spectrometer (Agilent, Santa Clara, CA, United States). A Phenomenex Gemini C18, 3 μm (100 × 3 mm) column was employed for the chromatographic separation: mobile phase H_2O/CH_3CN, gradient from 30% to 80% of CH_3CN in 8 min, 80% of CH_3CN until 22 min, and then up to 90% of CH_3CN in 2 min; flow rate 0.4 mL min^{-1} (Phenomenex, Torrance, CA, United States).

Chiral stationary phase (CSP)-HPLC analyses were performed on an Agilent Technologies Series 1200 instrument (Agilent, Santa Clara, CA, United States) using Daicel®chiral columns and *n*-hexane/2-propanol (*n*-Hex/IPA) mixtures (Daicel, Osaka, Japan).

Optical rotation measurements were performed on a polarimeter Schmidt+Haensch UniPol L1000 (Schmidt + Haensch GmbH & Co, Berlin, Germany).

Flash chromatography purifications were carried out using Merck silica gel 60 (230–400 mesh particle size). Thin layer chromatography was performed on Merck 60 F254 plates (Merck, Darmstadt, Germany).

Commercial reagents were used as received without additional purification, with exception of liquid aldehydes, which were distilled and stored under nitrogen atmosphere to avoid the formation of the corresponding acids. Dry methanol (Sure/Seal™ bottle) was used to ensure a reproducible water content.

The diastereomeric and enantiomeric compositions were checked on the crude products against the corresponding racemic products, obtained under the same reaction conditions using racemic proline.

3.2. Synthetic Procedures.

3.2.1. General Procedure for the Small-Scale Aldol Condensation Between Aldehydes **2** and Ketones **1**

The aldol reaction was carried out in a 2 mL vial. In a typical reaction, the vial was charged at room temperature with the reactants in the following order: (*S*)-proline (0.03 mmol), methanol (40 μL), water (10 μL), the selected ketone **1** (1.5 mmol), and the selected aldehyde **2** (0.3 mmol). The flask was capped with a stopper and sealed. Then, the reaction mixture was stirred at room temperature for the desired time. The conversion was monitored by TLC (Merck, Darmstadt, Germany) and ^{1}H-NMR (a small portion was taken, diluted, and immediately analyzed) (Agilent, Santa Clara, CA, United States). Then, the mixture was filtered on a short pad of silica with ethyl acetate and concentrated under reduced pressure.

The product conversion with respect to the limiting aldehyde and the diastereomeric ratio were determined by ^{1}H-NMR in $CDCl_3$ on the crude mixture. The enantiomeric excess was determined by chiral stationary phase (CSP)-HPLC (Agilent, Santa Clara, CA, United States) on the crude mixture.

The study of the solvent role (Tables 1–3), the study of the effects of ketone amount (Table 4), and the protocol application to other ketones (Table 5) were carried out following this general procedure.

3.2.2. Procedure for the Aldol Condensation Between Benzaldehyde **2d** and Cyclohexanone **1a** on 10 mmol Scale

The aldol reaction was conducted in a 25 mL flask. The flask was charged with (*S*)-proline (115 mg, 1 mmol), methanol (1.33 mL), water (330 μL), and cyclohexanone **1a** (5.18 mL, 50 mmol) and the mixture was allowed to stir for 10 min at room temperature. Then, the mixture was cooled at 0 °C and benzaldehyde **2d** (1.02 mL, 10 mmol) was slowly added by means of a syringe. The flask was capped with a stopper and sealed. The reaction mixture was stirred at room temperature for 30 h. Then, the mixture was filtered on a pad of silica with ethyl acetate and concentrated under reduced pressure. The conversion (85%) with respect to the limiting aldehyde and the diastereomeric ratio (90:10) were determined by ^{1}H-NMR in $CDCl_3$ on the crude mixture. The obtained residue was purified by column chromatography (ethyl acetate/cyclohexane = 2:8 as the eluent) to afford the product **3ad** in 78% yield. The enantiomeric excess (95% *ee*) was determined by CSP-HPLC on the pure product.

3.2.3. Procedure for the Aldol Condensation between Benzaldehyde **2d** and Cyclohexanone **1a** on 100 mmol Scale (Table 6)

The aldol reaction was conducted in a 250 mL flask. The flask was charged with (*S*)-proline (1.15 g, 10 mmol), methanol (13.33 mL), water (3.33 mL), and cyclohexanone **1a** (51.8 mL, 500 mmol) and the mixture was allowed to stir for 15 min at room temperature. Then, the mixture was cooled at 0 °C and benzaldehyde **2d** (10.2 mL, 100 mmol) was slowly added by means of (*i*) addition funnel (addition rate = 45 min), or (*ii*) syringe for slow addition (addition rate = 6 h). Then, the flask was capped with a stopper and sealed. The reaction mixture was stirred at room temperature. The reaction performance was monitored over time (a small portion was taken, diluted, and immediately analyzed); the product conversion with respect to the limiting aldehyde and the diastereomeric ratio were determined by ^{1}H-NMR in $CDCl_3$ on the crude mixture, and the enantiomeric excess was determined by CSP-HPLC on the crude mixture. After 49 h, the first reaction (addition rate = 45 min) was stopped, the reaction mixture (total volume = 79 mL) was divided in six portions, and they were treated as described in Table 8 (see below for details).

3.2.4. General Procedure for the Study of Reaction Outcome as a Function of Reaction Time (Table 7)

The aldol reaction was conducted in a 100 mL flask. The flask was charged with (*S*)-proline (575 mg, 5 mmol), methanol (6.67 mL), water (1.67 mL), and cyclohexanone **1a** (25.9 mL, 250 mmol) and the mixture was allowed to stir for 15 min at room temperature. Then, the mixture was cooled at 0 °C and the desired aldehyde **2** (50 mmol) was slowly added by means of an addition funnel. Then, the flask was capped with a stopper and sealed. The reaction mixture was stirred at room temperature. The reaction performance was monitored over time (a small portion was taken, diluted, and immediately analyzed); the product conversion with respect to the limiting aldehyde and the diastereomeric ratio were determined by ^{1}H-NMR in $CDCl_3$ on the crude mixture, and the enantiomeric excess was determined by CSP-HPLC on the crude mixture. At the reported time (Table 7), the mixture was filtered on a pad of silica with ethyl acetate and concentrated under reduced pressure. The obtained residue was purified by column chromatography (ethyl acetate/cyclohexane = 2:8 as the eluent) to afford the pure product (**3ad**: 76% yield, **3am**: 67% yield, **3ai**: 43% yield).

3.2.5. Procedure for the Aldol Condensation between Benzaldehyde **2d** (50 mmol) and Cyclohexanone **1a** (2 equivalents, 100 mmol)

The aldol reaction was conducted in a 100 mL flask. The flask was charged with (*S*)-proline (575 mg, 5 mmol), methanol (6.67 mL), water (1.67 mL), and cyclohexanone **1a** (10.36 mL, 100 mmol) and the mixture was allowed to stir for 15 min at room temperature. Then, the mixture was cooled at 0 °C and benzaldehyde **2d** (5.1 mL, 50 mmol) was slowly added by means of an addition funnel. Then, the flask was capped with a stopper and sealed. The reaction mixture was stirred at room temperature. The reaction performance was monitored over time (a small portion was taken, diluted, and immediately analyzed); the product conversion with respect to the limiting aldehyde and the diastereomeric ratio were determined by ^{1}H-NMR in $CDCl_3$ on the crude mixture, and the enantiomeric excess was determined by CSP-HPLC on the crude mixture. After 99 h, the reaction was stopped and it was filtered on a pad of silica with ethyl acetate and concentrated under reduced pressure. The obtained residue was purified by column chromatography (ethyl acetate/cyclohexane = 2:8 as the eluent) to afford the product **3ad** in 76% yield. The enantiomeric excess (91% *ee*) was determined by CSP-HPLC on the pure product.

3.3. *Work-Up Procedures*

3.3.1. Procedure for Filtration on a Silica-Pad (Method **A**, Table 8)

A portion (18 mL corresponding to 22.8 mmol) of the first reaction (addition rate = 45 min) carried out on 100 mmol of limiting aldehyde **2d** (see Section 3.2.3) was filtered on a silica-pad: 1.5 cm height, 9.6 cm diameter, gooch porosity = 4 (10–16 μm), mobile phase = EtOAc. In the last EtOAc portions (25 mL each), the presence of product was checked by TLC. The filtered reaction mixture (total volume = 242 mL) was concentrated under reduced pressure. The product conversion (87%) with respect to the limiting aldehyde and the diastereomeric ratio (84:16 = *anti*:/*syn*) were determined by ^{1}H-NMR in $CDCl_3$ on the obtained residue.

3.3.2. Procedure for Aqueous Work-Up Employing NH_4Cl (Method **B**, Table 8)

A portion (18 mL corresponding to 22.8 mmol) of the first reaction (addition rate = 45 min) carried out on 100 mmol of limiting aldehyde **2d** (see Section 3.2.3) was diluted with EtOAc (5 mL) and treated with an aqueous solution of NH_4Cl (242 mg, 2 equivalents with respect to proline, in 20 mL of H_2O). The two layers were separated and the aqueous phase was further extracted with EtOAc (2 x 20 mL, the complete product extraction was checked by TLC). The collected solution was dried with Na_2SO_4 (1.15 g), and then it was filtered and washed with EtOAc (3 x 9 mL, the complete product recovery was checked by TLC). The filtered reaction mixture (total volume = 90 mL) was concentrated under reduced pressure. The product conversion (86%) with respect to the limiting aldehyde and the diastereomeric ratio (83:17 = *anti*/*syn*) were determined by ^{1}H-NMR in $CDCl_3$ on the obtained residue.

3.3.3. Procedure for Dilution with EtOAc and Cooling (Method **C**, Table 8).

A portion (10 mL corresponding to 12.6 mmol) of the first reaction (addition rate = 45 min) carried out on 100 mmol of limiting aldehyde **2d** (see Section 3.2.3) was diluted with EtOAc (4 mL) and placed at −15 °C for 36 h. A white precipitate was formed, was filtered under vacuum, and washed with cold EtOAc (2 × 4 mL). Here, 86.2 mg of white solid was recovered. The obtained solution was placed at −15 °C for further 36 h. A second portion of white solid was filtered under vacuum and washed with cold EtOAc (2 × 3 mL). Here, 27.1 mg of white solid was recovered. The collected solution was dried with Na_2SO_4 (650 mg), and then it was filtered and washed with EtOAc (3 × 6 mL, the complete product recovery was checked by TLC). The filtered reaction mixture (total volume = 46 mL) was concentrated under reduced pressure. The product conversion (89%) with respect to the limiting aldehyde and the diastereomeric ratio (84:16 = *anti*/*syn*) were determined by ^{1}H-NMR in $CDCl_3$ on the obtained residue. Total recovered proline = 113.3 mg (78%). The nature of the white solid was confirmed by ^{1}H-NMR spectroscopy (see Supplementary Materials) and optical rotation measurement ($[\alpha]_D{}^{25} = -84$; c = 0.135, water) in comparison with the commercial compound ($[\alpha]_D{}^{25} = -86$; c = 0.133, water).

3.3.4. Procedure for Dilution with Et_2O and Cooling (Method **D**, Table 8).

A portion (10 mL corresponding to 12.6 mmol) of the first reaction (addition rate = 45 min) carried out on 100 mmol of limiting aldehyde **2d** (see Section 3.2.3) was diluted with Et_2O (4 mL) and placed at −15 °C for 36 h. A white precipitate was formed, it was filtered under vacuum and washed with cold Et_2O (2 × 4 mL). Here, 109.4 mg of white solid was recovered. The obtained solution was placed at −15 °C for further 36 h. A second portion of white solid was filtered under vacuum and washed with cold Et_2O (2 × 3 mL). Here, 15.4 mg of white solid was recovered. The collected solution was dried with Na_2SO_4 (650 mg), and then it was filtered and washed with Et_2O (3 × 8 mL, the complete product recovery was checked by TLC). The filtered reaction mixture (total volume = 52 mL) was concentrated under reduced pressure. The product conversion (90%) with respect to the limiting aldehyde and the diastereomeric ratio (85:15 = *anti*/*syn*) were determined by ^{1}H-NMR in $CDCl_3$ on the obtained residue. Total recovered proline = 124.8 mg (86%). The nature of the white solid was confirmed by ^{1}H-NMR spectroscopy (see Supplementary Materials) and optical rotation measurement ($[\alpha]_D{}^{25} = -80$; c = 0.131, water) in comparison with the commercial compound ($[\alpha]_D{}^{25} = -86$; c = 0.133, water).

3.3.5. Procedure for Dilution with DCM and Cooling (Method **E**, Table 8)

A portion (10 mL corresponding to 12.6 mmol) of the first reaction (addition rate = 45 min) carried out on 100 mmol of limiting aldehyde **2d** (see Section 3.2.3) was diluted with DCM (5 mL) and placed at −15 °C for 36 h. Two liquid phases were formed. The mixture was directly dried with Na_2SO_4 (650 mg), and then it was filtered and washed with DCM (3 × 7 mL, the complete product recovery was checked by TLC). The filtered reaction mixture (total volume = 36 mL) was concentrated under reduced pressure. The product conversion (88%) with respect to the limiting aldehyde and the diastereomeric ratio (85:15 = *anti*/*syn*) were determined by ^{1}H-NMR in $CDCl_3$ on the obtained residue.

3.3.6. Procedure for Dilution with n-Hexane and Cooling (Method **F**, Table 8)

A portion (10 mL corresponding to 12.6 mmol) of the first reaction (addition rate = 45 min) carried out on 100 mmol of limiting aldehyde **2d** (see Section 3.2.3) was diluted with *n*-hexane (5 mL) and placed at −15 °C for 36 h. Two liquid phases were formed. The mixture was directly dried with Na_2SO_4 (650 mg), and then it was filtered and washed with *n*-hexane (4 × 7 mL, the complete product recovery was checked by TLC). The filtered reaction mixture (total volume = 43 mL) was concentrated under reduced pressure. The product conversion (89%) with respect to the limiting aldehyde and the diastereomeric ratio (86:14 = *anti*:*syn*) were determined by ^{1}H-NMR in $CDCl_3$ on the obtained residue.

3.4. Products Characterization.

All the synthesized products were known compounds and the obtained data were in agreement with the published ones:

[134] for products **3aa**, **3ab**, **3ac**, **3ad**, **3af**, **3ag**, and **3ah**;
[135] for products **3ae**, **3ai**, **3ak**, and **3ba**;
[106] for product **3aj**;
[136] for product **3bd**;
[137] for product **3cd**;
[138] for products **3al** and **3am**;
[139] for product **3bg**;
[129] for product **3cg**.

As an example, the complete characterization of the most studied aldol product **3ad** (*anti* isomer) is reported in the Supplementary Materials. CSP-HPLC separation conditions and chromatograms of all the aldol products **3** are reported in the Supplementary Materials.

4. Conclusions

Since 2000, the time the first seminal publication by List, Lerner, and Barbas III on the intermolecular asymmetric aldol reaction catalyzed by proline appeared, a countless number of papers focused on enamine organocatalysis with the aim to solve a few critical issues inherent in the use of proline. Summarizing, high catalyst loading, long reaction times, solvent limitations owing to proline solubility, variable stereocontrol mainly dependent on the donor-acceptor aldol partners, difficult and/or expensive product isolation, and catalyst recovery characterize the proline-catalysed aldol protocol. On the other hand, advantages have been previously underlined such as low cost, no toxicity, no need for anhydrous solvents or controlled atmosphere, and process practicality.

Over these two decades, the greatest efforts have been dedicated to the design and synthesis of new catalysts, mostly sharing with proline the chiral pyrrolidine scaffold. These derivatives allow to enlarge the platform of solvent candidates, up to enabling the possibility of catalyst recycling. Reaction kinetics improve with shorter reaction times and lower catalyst loadings. If these improvements are beyond doubt, the costs coupled to their preparation are clearly a limiting factor. On the other hand, it is known that a number of common solvents have been questioned in recent years as their hazardous properties have come to light, for example, the environmental, safety, and health issues associated to the use of DCM, toluene, DMSO, and others.

The work presented here shows that very good results can be simply achieved using methanol/water mixtures as reaction medium. When only water is used, these reactions take place in a typical heterogeneous conditions (emulsions), where the interphase water has as many as about a quarter of the O–H bonds not being involved in hydrogen bonding. According to Jung and Marcus [140], the interactions of these unbound hydroxyl groups with organic reactants and, more importantly, with the transition states, lower the activation energies, enabling rate and yield enhancements. Faster reactions occur in pure methanol because of the homogeneous conditions, which allow all the amount of proline used to participate to catalysis, but this superior reactivity is characterized by a lower stereocontrol. Recent papers evidenced, by DFT calculations, the positive effects of co-additives such as water or methanol in stabilizing the transition states of the aldol reaction, with methanol displaying the larger effects [24,141–143]. These protic additives could directly participate in the reaction mechanism, acting as an active proton transfer relay between the proline carboxylic acid group and the incoming aldehyde. The amount and the nature of the protic additive could significantly change the reactivity and stereoselectivity of this transformation, as transition states with one, two, and even three molecules of the additive have been located and described.

If both methanol and water as pure solvents give largely unsatisfactory results that discouraged further investigations, we demonstrated that methanol/water mixtures provide the high reaction rates (good yields in short reaction times) typical of methanol and the high stereocontrol typical of water.

The efficient, simple, and cost-effective reaction protocol proposed, easily scaled up here up to the 100 mmol scale, as well as the safe handling of the methanol/water mixture, positively impact the overall efficiency and sustainability of this proline-catalysed aldol protocol. However, we have to observe that, also following this procedure, the recurring dependence of relative reaction rates and stereochemical outcome on the nature of the donor-acceptor pair has not been overcome. Thus, cyclohexanone is the best donor in terms of reactivity and stereocontrol, while cyclopentanone works faster, but with a much lower stereocontrol. Electron-rich aromatic aldehydes are the slowest reaction acceptors, requiring long reaction times, while electron-poor aldehydes are the best. Nevertheless, given that the usual relative behavior of ketones and aldehydes is confirmed, the aldol protocol in methanol/water can be considered a useful contribution, enabling the achievement of performance never obtained before (also for less reactive compounds), employing the smallest and cheapest organocatalytic species, proline.

Supplementary Materials: The following are available online at http://www.mdpi.com/2073-4344/10/6/649/s1, Table S1: Enantioselectivity variation as a function of reaction time; Figure S1: ^{1}H-NMR spectrum of commercial proline; Figure S2: ^{1}H-NMR spectrum of recovered proline employing work-up method C (Table 8); Figure S3: ^{1}H-NMR spectrum of recovered proline employing work-up method D (Table 8), CSP-HPLC separation conditions and chromatograms of aldols **3** (racemic and enantio-enriched), full characterization of *anti* aldol product **3ad** (CSP-HPLC chromatogram of enantio-enriched product, ^{1}H-NMR spectrum, ^{13}C-NMR spectrum, HPLC-MS chromatograms, ESI-MS spectrum).

Author Contributions: Conceptualization, M.L., A.Q., and C.T.; methodology, M.L., A.Q., and C.T.; validation, A.M. and A.Q.; investigation, M.G.E., A.T., and A.Q.; resources, M.L., A.Q., and C.T.; data curation, M.L., A.Q., and C.T.; writing—original draft preparation, A.Q.; writing—review and editing, A.M., M.L., A.Q., and C.T.; visualization, A.M. and A.Q.; supervision, C.T.; project administration, M.L. and A.Q.; funding acquisition, M.L., A.Q., and C.T.. All authors have read and agreed to the published version of the manuscript.

Funding: This research and the APC were funded by Università di Bologna (RFO) and MIUR (Rome).

Acknowledgments: We thank R. Miani for the execution of some experiments.

Conflicts of Interest: The authors declare no conflict of interest. The funders had no role in the design of the study; in the collection, analyses, or interpretation of data; in the writing of the manuscript; or in the decision to publish the results.

References

1. List, B.; Lerner, R.A.; Barbas, C.F. Proline-Catalyzed Direct Asymmetric Aldol Reactions. *J. Am. Chem. Soc.* **2000**, *122*, 2395–2396. [CrossRef]
2. Melchiorre, P.; Marigo, M.; Carlone, A.; Bartoli, G. Asymmetric Aminocatalysis-Gold Rush in Organic Chemistry. *Angew. Chem. Int. Ed.* **2008**, *47*, 6138–6171. [CrossRef] [PubMed]
3. Macmillan, D.W.C. The advent and development of organocatalysis. *Nature* **2008**, *455*, 304–308. [CrossRef] [PubMed]
4. Pihko, P.M.; Majander, I.; Erkkilä, A. Enamine Catalysis. *Top. Curr. Chem.* **2009**, *291*, 145–200. [CrossRef]
5. Schneider, J.F.; Ladd, C.L.; Bräse, S. *Chapter 5: Proline as an Asymmetric Organocatalyst*; RSC Publishing: Cambridge, UK, 2015; pp. 79–119. [CrossRef]
6. Yamashita, Y.; Yasukawa, T.; Yoo, W.-J.; Kitanosono, T.; Kobayashi, S. Catalytic enantioselective aldol reactions. *Chem. Soc. Rev.* **2018**, *47*, 4388–4480. [CrossRef] [PubMed]
7. Wang, L.; Liu, J. Recent Advances in Asymmetric Reactions Catalyzed by Proline and Its Derivatives. *Synthesis* **2016**, *49*, 960–972. [CrossRef]
8. List, B. Proline-catalyzed asymmetric reactions. *Tetrahedron* **2002**, *58*, 5573–5590. [CrossRef]
9. Barbas, C.F. Organocatalysis Lost: Modern Chemistry, Ancient Chemistry, and an Unseen Biosynthetic Apparatus. *Angew. Chem. Int. Ed.* **2008**, *47*, 42–47. [CrossRef]
10. Palomo, C.; Oiarbide, M.; García, J.M. Current progress in the asymmetric aldol addition reaction. *Chem. Soc. Rev.* **2004**, *33*, 65–75. [CrossRef] [PubMed]
11. Wennemers, H. Peptides as Asymmetric Catalysts for Aldol Reactions. *Chim. Int. J. Chem.* **2007**, *61*, 276–278. [CrossRef]

12. Notz, W.; Tanaka, F.; Barbas, C.F. Enamine-Based Organocatalysis with Proline and Diamines: The Development of Direct Catalytic Asymmetric Aldol, Mannich, Michael, and Diels–Alder Reactions. *Accounts Chem. Res.* **2004**, *37*, 580–591. [CrossRef] [PubMed]
13. Mukherjee, S.; Yang, J.W.; Hoffmann, S.; List, B. Asymmetric Enamine Catalysis. *Chem. Rev.* **2007**, *107*, 5471–5569. [CrossRef] [PubMed]
14. Alcaide, B.; Almendros, P. The Direct Catalytic Asymmetric Cross-Aldol Reaction of Aldehydes. *Angew. Chem. Int. Ed.* **2003**, *42*, 858–860. [CrossRef] [PubMed]
15. Alcaide, B.; Almendros, P. The Direct Catalytic Asymmetric Aldol Reaction. *Eur. J. Org. Chem.* **2002**, *2002*, 1595–1601. [CrossRef]
16. Guillena, G.; Najera, C.; Ramón, D.J. Enantioselective direct aldol reaction: The blossoming of modern organocatalysis. *Tetrahedron Asymmetry* **2007**, *18*, 2249–2293. [CrossRef]
17. Geary, L.M.; Hultin, P.G. The state of the art in asymmetric induction: The aldol reaction as a case study. *Tetrahedron Asymmetry* **2009**, *20*, 131–173. [CrossRef]
18. Schetter, B.; Mahrwald, R. Modern Aldol Methods for the Total Synthesis of Polyketides. *Angew. Chem. Int. Ed.* **2006**, *45*, 7506–7525. [CrossRef] [PubMed]
19. Mestres, R. A green look at the aldol reaction. *Green Chem.* **2004**, *6*, 583. [CrossRef]
20. Bahmanyar, S.; Houk, K.N.; Martin, H.J.; List, B. Quantum Mechanical Predictions of the Stereoselectivities of Proline-Catalyzed Asymmetric Intermolecular Aldol Reactions. *J. Am. Chem. Soc.* **2003**, *125*, 2475–2479. [CrossRef] [PubMed]
21. Seebach, D.; Beck, A.K.; Badine, D.M.; Limbach, M.; Eschenmoser, A.; Treasurywala, A.M.; Hobi, R.; Prikoszovich, W.; Linder, B. Are Oxazolidinones Really Unproductive, Parasitic Species in Proline Catalysis?—Thoughts and Experiments Pointing to an Alternative View. *Helvetica Chim. Acta* **2007**, *90*, 425–471. [CrossRef]
22. Zotova, N.; Franzke, A.; Armstrong, A.; Blackmond, D.G. Clarification of the Role of Water in Proline-Mediated Aldol Reactions. *J. Am. Chem. Soc.* **2007**, *129*, 15100–15101. [CrossRef] [PubMed]
23. Zotova, N.; Broadbelt, L.J.; Armstrong, A.; Blackmond, D.G. Kinetic and mechanistic studies of proline-mediated direct intermolecular aldol reactions. *Bioorganic Med. Chem. Lett.* **2009**, *19*, 3934–3937. [CrossRef] [PubMed]
24. Sharma, A.K.; Sunoj, R.B. Enamine versus Oxazolidinone: What Controls Stereoselectivity in Proline-Catalyzed Asymmetric Aldol Reactions? *Angew. Chem. Int. Ed.* **2010**, *49*, 6373–6377. [CrossRef] [PubMed]
25. Orlandi, M.; Ceotto, M.; Benaglia, M. Kinetics versus thermodynamics in the proline catalyzed aldol reaction. *Chem. Sci.* **2016**, *7*, 5421–5427. [CrossRef] [PubMed]
26. Haindl, M.H.; Hioe, J.; Gschwind, R.M. The Proline Enamine Formation Pathway Revisited in Dimethyl Sulfoxide: Rate Constants Determined via NMR. *J. Am. Chem. Soc.* **2015**, *137*, 12835–12842. [CrossRef] [PubMed]
27. Renzi, P.; Hioe, J.; Gschwind, R.M. Enamine/Dienamine and Brønsted Acid Catalysis: Elusive Intermediates, Reaction Mechanisms, and Stereoinduction Modes Based on in Situ NMR Spectroscopy and Computational Studies. *Accounts Chem. Res.* **2017**, *50*, 2936–2948. [CrossRef] [PubMed]
28. Notz, W.; List, B. Catalytic Asymmetric Synthesis ofanti-1,2-Diols. *J. Am. Chem. Soc.* **2000**, *122*, 7386–7387. [CrossRef]
29. List, B.; Pojarliev, P.; Castello, C. Proline-Catalyzed Asymmetric Aldol Reactions between Ketones and α-Unsubstituted Aldehydes. *Org. Lett.* **2001**, *3*, 573–575. [CrossRef] [PubMed]
30. Sakthivel, K.; Notz, W.; Bui, T.; Barbas, C.F. Amino acid catalyzed direct asymmetric aldol reactions: A bioorganic approach to catalytic asymmetric carbon-carbon bond-forming reactions. *J. Am. Chem. Soc.* **2001**, *123*, 5260–5267. [CrossRef] [PubMed]
31. Córdova, A.; Notz, W.; Barbas, C.F. Direct organocatalytic aldol reactions in buffered aqueous media. *Chem. Commun.* **2002**, *24*, 3024–3025. [CrossRef] [PubMed]
32. Pihko, P.M.; Nyberg, A.I.; Usano, A. Proline-Catalyzed Ketone-Aldehyde Aldol Reactions are Accelerated by Water. *Synlett* **2004**, *2004*, 1891–1896. [CrossRef]
33. Peng, Y.-Y.; Ding, Q.-P.; Li, Z.; Wang, P.G.; Cheng, J.-P. Proline catalyzed aldol reactions in aqueous micelles: An environmentally friendly reaction system. *Tetrahedron Lett.* **2003**, *44*, 3871–3875. [CrossRef]
34. Darbre, T.; Machuqueiro, M. Zn-Proline catalyzed direct aldol reaction in aqueous media. *Chem. Commun.* **2003**, 1090–1091. [CrossRef]

35. Hayashi, Y.; Tsuboi, W.; Shoji, M.; Suzuki, N. Application of high pressure, induced by water freezing, to the direct asymmetric aldol reaction. *Tetrahedron Lett.* **2004**, *45*, 4353–4356. [CrossRef]
36. Wu, Y.-S.; Shao, W.-Y.; Zheng, C.-Q.; Huang, Z.-L.; Cai, J.; Deng, Q.-Y. Studies on Direct Stereoselective Aldol Reactions in Aqueous Media. *Helvetica Chim. Acta* **2004**, *87*, 1377–1384. [CrossRef]
37. Trost, B.M.; Brindle, C.S. The direct catalytic asymmetric aldol reaction. *Chem. Soc. Rev.* **2010**, *39*, 1600–1632. [CrossRef] [PubMed]
38. Li, Z.-Y.; Chen, Y.; Zheng, C.-Q.; Yin, Y.; Wang, L.; Sun, X.-Q. Highly enantioselective aldol reactions catalyzed by reusable upper rim-functionalized calix[4]arene-based l -proline organocatalyst in aqueous conditions. *Tetrahedron* **2017**, *73*, 78–85. [CrossRef]
39. Benaglia, M. *Recoverable and Recyclable Catalysts*, 1st ed.; Wiley & Sons: Chichester, UK, 2009.
40. Liu, Y.-X.; Sun, Y.-N.; Tan, H.-H.; Tao, J.-C. Asymmetric Aldol Reaction Catalyzed by New Recyclable Polystyrene-supported l-proline in the Presence of Water. *Catal. Lett.* **2007**, *120*, 281–287. [CrossRef]
41. Liu, Y.-X.; Sun, Y.-N.; Tan, H.-H.; Liu, W.; Tao, J.-C. Linear polystyrene anchored l-proline, new recyclable organocatalysts for the aldol reaction in the presence of water. *Tetrahedron Asymmetry* **2007**, *18*, 2649–2656. [CrossRef]
42. Gruttadauria, M.; Giacalone, F.; Marculescu, A.M.; Meo, P.L.; Rielaa, S.; Noto, R. Hydrophobically Directed Aldol Reactions: Polystyrene-SupportedL-Proline as a Recyclable Catalyst for Direct Asymmetric Aldol Reactions in the Presence of Water. *Eur. J. Org. Chem.* **2007**, *2007*, 4688–4698. [CrossRef]
43. Lu, A.; Smart, T.P.; Epps, T.H.; Longbottom, D.A.; O'Reilly, R.K. l-Proline Functionalized Polymers Prepared by RAFT Polymerization and Their Assemblies as Supported Organocatalysts. *Macromolecules* **2011**, *44*, 7233–7241. [CrossRef] [PubMed]
44. Benaglia, M.; Cinquini, M.; Cozzi, F.; Puglisi, A.; Celentano, G. Poly(Ethylene Glycol)-Supported Proline: A Versatile Catalyst for the Enantioselective Aldol and Iminoaldol Reactions. *Adv. Synth. Catal.* **2002**, *344*, 533–542. [CrossRef]
45. Benaglia, M.; Celentano, G.; Cozzi, F. Enantioselective Aldol Condensations Catalyzed by Poly(ethylene glycol)-Supported Proline. *Adv. Synth. Catal.* **2001**, *343*, 171–173. [CrossRef]
46. Calderón, F.; Fernández, R.; Sánchez, F.; Fernández-Mayoralas, A. Asymmetric Aldol Reaction Using Immobilized Proline on Mesoporous Support. *Adv. Synth. Catal.* **2005**, *347*, 1395–1403. [CrossRef]
47. Giacalone, F.; Gruttadauria, M.; Marculescu, A.M.; Noto, R. Polystyrene-supported proline and prolinamide. Versatile heterogeneous organocatalysts both for asymmetric aldol reaction in water and α-selenenylation of aldehydes. *Tetrahedron Lett.* **2007**, *48*, 255–259. [CrossRef]
48. Font, D.; Jimeno, C.; Pericàs, M.A. Polystyrene-Supported Hydroxyproline: An Insoluble, Recyclable Organocatalyst for the Asymmetric Aldol Reaction in Water. *Org. Lett.* **2006**, *8*, 4653–4655. [CrossRef] [PubMed]
49. Yang, H.; Zhang, X.; Li, S.; Wang, X.; Ma, J. The high catalytic activity and reusability of the proline functionalized cage-like mesoporous material SBA-16 for the asymmetric aldol reaction proceeding in methanol–water mixed solvent. *RSC Adv.* **2014**, *4*, 9292–9299. [CrossRef]
50. Angeloni, M.; Piermatti, O.; Pizzo, F.; Vaccaro, L. Synthesis of Zirconium Phosphonate SupportedL-Proline as an Effective Organocatalyst for Direct Asymmetric Aldol Addition. *Eur. J. Org. Chem.* **2014**, *2014*, 1716–1726. [CrossRef]
51. Aghahosseini, H.; Ramazani, A.; Ślepokura, K.; Lis, T. The first protection-free synthesis of magnetic bifunctional l-proline as a highly active and versatile artificial enzyme: Synthesis of imidazole derivatives. *J. Colloid Interface Sci.* **2018**, *511*, 222–232. [CrossRef] [PubMed]
52. Doyagüez, E.G.; Calderón, F.; Sánchez, F.; Fernández-Mayoralas, A. Asymmetric Aldol Reaction Catalyzed by a Heterogenized Proline on a Mesoporous Support. The Role of the Nature of Solvents. *J. Org. Chem.* **2007**, *72*, 9353–9356. [CrossRef] [PubMed]
53. Campisciano, V.; Giacalone, F.; Gruttadauria, M. Supported Ionic Liquids: A Versatile and Useful Class of Materials. *Chem. Rec.* **2017**, *17*, 918–938. [CrossRef] [PubMed]
54. Kong, Y.; Tan, R.; Zhao, L.; Yin, D. l-Proline supported on ionic liquid-modified magnetic nanoparticles as a highly efficient and reusable organocatalyst for direct asymmetric aldol reaction in water. *Green Chem.* **2013**, *15*, 2422. [CrossRef]
55. Ferré, M.; Pleixats, R.; Man, M.W.C.; Cattoën, X. Recyclable organocatalysts based on hybrid silicas. *Green Chem.* **2016**, *18*, 881–922. [CrossRef]

56. Aprile, C.; Giacalone, F.; Gruttadauria, M.; Marculescu, A.M.; Noto, R.; Revell, J.D.; Wennemers, H. New ionic liquid-modified silica gels as recyclable materials for l-proline- or H–Pro–Pro–Asp–NH2-catalyzed aldol reaction. *Green Chem.* **2007**, *9*, 1328. [CrossRef]
57. Lombardo, M.; Trombini, C. Ionic Tags in Catalyst Optimization: Beyond Catalyst Recycling. *ChemCatChem* **2010**, *2*, 135–145. [CrossRef]
58. Lombardo, M.; Quintavalla, A.; Chiarucci, M.; Trombini, C. Multiphase Homogeneous Catalysis: Common Procedures and Recent Applications. *Synlett* **2010**, *2010*, 1746–1765. [CrossRef]
59. Lombardo, M.; Pasi, F.; Easwar, S.; Trombini, C. An Improved Protocol for the Direct Asymmetric Aldol Reaction in Ionic Liquids, Catalysed by Onium Ion-Tagged Prolines. *Adv. Synth. Catal.* **2007**, *349*, 2061–2065. [CrossRef]
60. Lombardo, M.; Pasi, F.; Easwar, S.; Trombini, C. Direct Asymmetric Aldol Reaction Catalyzed by an Imidazolium-Tagged trans-4-Hydroxy-l-proline under Aqueous Biphasic Conditions. *Synlett* **2008**, 2471–2474. [CrossRef]
61. Lombardo, M.; Easwar, S.; Pasi, F.; Trombini, C. The Ion Tag Strategy as a Route to Highly Efficient Organocatalysts for the Direct Asymmetric Aldol Reaction. *Adv. Synth. Catal.* **2009**, *351*, 276–282. [CrossRef]
62. Bhati, M.; Upadhyay, S.; Easwar, S.; Srinivasan, E. Exploring "Through-Bond" Proximity between the Ion Tag and Reaction Site of an Imidazolium-Proline Catalyst for the Direct Asymmetric Aldol Reaction. *Eur. J. Org. Chem.* **2017**, 1788–1793. [CrossRef]
63. Rosso, C.; Emma, M.G.; Martinelli, A.; Lombardo, M.; Quintavalla, A.; Trombini, C.; Syrgiannis, Z.; Prato, M. A Recyclable Chiral 2-(Triphenylmethyl)pyrrolidine Organocatalyst Anchored to [60]Fullerene. *Adv. Synth. Catal.* **2019**, *361*, 2936–2944. [CrossRef]
64. Bottoni, A.; Lombardo, M.; Miscione, G.P.; Montroni, E.; Quintavalla, A.; Trombini, C. Electrosteric Activation by using Ion-Tagged Prolines: A Combined Experimental and Computational Investigation. *ChemCatChem* **2013**, *5*, 2913–2924. [CrossRef]
65. Montroni, E.; Lombardo, M.; Quintavalla, A.; Trombini, C.; Gruttadauria, M.; Giacalone, F. A Liquid-Liquid Biphasic Homogeneous Organocatalytic Aldol Protocol Based on the Use of a Silica Gel Bound Multilayered Ionic Liquid Phase. *ChemCatChem* **2012**, *4*, 1000–1006. [CrossRef]
66. Montroni, E.; Sanap, S.P.; Trombini, C.; Dhavale, D.D.; Lombardo, M.; Quintavalla, A. A New Robust and Efficient Ion?Tagged Proline Catalyst Carrying an Amide Spacer for the Asymmetric Aldol Reaction. *Adv. Synth. Catal.* **2011**, *353*, 3234–3240. [CrossRef]
67. Loh, T.-P.; Feng, L.-C.; Yang, H.-Y.; Yang, J.-Y. l-Proline in an ionic liquid as an efficient and reusable catalyst for direct asymmetric aldol reactions. *Tetrahedron Lett.* **2002**, *43*, 8741–8743. [CrossRef]
68. Kotrusz, P.; Kmentová, I.; Gotov, B.; Toma, Š.; Solčániová, E. Proline-catalysed asymmetric aldol reaction in the room temperature ionic liquid [bmim]PF6. *Chem. Commun.* **2002**, 2510–2511. [CrossRef]
69. Gruttadauria, M.; Rielaa, S.; Aprile, C.; Meo, P.L.; D'Anna, F.; Noto, R. Supported Ionic Liquids. New Recyclable Materials for theL-Proline-Catalyzed Aldol Reaction. *Adv. Synth. Catal.* **2006**, *348*, 82–92. [CrossRef]
70. Kitazume, T.; Jiang, Z.; Kasai, K.; Mihara, Y.; Suzuki, M. Synthesis of fluorinated materials catalyzed by proline or antibody 38C2 in ionic liquid. *J. Fluor. Chem.* **2003**, *121*, 205–212. [CrossRef]
71. Córdova, A. Direct catalytic asymmetric cross-aldol reactions in ionic liquid media. *Tetrahedron Lett.* **2004**, *45*, 3949–3952. [CrossRef]
72. Miao, W.; Chan, T.-H. Ionic-Liquid-Supported Organocatalyst: Efficient and Recyclable Ionic-Liquid-Anchored Proline for Asymmetric Aldol Reaction. *Adv. Synth. Catal.* **2006**, *348*, 1711–1718. [CrossRef]
73. List, B.; Martínez, A.; Zumbansen, K.; Döhring, A.; Van Gemmeren, M. Improved Conditions for the Proline-Catalyzed Aldol Reaction of Acetone with Aliphatic Aldehydes. *Synlett* **2014**, *25*, 932–934. [CrossRef]
74. Trajković, J.M.; Milanovic, V.D.; Ferjančić, Z.; Saicic, R.N. On the Asymmetric Induction in Proline-Catalyzed Aldol Reactions: Reagent-Controlled Addition Reactions of 2,2-Dimethyl-1,3-dioxane-5-one to Acyclic Chiral α-Branched Aldehydes. *Eur. J. Org. Chem.* **2017**, 6146–6153. [CrossRef]
75. Rougeot, C.; Situ, H.; Cao, B.H.; Vlachos, V.; Hein, J.E. Automated reaction progress monitoring of heterogeneous reactions: Crystallization-induced stereoselectivity in amine-catalyzed aldol reactions. *React. Chem. Eng.* **2017**, *2*, 226–231. [CrossRef]

76. Brenna, D.; Massolo, E.; Puglisi, A.; Rossi, S.; Celentano, G.; Benaglia, M.; Capriati, V. Towards the development of continuous, organocatalytic, and stereoselective reactions in deep eutectic solvents. *Beilstein J. Org. Chem.* **2016**, *12*, 2620–2626. [CrossRef] [PubMed]
77. Martínez, R.; Berbegal, L.; Guillena, G.; Ramón, D.J. Bio-renewable enantioselective aldol reaction in natural deep eutectic solvents. *Green Chem.* **2016**, *18*, 1724–1730. [CrossRef]
78. Rodriguez, B.; Bruckmann, A.; Bolm, C. A Highly Efficient Asymmetric Organocatalytic Aldol Reaction in a Ball Mill. *Chem. A Eur. J.* **2007**, *13*, 4710–4722. [CrossRef] [PubMed]
79. Bruckmann, A.; Krebs, A.; Bolm, C. Organocatalytic reactions: Effects of ball milling, microwave and ultrasound irradiation. *Green Chem.* **2008**, *10*, 1131. [CrossRef]
80. Veverková, E.; Modrocká, V.; Sebesta, R. Organocatalyst Efficiency in the α-Aminoxylation and α-Hydrazination of Carbonyl Derivatives in Aqueous Media or in a Ball-Mill. *Eur. J. Org. Chem.* **2017**, 1191–1195. [CrossRef]
81. Kumar, A.; Gupta, M.K.; Kumar, M. l-Proline catalysed multicomponent synthesis of 3-amino alkylated indolesvia a Mannich-type reaction under solvent-free conditions. *Green Chem.* **2012**, *14*, 290–295. [CrossRef]
82. Curzons, A.D.; Constable, D.J.C.; Cunningham, V.L. Expanding GSK's Solvent Selection Guide?application of life cycle assessment to enhance solvent selections. *Clean Technol. Environ. Policy* **2004**, *7*, 42–50. [CrossRef]
83. Byrne, F.; Jin, S.; Paggiola, G.; Petchey, T.H.M.; Clark, J.H.; Farmer, T.J.; Hunt, A.J.; McElroy, C.R.; Sherwood, J. Tools and techniques for solvent selection: Green solvent selection guides. *Sustain. Chem. Process.* **2016**, *4*, 1034. [CrossRef]
84. Henderson, R.K.; Jiménez-González, C.; Constable, D.J.C.; Alston, S.R.; Inglis, G.G.A.; Fisher, G.; Sherwood, J.; Binks, S.P.; Curzons, A.D. Expanding GSK's solvent selection guide – embedding sustainability into solvent selection starting at medicinal chemistry. *Green Chem.* **2011**, *13*, 854. [CrossRef]
85. Pihko, P.M.; Laurikainen, K.M.; Usano, A.; Nyberg, A.I.; Kaavi, J.A. Effect of additives on the proline-catalyzed ketone–aldehyde aldol reactions. *Tetrahedron* **2006**, *62*, 317–328. [CrossRef]
86. Kaur, N.; Kishore, D. Synthetic Strategies Applicable in the Synthesis of Privileged Scaffold: 1,4-Benzodiazepine. *Synth. Commun.* **2014**, *44*, 1375–1413. [CrossRef]
87. Ward, D.; Jheengut, V. Proline-catalyzed asymmetric aldol reactions of tetrahydro-4H-thiopyran-4-one with aldehydes. *Tetrahedron Lett.* **2004**, *45*, 8347–8350. [CrossRef]
88. Cai, J.; Wu, Y.-S.; Chen, Y.; Deng, D.-S. Proline-Catalyzed Asymmetric Direct Aldol Reaction Assisted by d-Camphorsulfonic Acid in Aqueous Media. *Synlett* **2005**, 1627–1629. [CrossRef]
89. Zhou, Y.; Shan, Z. (R)- or (S)-Bi-2-naphthol assisted, l-proline catalyzed direct aldol reaction. *Tetrahedron Asymmetry* **2006**, *17*, 1671–1677. [CrossRef]
90. Zhou, Y.; Shan, Z. Chiral Diols: A New Class of Additives for Direct Aldol Reaction Catalyzed byl-Proline. *J. Org. Chem.* **2006**, *71*, 9510–9512. [CrossRef] [PubMed]
91. Reis, O.; Eymur, S.; Reis, B.; Demir, A.S. Direct enantioselective aldol reactions catalyzed by a proline–thiourea host–guest complex. *Chem. Commun.* **2009**, 1088. [CrossRef] [PubMed]
92. El-Hamdouni, N.; Companyó, X.; Rios, R.; Moyano, A. Substrate-Dependent Nonlinear Effects in Proline-Thiourea-Catalyzed Aldol Reactions: Unraveling the Role of the Thiourea Co-Catalyst. *Chem. A Eur. J.* **2010**, *16*, 1142–1148. [CrossRef] [PubMed]
93. Demir, A.S.; Basceken, S. Study of asymmetric aldol and Mannich reactions catalyzed by proline–thiourea host–guest complexes in nonpolar solvents. *Tetrahedron Asymmetry* **2013**, *24*, 515–525. [CrossRef]
94. Martinez-Castaneda, A.; Rodriguez-Solla, H.; Concellón, C.; Del Amo, V. Switching Diastereoselectivity in Proline-Catalyzed Aldol Reactions. *J. Org. Chem.* **2012**, *77*, 10375–10381. [CrossRef] [PubMed]
95. Xu, Z.; Daka, P.; Wang, H. Primary amine-metal Lewis acid bifunctional catalysts: The application to asymmetric direct aldol reactions. *Chem. Commun.* **2009**, 6825. [CrossRef] [PubMed]
96. Daka, P.; Xu, Z.; Alexa, A.; Wang, H. Primary amine-metal Lewis acid bifunctional catalysts based on a simple bidentate ligand: Direct asymmetric aldol reaction. *Chem. Commun.* **2011**, *47*, 224–226. [CrossRef] [PubMed]
97. Paradowska, J.; Stodulski, M.; Mlynarski, J. Direct Catalytic Asymmetric Aldol Reactions Assisted by Zinc Complex in the Presence of Water. *Adv. Synth. Catal.* **2007**, *349*, 1041–1046. [CrossRef]
98. Lu, Z.; Mei, H.; Han, J.; Pan, Y. The Mimic of Type II Aldolases Chemistry: Asymmetric Synthesis of β-Hydroxy Ketones by Direct Aldol Reaction. *Chem. Boil. Drug Des.* **2010**, *76*, 181–186. [CrossRef] [PubMed]

99. Itoh, S.; Kitamura, M.; Yamada, Y.; Aoki, S. Chiral Catalysts Dually Functionalized with Amino Acid and Zn2+Complex Components for Enantioselective Direct Aldol Reactions Inspired by Natural Aldolases: Design, Synthesis, Complexation Properties, Catalytic Activities, and Mechanistic Study. *Chem. A Eur. J.* **2009**, *15*, 10570–10584. [CrossRef]
100. Kofoed, J.; Darbre, T.; Reymond, J. Dual mechanism of zinc-proline catalyzed aldol reactions in water. *Chem. Commun.* **2006**, 1482. [CrossRef] [PubMed]
101. Kofoed, J.; Machuqueiro, M.; Reymond, J.; Darbre, T. Zinc–proline catalyzed pathway for the formation of sugars. *Chem. Commun.* **2004**, 1540. [CrossRef] [PubMed]
102. Fernandez-Lopez, R.; Kofoed, J.; Machuqueiro, M.; Darbre, T. A Selective Direct Aldol Reaction in Aqueous Media Catalyzed byZinc-Proline. *Eur. J. Org. Chem.* **2005**, 5268–5276. [CrossRef]
103. Kofoed, J.; Reymond, J.; Darbre, T. Prebiotic carbohydrate synthesis: Zinc–proline catalyzes direct aqueous aldol reactions of α-hydroxy aldehydes and ketones. *Org. Biomol. Chem.* **2005**, *3*, 1850. [CrossRef] [PubMed]
104. Akagawa, K.; Sakamoto, S.; Kudo, K. Direct asymmetric aldol reaction in aqueous media using polymer-supported peptide. *Tetrahedron Lett.* **2005**, *46*, 8185–8187. [CrossRef]
105. Penhoat, M.; Barbry, D.; Rolando, C. Direct asymmetric aldol reaction co-catalyzed by l-proline and group 12 elements Lewis acids in the presence of water. *Tetrahedron Lett.* **2011**, *52*, 159–162. [CrossRef]
106. Karmakar, A.; Maji, T.; Wittmann, S.; Reiser, O. L-Proline/CoCl2-Catalyzed Highly Diastereo- and Enantioselective Direct Aldol Reactions. *Chem. A Eur. J.* **2011**, *17*, 11024–11029. [CrossRef] [PubMed]
107. Zhang, Q.; Hou, Q.; Huang, G.; Fan, Q. Removal of heavy metals in aquatic environment by graphene oxide composites: A review. *Environ. Sci. Pollut. Res.* **2019**, *27*, 190–209. [CrossRef] [PubMed]
108. Singh, S.; Kumar, V.; Datta, S.; Dhanjal, D.S.; Sharma, K.; Samuel, J.; Singh, J. Current advancement and future prospect of biosorbents for bioremediation. *Sci. Total. Environ.* **2020**, *709*, 135895. [CrossRef] [PubMed]
109. Vareda, J.P.; Valente, A.J.; Durães, L. Assessment of heavy metal pollution from anthropogenic activities and remediation strategies: A review. *J. Environ. Manag.* **2019**, *246*, 101–118. [CrossRef] [PubMed]
110. Lindström, U.M. Stereoselective Organic Reactions in Water. *Chem. Rev.* **2002**, *102*, 2751–2772. [CrossRef] [PubMed]
111. Sinou, D. Asymmetric Organometallic-Catalyzed Reactions in Aqueous Media. *Adv. Synth. Catal.* **2002**, *344*, 221–237. [CrossRef]
112. Prat, D.; Pardigon, O.; Flemming, H.-W.; Letestu, S.; Ducandas, V.; Isnard, P.; Guntrum, E.; Senac, T.; Ruisseau, S.; Cruciani, P.; et al. Sanofi's Solvent Selection Guide: A Step Toward More Sustainable Processes. *Org. Process. Res. Dev.* **2013**, *17*, 1517–1525. [CrossRef]
113. Capello, C.; Fischer, U.; Hungerbühler, K. What is a green solvent? A comprehensive framework for the environmental assessment of solvents. *Green Chem.* **2007**, *9*, 927–934. [CrossRef]
114. Clegg, W.; Harrington, R.W.; North, M.; Pizzato, F.; Villuendas, P. Cyclic carbonates as sustainable solvents for proline-catalysed aldol reactions. *Tetrahedron Asymmetry* **2010**, *21*, 1262–1271. [CrossRef]
115. Tan, R.; Li, C.; Luo, J.; Kong, Y.; Zheng, W.; Yin, D. An effective heterogeneous l-proline catalyst for the direct asymmetric aldol reaction using graphene oxide as support. *J. Catal.* **2013**, *298*, 138–147. [CrossRef]
116. Qian, Y.; Zheng, X.; Wang, X.; Xiao, S.; Wang, Y. An Efficient Ionic Liquid Additive for Proline-catalyzed Direct Asymmetric Aldol Reactions between Cyclic Ketones and Aromatic Aldehydes. *Chem. Lett.* **2009**, *38*, 576–577. [CrossRef]
117. Rodriguez, B.; Rantanen, T.; Bolm, C. Solvent-Free Asymmetric Organocatalysis in a Ball Mill. *Angew. Chem. Int. Ed.* **2006**, *45*, 6924–6926. [CrossRef] [PubMed]
118. Obregon, A.; Milán, M.; Juaristi, E. Improving the Catalytic Performance of (S)-Proline as Organocatalyst in Asymmetric Aldol Reactions in the Presence of Solvate Ionic Liquids: Involvement of a Supramolecular Aggregate. *Org. Lett.* **2017**, *19*, 1108–1111. [CrossRef] [PubMed]
119. Martínez-Castañeda, A.; Poladura, B.; Rodriguez-Solla, H.; Concellón, C.; Del Amo, V. Direct Aldol Reactions Catalyzed by a Heterogeneous Guanidinium Salt/Proline System under Solvent-Free Conditions‡. *Org. Lett.* **2011**, *13*, 3032–3035. [CrossRef] [PubMed]
120. Szőllősi, G.; Fekete, M.; Gurka, A.A.; Bartók, M. Reversal of Enantioselectivity in Aldol Reaction: New Data on Proline/γ-Alumina Organic–Inorganic Hybrid Catalysts. *Catal. Lett.* **2013**, *144*, 478–486. [CrossRef]
121. North, M.; Villuendas, P. A Chiral Solvent Effect in Asymmetric Organocatalysis. *Org. Lett.* **2010**, *12*, 2378–2381. [CrossRef] [PubMed]

122. Hayashi, Y.; Aratake, S.; Itoh, T.; Okano, T.; Sumiya, T.; Shoji, M. Dry and wet prolines for asymmetric organic solvent-free aldehyde–aldehyde and aldehyde–ketone aldol reactions. *Chem. Commun.* **2007**, 957–959. [CrossRef] [PubMed]
123. Guo, G.; Wu, Y.; Zhao, X.; Wang, J.; Zhang, L.; Cui, Y. Polymerization of l-proline functionalized styrene and its catalytic performance as a supported organocatalyst for direct enantioselective aldol reaction. *Tetrahedron Asymmetry* **2016**, *27*, 740–746. [CrossRef]
124. Zhang, X.; Wang, L.; Wang, Z.; Yan, J. Merrifield Resin Supported Ionic Liquids/l-Proline as Efficient and Recyclable Catalyst Systems for Asymmetric Aldol Reaction. *Synthesis* **2009**, 3744–3750. [CrossRef]
125. Yang, H.; Li, S.; Wang, X.; Zhang, F.; Zhong, X.; Dong, Z.; Ma, J. Core–shell silica magnetic microspheres supported proline as a recyclable organocatalyst for the asymmetric aldol reaction. *J. Mol. Catal. A Chem.* **2012**, *363*, 404–410. [CrossRef]
126. Yacob, Z.; Nan, A.; Liebscher, J. Proline-Functionalized Magnetic Core-Shell Nanoparticles as Efficient and Recyclable Organocatalysts for Aldol Reactions. *Adv. Synth. Catal.* **2012**, *354*, 3259–3264. [CrossRef]
127. Liebscher, J.; Shah, J.; Khan, S.; Blumenthal, H. 1,2,3-Triazolium-Tagged Prolines and Their Application in Asymmetric Aldol and Michael Reactions. *Synthesis* **2009**, 3975–3982. [CrossRef]
128. Kucherenko, A.S.; Struchkova, M.I.; Zlotin, S.G. The (S)-Proline/Polyelectrolyte System: An Efficient, Heterogeneous, Reusable Catalyst for Direct Asymmetric Aldol Reactions. *Eur. J. Org. Chem.* **2006**, 2000–2004. [CrossRef]
129. Ibrahem, I.; Zou, W.; Xu, Y.; Córdova, A. Amino Acid-Catalyzed Asymmetric Carbohydrate Formation: Organocatalytic One-StepDe Novo Synthesis of Keto and Amino Sugars. *Adv. Synth. Catal.* **2006**, *348*, 211–222. [CrossRef]
130. Suri, J.T.; Mitsumori, S.; Albertshofer, K.; Tanaka, F.; Barbas, C.F. Dihydroxyacetone Variants in the Organocatalytic Construction of Carbohydrates: Mimicking Tagatose and Fuculose Aldolases. *J. Org. Chem.* **2006**, *71*, 3822–3828. [CrossRef] [PubMed]
131. Grondal, C.; Enders, D. Direct asymmetric organocatalytic de novo synthesis of carbohydrates. *Tetrahedron* **2006**, *62*, 329–337. [CrossRef]
132. Majewski, M.; Niewczas, I.; Palyam, N. Acids as Proline Co-catalysts in the Aldol Reaction of 1,3-Dioxan-5-ones. *Synlett* **2006**, 2387–2390. [CrossRef]
133. Ibrahem, I.; Córdova, A. Amino acid catalyzed direct enantioselective formation of carbohydrates: One-step de novo synthesis of ketoses. *Tetrahedron Lett.* **2005**, *46*, 3363–3367. [CrossRef]
134. Gong, Z.; Wei, C.; Shi, Y.; Zheng, Q.; Song, Z.; Liu, Z. Novel chiral bifunctional l-thiazoline-amide derivatives: Design and application in the direct enantioselective aldol reactions. *Tetrahedron* **2014**, *70*, 1827–1835. [CrossRef]
135. Miura, T.; Kasuga, H.; Imai, K.; Ina, M.; Tada, N.; Imai, N.; Itoh, A. Highly efficient asymmetric aldol reaction in brine using a fluorous sulfonamide organocatalyst. *Org. Biomol. Chem.* **2012**, *10*, 2209. [CrossRef] [PubMed]
136. Han, X.; Wang, Y.; Gai, X.; Zeng, X. Highly Enantio- and Diastereoselective l-Proline Derived Acetylglucose Amide Catalyzed Aldol Reaction of Ketones to Aldehydes under Solvent-Free Conditions. *Synlett* **2015**, *26*, 2858–2862. [CrossRef]
137. Majewski, M.; Gleave, D.M.; Nowak, P. 1,3-Dioxan-5-ones: Synthesis, deprotonation, and reactions of their lithium enolates. *Can. J. Chem.* **1995**, *73*, 1616–1626. [CrossRef]
138. Ying, A.; Liu, S.; Li, Z.; Chen, G.; Yang, J.; Yan, H.; Xu, S. Magnetic Nanoparticles-Supported Chiral Catalyst with an Imidazolium Ionic Moiety: An Efficient and Recyclable Catalyst for Asymmetric Michael and Aldol Reactions. *Adv. Synth. Catal.* **2016**, *358*, 2116–2125. [CrossRef]
139. Sai, M.; Yamamoto, H. Chiral Brønsted Acid as a True Catalyst: Asymmetric Mukaiyama Aldol and Hosomi–Sakurai Allylation Reactions. *J. Am. Chem. Soc.* **2015**, *137*, 7091–7094. [CrossRef] [PubMed]
140. Jung, Y.; Marcus, R.A. On the Theory of Organic Catalysis "on Water". *J. Am. Chem. Soc.* **2007**, *129*, 5492–5502. [CrossRef] [PubMed]
141. Armstrong, A.; Boto, R.A.; Dingwall, P.; Contreras-Garcia, J.; Harvey, M.J.; Mason, N.; Rzepa, H. The Houk–List transition states for organocatalytic mechanisms revisited. *Chem. Sci.* **2014**, *5*, 2057–2071. [CrossRef]
142. Patil, M.P.; Sunoj, R.B. Insights on Co-Catalyst-Promoted Enamine Formation between Dimethylamine and Propanal through Ab Initio and Density Functional Theory Study. *J. Org. Chem.* **2007**, *72*, 8202–8215. [CrossRef] [PubMed]

143. Wheeler, S.E.; Seguin, T.J.; Guan, Y.; Doney, A. Noncovalent Interactions in Organocatalysis and the Prospect of Computational Catalyst Design. *Acc. Chem. Res.* **2016**, *49*, 1061–1069. [CrossRef] [PubMed]

catalysts

MDPI

Review

Recent Advances in Asymmetric Catalytic Electrosynthesis

Cristiana Margarita and Helena Lundberg *

Department of Chemistry, KTH Royal Institute of Technology, SE-100 44 Stockholm, Sweden; mcristia@kth.se
* Correspondence: hellundb@kth.se; Tel.: +46-8-790-8125

Received: 31 July 2020; Accepted: 26 August 2020; Published: 1 September 2020

Abstract: The renewed interest in electrosynthesis demonstrated by organic chemists in the last years has allowed for rapid development of new methodologies. In this review, advances in enantioselective electrosynthesis that rely on catalytic amounts of organic or metal-based chiral mediators are highlighted with focus on the most recent developments up to July 2020. Examples of C-H functionalization, alkene functionalization, carboxylation and cross-electrophile couplings are discussed, along with their related mechanistic aspects.

Keywords: organic electrosynthesis; asymmetric catalysis; electrochemistry; enantioselectivity; organocatalysis; transition-metal catalysis

1. Introduction

The stereochemistry of organic compounds can dramatically influence their properties, such as biological activity of pharmaceuticals or macroscopic physical characteristics of polymers. Thus, the ability to control the stereochemical outcome in organic synthesis can be of great importance, and asymmetric catalysis is an indispensable tool in this endeavor. In asymmetric catalysis, racemic or prochiral substrates are converted to stereogenically enriched products with the aid of a chiral mediator in sub-stoichiometric amounts. As the catalyst is continually being reused in the process, the strategy presents great opportunities for highly atom-efficient processes. Traditionally, asymmetric catalysis has focused on two-electron transformations [1,2]. However, the toolbox of the organic chemist has been expanded tremendously in the last decade with the introduction of methods for asymmetric transformations with open-shell radical intermediate species using chemical, photochemical and electrochemical strategies [3–10].

In electrosynthesis, chemical redox reagents are replaced with electricity for the transformation of organic molecules. While dating back to the 19th century, the topic is currently receiving renewed interest due to its enabling potential for selective and sustainable synthesis [11–19]. In an electrochemical reaction, oxidation occurs at the anode and reduction at the cathode in a conductive medium. The redox processes can take place at constant current or constant potential in an undivided or divided electrochemical cell, where the latter separates the anodic and cathodic compartments by a conductive membrane [20]. The redox event that ultimately converts starting material to product in electrosynthesis can occur via different paths. As exemplified for a net-oxidative process, the electron transfer from a substrate to the anode can occur directly at the electrode surface (Figure 1a) or take place in solution with the aid of a homogeneous redox mediator. The latter represents indirect electrolysis (Figure 1b), where the mediator effectively acts as a catalyst for the electron transfer. The use of such redox mediators can facilitate the redox event from electrode to substrate and thereby reduce energy consumption and enable milder conditions with higher chemoselectivities [15,21]. In addition, electricity can be used to drive product formation by (re)generation of catalytically active species (Figure 1c). For clarity, we use the word "redox mediator" for electron transfer catalysts in this review, whereas "catalyst" denotes compounds added in catalytic amounts to mediate a transformation by other mechanistic action. In the context of asymmetric electrosynthesis, both redox mediators and catalysts can be chiral and induce stereoselectivity in the reaction in which they are used. Mechanistic insight can be crucial for rational development of new catalysts and synthetic protocols. To probe the reaction mechanism of electrosynthetic reactions, electroanalytical techniques such as cyclic voltammetry (CV) in combination with classic chemical approaches are commonly used [22].

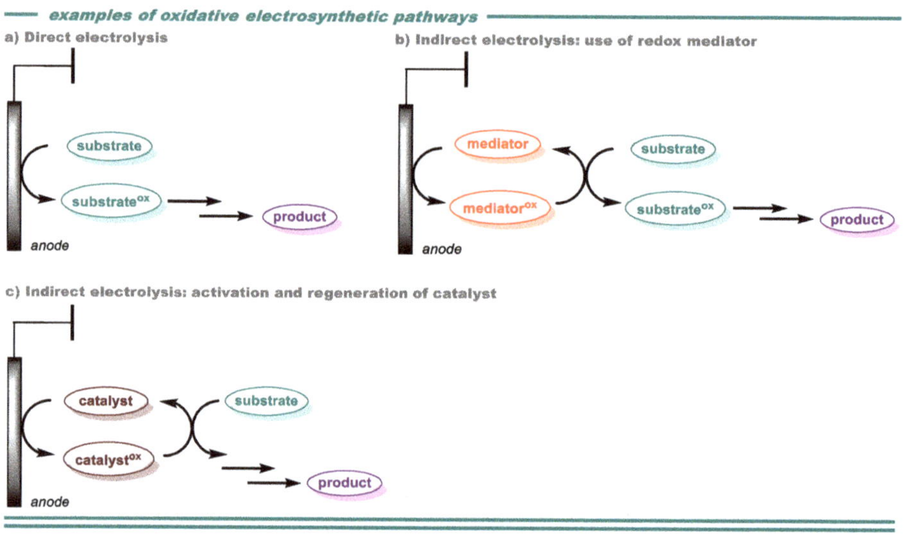

Figure 1. Examples of electrosynthetic pathways for substrate oxidation (cathodic reaction not shown) (**a**) direct anodic oxidation; (**b**) indirect anodic oxidation; (**c**) indirect oxidation by anodic activation and regeneration of catalyst.

In this review, electrosynthetic protocols that utilize catalytic amounts of small organic or metal-based chiral mediators to afford asymmetric induction in C–H functionalization, alkene functionalization, carboxylation and cross-electrophile couplings are discussed, along with mechanistic aspects of the transformations. Other strategies for asymmetric electrosynthesis, e.g., biocatalysis, chiral auxiliaries, pre-functionalized chiral electrodes and chiral media have recently been covered elsewhere and will not be discussed here [3–6,23–25]. This review is divided into oxidative and reductive transformations and highlights recent developments up to July 2020.

2. Oxidative Transformations

2.1. C-H Functionalization

Asymmetric C–H functionalization of α-positions in carbonyl compounds through enamine intermediates is a well-explored topic in organocatalysis, often employing proline-derived catalysts [26]. In an electrochemical setting, Jang and co-workers reported on α-oxyamination of aldehydes in the presence of the prolinol derivative (S)-α,α-diphenyl-2-pyrrolidine methanol trimethylsilyl ether in catalytic amounts using platinum electrodes under constant current in an undivided cell (Scheme 1a) [27]. Supported by control experiments and CV, the authors suggested that the reaction proceeds via single-electron oxidation of the intermediate enamine, formed by dehydrative condensation between the substrate and pyrrolidine catalyst. The resulting cation radicals were trapped by TEMPO to form the oxyaminated products in 60–70% *ee*, in accordance with the design principle in Figure 2a. As such, this study was the first to demonstrate that anodic oxidation could be used to promote enamine-mediated organocatalytic reactions via cationic radical enamine intermediates. A similar strategy was used by the same group for α-alkylation of aldehydes with xanthene using chiral pyrrolidine catalysts in an undivided cell with platinum electrodes under constant current conditions (Scheme 1b) [28]. Supported by CV analysis and control experiments, the mechanism was proposed to proceed via single-electron oxidation of both the enamine and the xanthene, followed by radical coupling to afford the product in up to 74% yield and 70% *ee* (Figure 2a). However, as homocoupled products were not observed, a mechanism in accordance with the design principle of Figure 2b proceeding via enamine attack of an electrochemically formed xanthene cation could not be excluded. This alternative activation mode was proposed by Jørgensen and co-workers for α-arylation of aldehydes with N-tosyl *p*-anisidine to form enantioenriched alcohols with *ee*'s up to 96% using (S)-α,α-diphenyl-2-pyrrolidine methanol trimethylsilyl ether as catalyst (10 mol%) in an undivided cell with carbon anode and platinum cathode under constant current conditions (Scheme 1c) [29]. The authors suggested that electrochemical oxidation of the phenolic compound results in an electrophilic intermediate that is attacked by the chiral enamine, formed from the catalyst and the aldehyde substrate. Following the mechanistic rationale of traditional enamine catalysis, the resulting iminium ion is hydrolyzed to release the catalyst, which enters a new cycle, whereas the product would rearomatize and eventually cyclize to the corresponding dihydrobenzofuran.

Building on this initial work [27–29], Luo and co-workers utilized the enamine strategy for electrooxidative coupling of tertiary amines with ketones using a chiral diamine catalyst to obtain enantioenriched alkylated tetrahydroisoquinolines, using a graphite anode and Pt cathode in an undivided cell at constant potential (Scheme 2) [30]. No reaction was observed in the absence of either catalyst or current, and under optimized conditions the reaction proceeded with good to excellent enantioselectivities (up to 95% *ee*) and good yields for several N-arylated tetrahydroisoquinolines and ketones. The transformation was proposed to proceed via anodic oxidation of the benzylic position in the N-aryl tetrahydroisoquinoline substrate to form an intermediate iminium ion (Scheme 2), in accordance with the design principle in Figure 2b. Subsequent attack by the chiral enamine, formed by dehydrative condensation of the catalyst and ketone substrate, results in the enantioenriched product in high yields and *ee*'s after hydrolysis with release of the catalyst. The presence of a protonic additive (CF_3CH_2OH) was found to facilitate the C–C bond formation, supposedly by capturing the iminium ion as a more stable hemiaminal intermediate. The proposed mechanism was supported by control experiments, where pre-oxidation of the tetrahydroisoquinoline substrate to the corresponding iminium ion and subsequent addition of catalyst/TfOH and ketone gave close to identical results compared to standard conditions. Furthermore, electroanalytic studies confirmed that preferential oxidation of the tetrahydroisoquinoline occurred in the reaction mixture.

The use of TEMPO and analogous N-oxide compounds as redox mediators in electrochemical oxidation of organic substrates is well-established. In several examples, chiral N-oxide derivatives were developed and afforded asymmetric electrochemical transformations such as the oxidative kinetic resolution of secondary alcohols and amines [31–39]. Similarly, asymmetric Cu-catalysis under electrooxidative conditions has been reported for kinetic resolution of 1, 2 diols, aminoalcohols and aminoaldehydes [40]. Mei and co-workers reported on a co-catalytic Cu/TEMPO system for oxidative substitution of N-aryl and N-carbamate tetrahydroisoquinolines with terminal alkynes [41]. The use of TEMPO as redox mediator (20 mol%) and a Cu^{II}-species (10 mol%) with a chiral bisoxazoline ligand enabled good yields and up to 98% *ee* of the C1-alkynylated products under constant current conditions in an undivided cell using Pt electrodes (Scheme 3). The broad substrate scope demonstrated good functional group tolerance of the process and control experiments indicated that each reaction component was necessary for converting starting material into product. Based on CV measurements, the authors proposed that TEMPO acts as a redox mediator for Shono oxidation [42] of the tetrahydroisoquinoline to form the corresponding iminium species in benzylic position. Addition of the chiral copper acetylide, supposedly formed in situ from the Cu^{I} catalyst and the alkyne in the presence of base, results in the enantioenriched product. As such, the authors propose a mechanism following the general principles of Figures 1b and 2b, although alternative cooperative mechanisms involving both Cu and TEMPO were not ruled out. The active Cu^{I} was suggested to form by interaction of Cu^{II} with the reduced TEMPO mediator (TEMPO-H) or by cathodic reduction. A kinetic isotope effect study for the benzylic α-hydrogens of the tetrahydroisoquinoline indicated C–H cleavage is not rate-determining in the process.

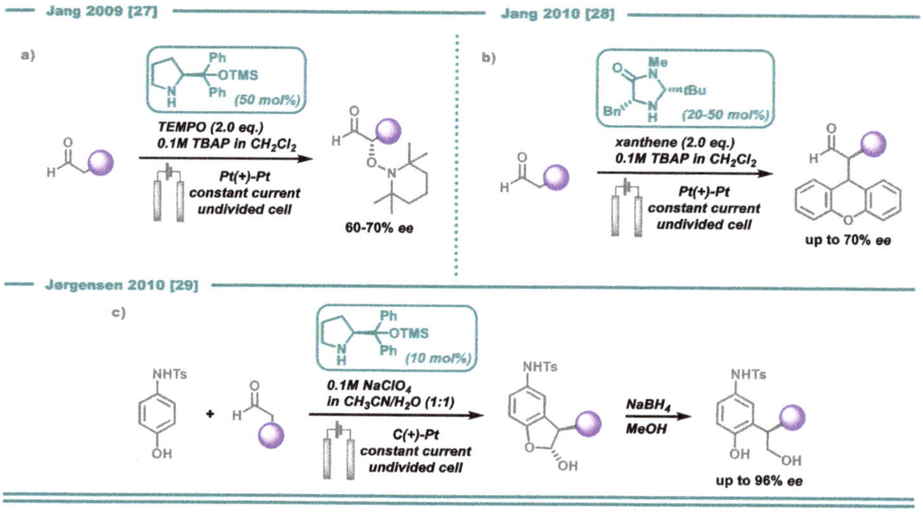

Scheme 1. (**a**) electrochemical organocatalytic α-oxyamination of aldehydes; (**b**) electrochemical organocatalytic α-alkylation of aldehydes; (**c**) electrochemical organocatalytic α-arylation of aldehydes.

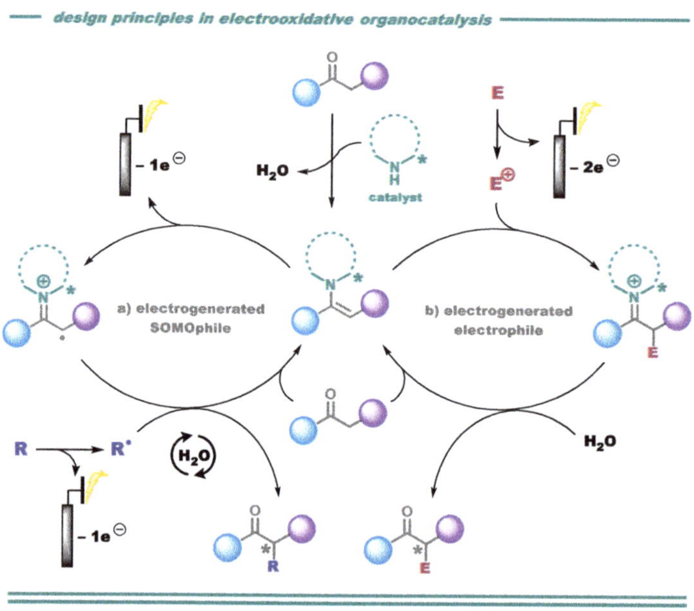

Figure 2. Two design principles for α-functionalization of carbonyl compounds in electrooxidative enamine catalysis: (**a**) electrogenerated SOMOphile; (**b**) electrogenerated electrophile.

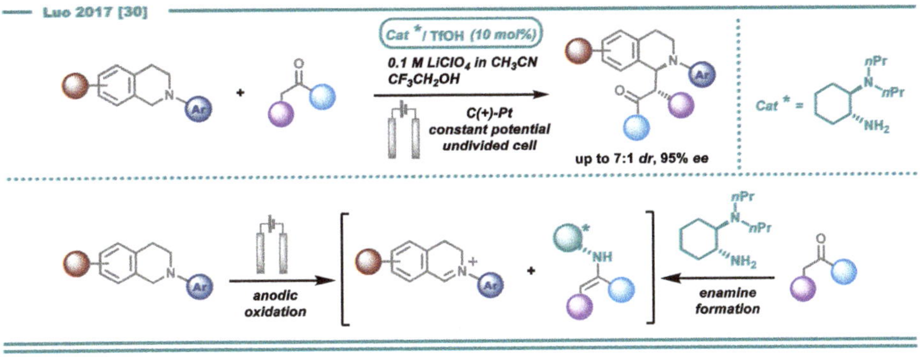

Scheme 2. Enantioselective organocatalytic coupling of tetrahydroisoquinolines and ketones under electrooxidative conditions.

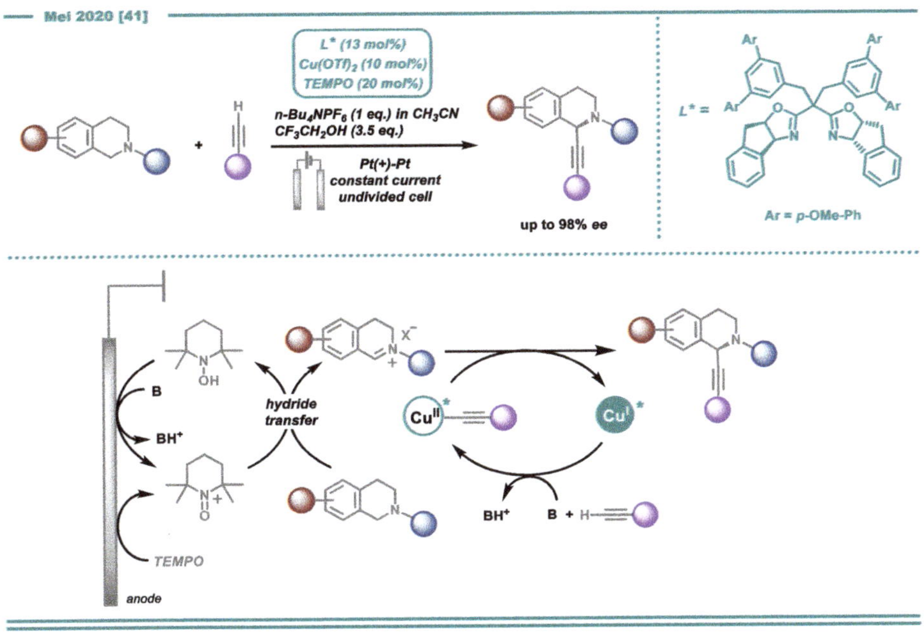

Scheme 3. Cu^{II}/TEMPO-catalyzed enantioselective electrooxidative alkynylation of tetrahydroisoquinolines.

Guan, He and co-workers reported on the proline-catalyzed enantioselective synthesis of C2-quaternary indolin-3-ones from 2-arylindoles and ketones under electrochemical conditions (Scheme 4) [43]. The transformation was carried out in an undivided cell at constant current with Pt electrodes and resulted in moderate to good yields and excellent diastereo-and enantioselectivities (up to 99% *ee*) of 2,2-disubstituted indol-3-one products bearing a quaternary stereocenter, an interesting compound class found in naturally occurring biologically active substances. The procedure was mostly found to be efficient on cyclic ketones, but two acyclic examples were also reported and could afford high *ee*'s. Similar to the findings of Luo and co-workers [30], no reaction occurred in the absence of proline or electric current. In addition, no reaction was observed when starting from N-methyl-substituted indole. The reaction was proposed to proceed in accordance with the design principle in Figure 1b via indirect oxidation of the indole to the corresponding radical cation, using TEMPO as redox mediator. Upon loss of a proton, the resulting indole radical was proposed to react with atmospheric oxygen after a hydrogen atom transfer (HAT). Subsequent attack of the imine carbon by the enamine, formed by the proline catalyst and the ketone substrate, results in the product after hydrolysis. Control experiments using isolated imine in the presence of the proline catalyst and ketone substrate under non-electrochemical conditions resulted in excellent yields and enantioselectivity and provided support for the proposed mechanism. Furthermore, the addition of BHT to standard conditions resulted in lower yields of the product as well as in the formation of a BHT adduct of the indole, detected by HRMS, thus indicating that a radical mechanism is operating. The use of ^{18}O-labelled O_2 and H_2O indicated that the former was indeed the source of the incorporated oxygen, a hypothesis that was further strengthened by the greatly reduced yields observed when the reaction was performed under an Ar atmosphere. Benzoic acid was used as weakly acidic additive, which aided to increase the yield by facilitating enamine formation, an effect reported in various examples of asymmetric enamine catalysis [44]. The reaction was found to proceed in the absence of TEMPO with the same enantioselectivity but reaching lower yields. Based on CV analysis, the authors proposed that direct oxidation of the indole at the anode is occurring in the absence of TEMPO.

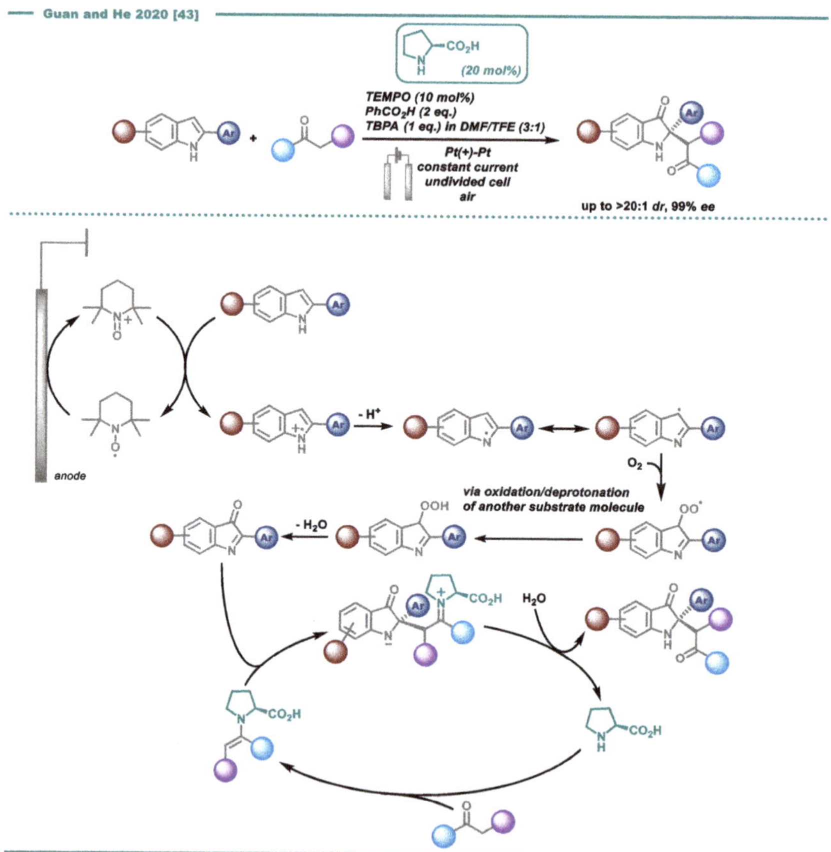

Scheme 4. Enantioselective proline-catalyzed electrosynthesis of C2-quaternary indolin-3-ones.

Luo and co-workers reported on the first asymmetric enamine-benzyne coupling (Scheme 5) [45]. Using a chiral diamine catalyst in an undivided cell with Pt electrodes and constant current conditions, α-arylation of cyclic α-ketocarbonyls with anodically formed benzyne occurred to form products with quaternary carbon stereocenters in good yields and with *ee*'s up to 99%. Under the electrochemical conditions, 1-aminobenzotriazole could be used as benzyne precursor in the absence of the toxic oxidant $Pb(OAc)_4$, typically required under non-electrochemical conditions [46–50]. In addition, the strategy was extended to the generation and use of cyclohexyne via the corresponding triazole precursor. Trapping experiments with tetraphenylcyclopentadienone under standard conditions resulted in good yields of the benzyne and cyclohexyne adducts but only trace amounts of the benchmark products, thus confirming the existence of the elusive intermediates. The substrate scope was limited to cyclic ketoesters and products were in some cases obtained as regioisomers, however all with high *ee*'s and in moderate to good yields. No reaction was observed in the absence of the aminocatalyst, whereas results in the absence of electric current was not explicitly discussed. Experimentally, it was found that $Co(OAc)_2{\cdot}4H_2O$ was beneficial as an additive (20 mol%) with a major effect on yield (71% vs. 28%) and a minor effect on *ee* (94% *ee* vs. 88% *ee*) of the product. With the use of CV, the authors found that the redox potentials of the Co salt and 1-aminobenzotriazole were comparable (E_{ox} = 0.83 vs. 0.84 V), hence suggesting that the Co salt was likely not functioning as a redox mediator for oxidative benzyne generation. However, control experiments with benzyne quenching reagents in the absence of α-ketocarbonyl substrates resulted in rapid decomposition of the benzotriazole but only trace amounts of quenching adducts in the absence of the Co salt, whereas considerably higher yields were obtained in the presence of Co. Based on these findings, the authors proposed that the Co salt stabilizes the intermediate arynes by binding to the triple bond, thereby enhancing the propensity for enamine coupling. This hypothesis was supported by DTF calculations, indicating that coordination to Co acetate stabilizes benzyne by 18.6 kcal/mol.

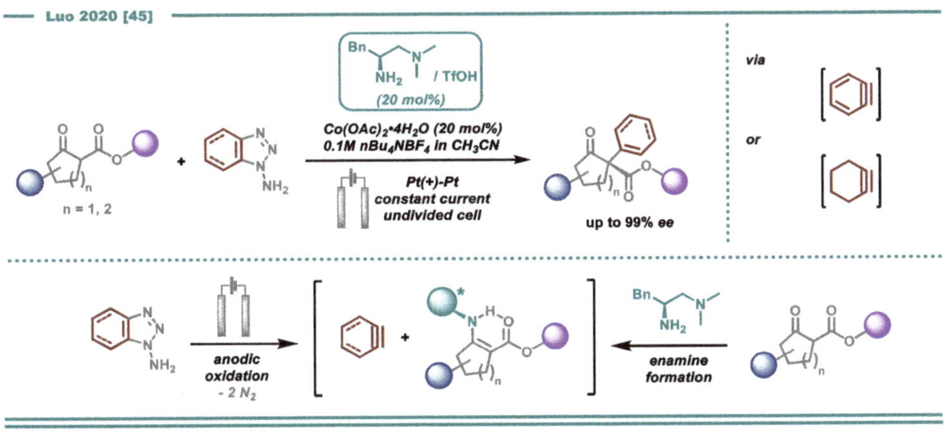

Scheme 5. Asymmetric catalytic α-arylation of cyclic α-ketocarbonyls with anodically formed benzyne or cyclohexyne.

Meggers and co-workers described an oxidative metal-catalyzed asymmetric C–H functionalization of the α-position of 2-acyl imidazoles in an undivided cell equipped with a C anode and Pt cathode under constant current conditions (Scheme 6) [51]. Using a chiral Lewis acidic Rh complex, 2-acyl imidazoles and silyl enol ethers were coupled to provide enantioenriched 1,4-dicarbonyls in up to 91% yield and >99% *ee*, including >10 examples of the formation of all-carbon quaternary stereocenters as well as two examples of complex natural product derivatization. Considerably lower yields were observed in the absence of a base, whereas only homocoupling of the silyl enol ether was obtained in the absence of the Rh catalyst. Mechanistically, the reaction was proposed to proceed via initial coordination of the 2-acyl imidazole substrate to the metal catalyst, followed by deprotonation by the external base (2,6-lutidine). The resulting Rh-coordinated enolate was envisioned to undergo anodic oxidation to afford a C-centered radical in α-position that would undergo stereocontrolled C–C bond formation with the silyl enol ether and form a TMS-protected ketyl radical. A second anodic oxidation, followed by subsequent desilylation would close the catalytic cycle (Scheme 6). As presented, the anodic oxidations of the catalyst-bound substrate were envisioned as direct electrolysis events in accordance with the general principle in Figure 1a. The mechanistic proposal was supported by CV analysis, indicating that the Rh-bound 2-acyl imidazole has a considerably lower oxidation potential compared to the silyl enol ether and the unbound 2-acyl imidazole. Furthermore, control experiments using TEMPO as a radical quenching agent resulted in high yields of TEMPO-functionalized 2-acyl imidazole in α-position, thus indicating that this carbon indeed hosts an intermediate radical.

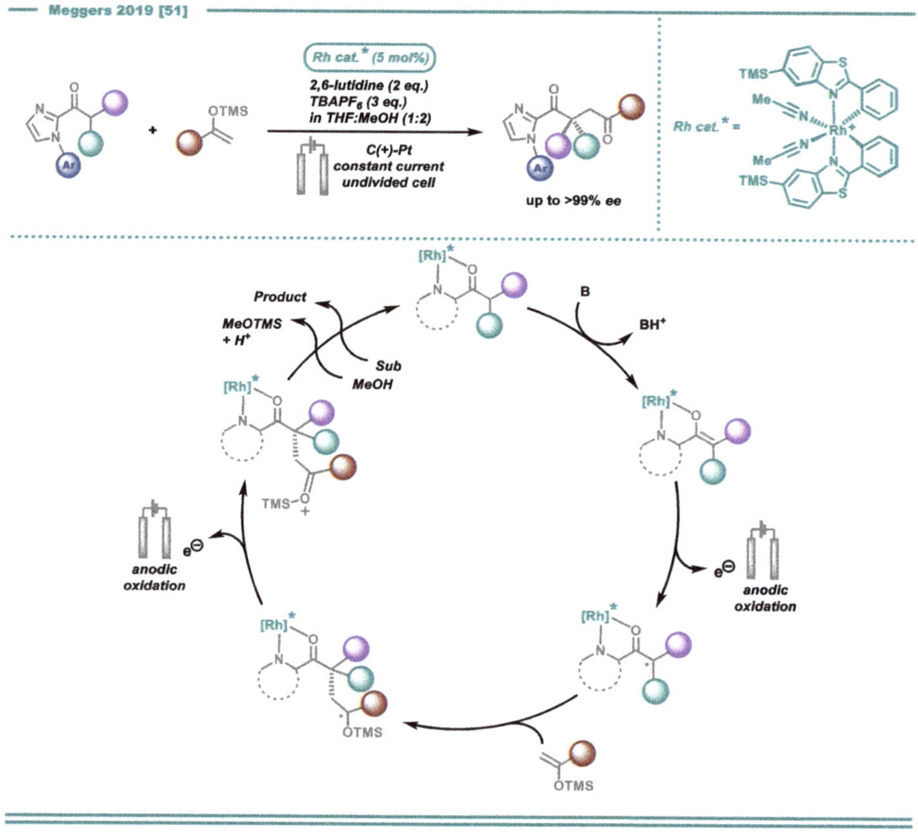

Scheme 6. Electrooxidative Lewis acid-catalyzed asymmetric cross-coupling of enolates.

Similar to the work of Meggers, Guo and co-workers reported the use of a Lewis acidic chiral Ni catalyst for enantioselective α-benzylation of 2-acyl imidazoles with substituted hydroxytoluenes (Scheme 7) [52]. In an undivided cell at 0 °C using C electrodes at constant potential, electrochemical benzylation was carried out with good yields and excellent enantioselectivities (up to 97% *ee*) in the presence of $Ni(OAc)_2$, a chiral diamine ligand and base (quinuclidine). No product formed in the absence of either Ni, diamine ligand or current, whereas the absence of base resulted in reduced yields but similar *ee*. Trapping experiments afforded 11% of the TEMPO adduct in α-position to the 2-acyl imidazole, suggesting that a radical can form at this carbon. Furthermore, dimerization of the 1,4,6-trimethylphenol under different conditions indicated that an intermediate benzylic radical may also form. Based on the combined results from these studies as well as CV and electron paramagnetic resonance (EPR) spectroscopy, the authors proposed the mechanism found in Scheme 7.

Initial coordination of the 2-acyl imidazole to the chiral Ni catalyst followed by deprotonation by the base affords the Ni-bound enolate. Anodic oxidation of this species affords the C-centered radical in α-position to the coordinated carbonyl, which undergoes coupling with a benzylic radical, formed via anodic oxidation via a separate route. As such, a mechanism similar to that of electrogenerated SOMOphiles in enamine catalysis (Figure 2a) was proposed. The possibility of direct interaction between the benzylic radical and the Ni center that could result in product formation after reductive elimination was not discussed by the authors. While quinuclidine is a known HAT catalyst in photoredox catalysis [53–58], additional mechanistic roles of this base, e.g., assistance in the formation of the postulated benzylic radical coupling partner, were not proposed.

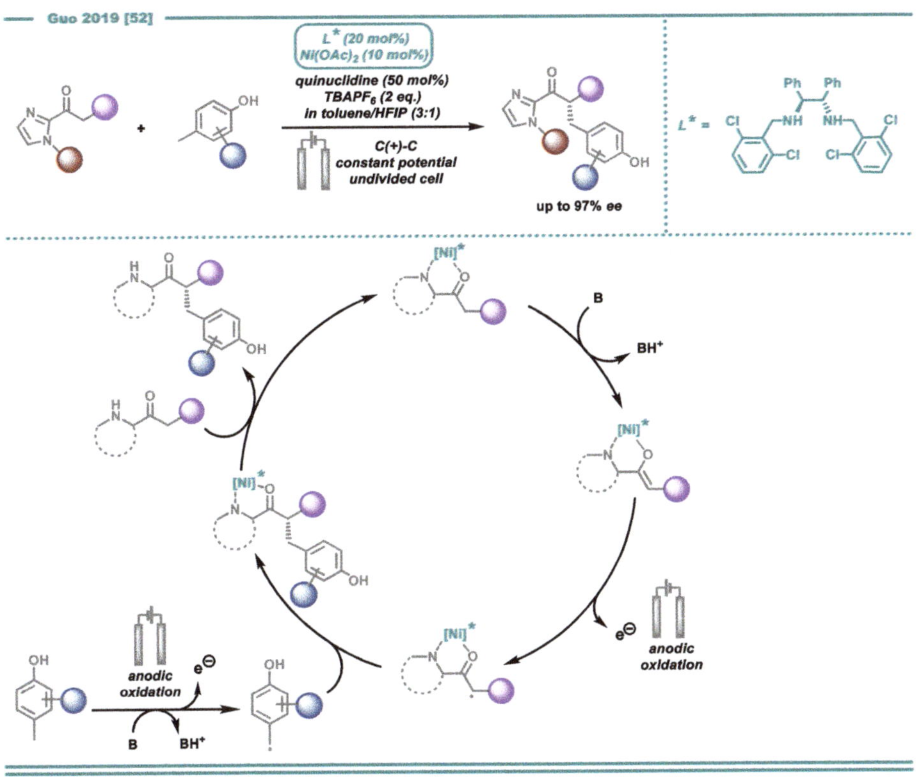

Scheme 7. Lewis acid-catalyzed electrooxidative enantioselective α-benzylation of 2-acyl imidazoles.

Along the same lines, the same group developed an enantioselective bifunctional squaramide-catalyzed detrifluoroacetylative alkylation reaction using substituted hydroxytoluenes under electrochemical conditions [59]. Using carbon electrodes under constant current conditions at 70 °C, fluorine-containing compounds bearing stereocenters at the C-F bond were formed in good yields and *ee*'s up to 95%. As suggested by the authors, the trifluoromethyl α-fluorinated β-keto *gem*-diol substrate is activated by the squaramide catalyst and reacts with the electrochemically formed *p*-quinone methide (Scheme 8), as such resembling the principles of Figure 2b with anodic electrophile formation. Base-induced detrifluoroacetylation followed by stereoselective proton transfer generates the product as the final step in the absence of electricity.

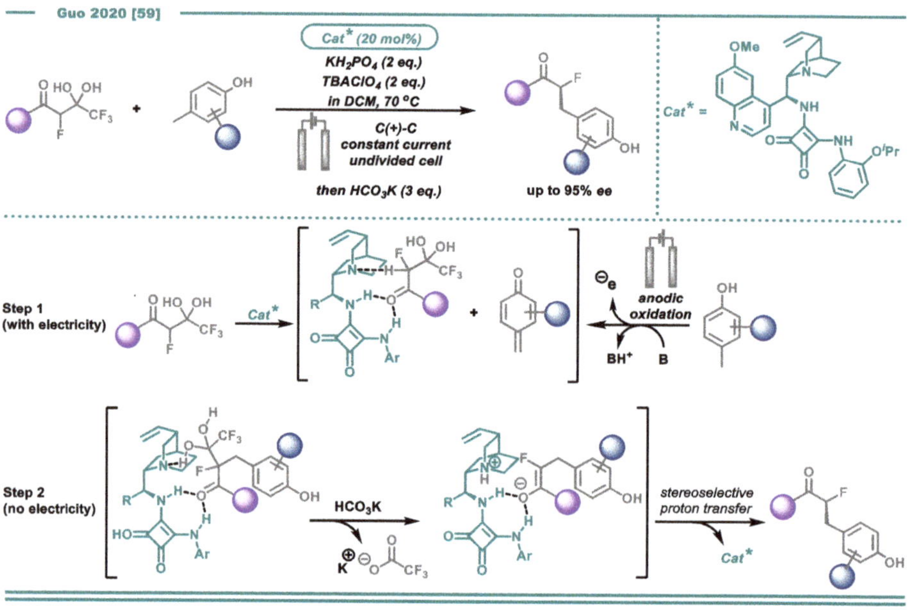

Scheme 8. Squaramide-catalyzed electrooxidative enantioselective α-benzylation/detrifluoroacetylation of trifluoromethyl α-fluorinated β-keto *gem*-diols.

Ackermann and co-workers demonstrated the first asymmetric electrooxidative C–H activation process using Pd-catalysis and L-*tert*-leucine as transient directing group to provide access to axially chiral biaryls (Scheme 9) [60]. In an undivided cell with graphite felt anode and Pt plate cathode and constant current at 60 °C, various aldehyde-substituted biaryls were alkenylated to form axially chiral products in good yields with excellent *ee*'s (up to 99%) and high position- and diastereocontrol. The substrate scope also included examples of N–C axially chiral N-aryl pyrroles and the use of perfluorinated alkenes. Furthermore, it was demonstrated how the axially chiral products could be converted into enantioenriched [5]-and [6]-helicenes, dicarboxylic acids and BINOL derivatives in high yields and optical purity. Control experiments revealed that no reaction occurred in the absence of either Pd catalyst or L-*tert*-leucine, whereas the absence of current resulted in considerably lower yield of the benchmark substrate (25% vs. 71%). Labelling experiments indicated that C–H activation is the rate-limiting step and that no H/D scrambling occurs between the substrate and the acetic acid solvent. Furthermore, a non-linear-effect (NLE) was not observed, indicating that the reaction proceeds with a metal to ligand ratio of 1:1 in the enantiodetermining step.

With the aid of DFT calculations, the authors proposed a Pd^{II}-mechanism where initial imine formation between the leucine ligand and the aldehyde substrate results in a transient directing group that facilitates the asymmetric Pd-catalyzed $C(sp^2)$–H activation. The nature of the transition state for this step was not discussed by the authors. Based on the presented geometry-optimized structures of intermediates, it does however appear as if the C–H bond breakage is envisioned to occur with the aid of an acetate base. The resulting Pd^{II} species coordinates an alkene reactant, followed by its insertion into the Pd–C bond. Although not explicitly mentioned, it can be envisioned that β-hydride elimination followed by hydrolysis of the transient imine releases the product. The role of the electricity was not specifically discussed by the authors. However, as the control reaction without current resulted in a product yield corresponding to more than two catalyst turnovers, it appears as if electricity is accelerating product formation rather than enabling it.

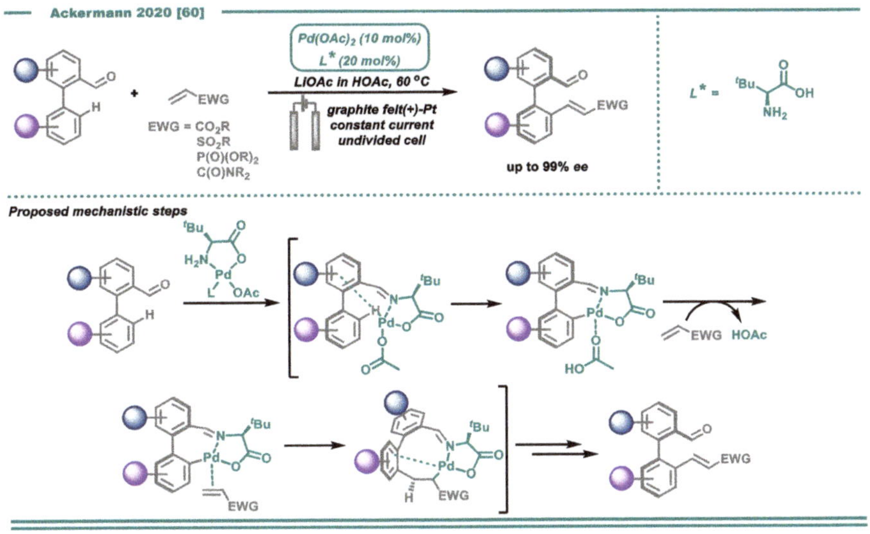

Scheme 9. Palladium-catalyzed C–H activation with a transient directing group under electrochemical conditions.

2.2. Alkene Functionalization

The prochiral nature of alkenes makes them excellent starting materials for asymmetric synthesis. A classic asymmetric oxidative transformation of alkenes is the Sharpless dihydroxylation using osmium catalysis in conjunction with chiral amine ligands and a stoichiometric oxidant [61–63]. As reported by Tsuji and Sharpless in the early 1990's, potassium ferricyanide [$K_3Fe(CN)_6$] can efficiently reoxidize an Os catalyst back to the active +8 state [64–66]. In the original Sharpless procedure, 3 equivalents of the ferricyanide salt were added to enable good turnover for the asymmetric dihydroxylation with excellent *ee*'s. Making use of anodic regeneration, Amundsen and Balko were able to reduce the stoichiometry to 0.4 equivalents using coupled redox cycles (Figure 3) in a divided cell with Pt electrodes at constant potential [67]. Similar to the Sharpless protocol, the authors used a biphasic mixture with the Fe salt in the aqueous phase and the organic transformation occurring in the organic phase, with re-oxidation of the Os catalyst taking place at the solvent interface. With an analogous approach, Torii et al. were able to further reduce the loading of both Os and $K_3Fe(CN)_6$ using an undivided cell with Pt electrodes at constant current conditions at 0 °C [68]. In addition, Torii et al. developed an electrochemical iodine-assisted asymmetric dihydroxylation under similar electrochemical conditions, replacing ferricyanide with substoichiometric I_2 as co-oxidant. As pointed out by the authors, although the Os^{IV} complex could be directly oxidized at the anode, mediated electrolysis in solution is preferred as direct oxidation can lead to catalyst adsorption on the electrode surface that may inhibit catalysis [69].

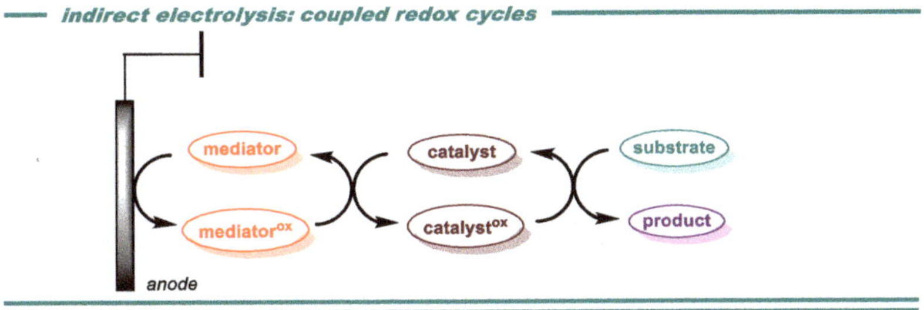

Figure 3. Principle for indirect oxidative electrolysis using coupled redox cycles.

Using the same principle of electrochemically driven coupled redox cycles, Torii and co-workers developed a protocol using anodically generated chloronium oxidants for olefin epoxidation [70], inspired by the Katsuki-Jacobsen epoxidation [71–73]. With an optically active Mn-salen complex (5 mol%) in an undivided cell and Pt electrodes at 0 °C under constant current conditions in biphasic media, a handful of substrates were converted to the corresponding epoxides in up to 93% yield and 87% *ee*. According to the authors, chloride ions were oxidized in the aqueous phase to chloronium species that subsequently entered the organic phase to oxidize the Mn catalyst, thereby driving the oxidation process forward. Along the same lines, Bethell and co-workers utilized electrogenerated percarbonate and persulfate oxidants for iminium catalyzed epoxidation of alkenes with moderate enantiomeric excess (up to 64% *ee*), using a boron-doped diamond anode and Pt wire cathode in an undivided cell under constant current conditions [74]. In recent years, asymmetric electrochemical Os-catalyzed dihydroxylation and Mn-catalyzed epoxidation have been demonstrated to work well on a photovoltaic platform for improved process sustainability (Scheme 10) [75,76].

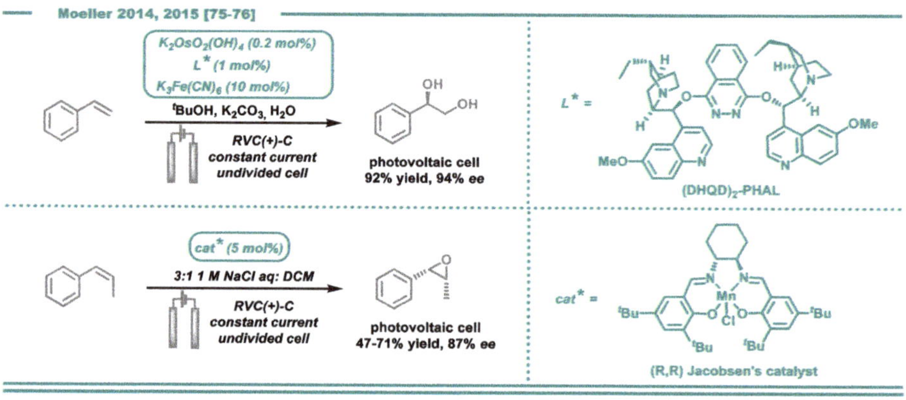

Scheme 10. Indirect electrooxidation of alkenes using coupled redox cycles in a photovoltaic system.

Lin and co-workers described enantioselective cyanophosphinoylation and cyanosulfinylation of alkenes in an anodically driven electrochemical process, using a copper catalyst with a newly developed serine-derived bisoxazoline ligand (Scheme 11) [77]. With a carbon felt anode and Pt cathode in an undivided cell under constant current conditions at 0 °C, a variety of styrene derivatives were functionalized in moderate to good yields with *ee*'s up to 95% (cyanophosphinoylation) and 98% (cyanosulfinylation), including heterocyclic substrates and substrates with functional groups sensitive to oxidation such as aldehyde and sulfide. The authors proposed that the electric current is required to oxidize the Cu catalyst from oxidation state +1 to its catalytically active +2 state. In its activated state, the copper catalyst was claimed to have a dual role (Scheme 11). Initially, the Cu^{II} catalyst was envisioned to oxidize a secondary phosphine oxide to its corresponding radical. This would in turn attack the alkene substrate to form an intermediate with the resulting radical situated in the stabilized benzylic position. Secondly, the reduced Cu catalyst would undergo another anodic oxidation event to form once more the active Cu^{II} complex. The active complex would thereafter accommodate the benzylic radical intermediate and form the enantiomerically enriched product upon reductive elimination. As such, the Cu catalyst is envisioned to act as a redox mediator in the first step in accordance with the general principle in Figure 1b, whereas it acts as an enantioselective cross-coupling catalyst in an electrochemically driven redox cycle in the second step (Figure 1c). Based on the assumption of a Cu^{III} intermediate, optimization of the ligand structure was carried out by adding ancillary ester groups to the bisoxazoline (BOX) scaffold to stabilize the high-valent intermediate prior to reductive elimination. In addition, it was hypothesized that this modification would allow for a more rigid structure that could improve selectivity in the enantio-determining step, as well as prevent cathodic demetalation of the catalyst. With the new ligand, an increase in *ee* from 84% to 95% was observed for the benchmark substrate.

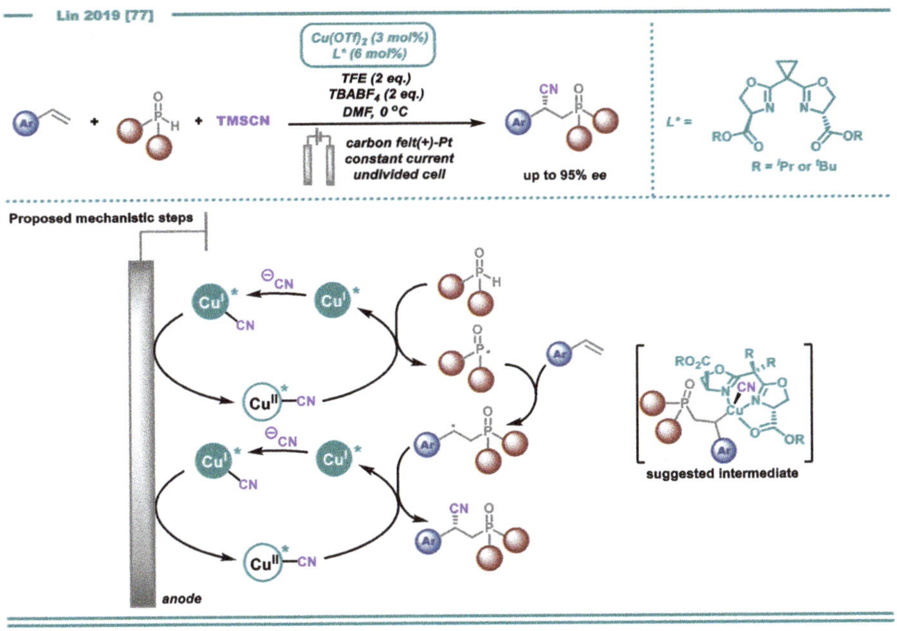

Scheme 11. Copper-catalyzed asymmetric oxidative difunctionalization of alkenes.

Along the same lines, the Lin group published a second anodically driven enantioselective functionalization of alkenes [78]. In conjunction with a cobalt(salen) complex, the use of the same type of asymmetric Cu catalyst as in their prior work enabled hydrocyanation of differently substituted olefins using $PhSiH_3$ and TMSCN in an undivided cell at 0 °C and constant potential, equipped with a carbon anode and Pt cathode (Scheme 12). A range of terminal and internal alkenes were transformed into their corresponding hydrocyanated products with high enantioselectivity (80–95% *ee*). Dienes, enynes and allenes were also compatible substrates, with terminal bonds being favored sites for hydrocyanation. In the case of allenes, excess reagents allowed for dicyanation. A comparison between the developed electrochemical protocol and the use of several chemical oxidants indicated that anodic oxidation resulted in the best yields and enantioselectivities under the examined conditions. Acetic acid was used as an additive in the transformation, rationalized as preventing undesired cathodic reduction of the Cu catalyst.

The suggested mechanism for the transformation resembles that proposed for the Lin group's preceding work on Cu-catalyzed alkene functionalization in Scheme 11 [77]. The first anodic event leads to the formation of a C-centered radical and the second anodic event produces the active asymmetric Cu^{II} cross-coupling catalyst that reacts with the C-centered radical and forms product upon reductive elimination. For the hydrocyanation reaction, the initial anodic oxidation step was proposed to involve electrochemical oxidation of the Co^{II} catalyst, followed by the formation of a cobalt hydride species upon reaction with the hydrosilane reagent. Hydrogen atom transfer of this metal hydride to the alkene would result in a C-centered radical that in turn reacts with the asymmetric Cu-catalyst and eventually forms product. As such, both electrochemical processes appear to follow the general principle of Figure 1c. A radical rearrangement experiment provided support for the notion of intermediate radicals, whereas DFT calculations suggested that the enantio-determining C–CN bond formation was the turnover-limiting step of the process.

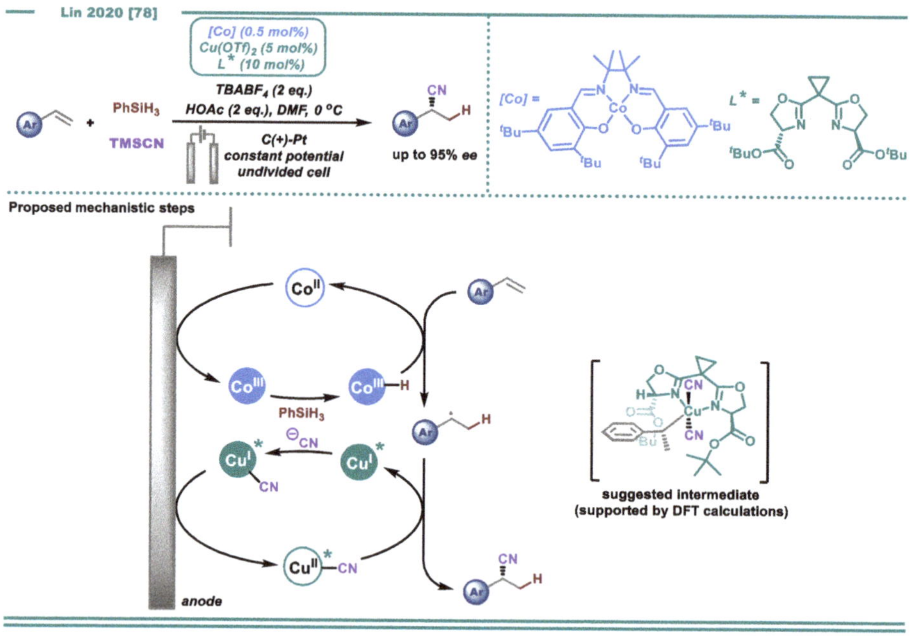

Scheme 12. Anodically driven asymmetric hydrocyanation of alkenes by Co/Cu catalysis.

3. Reductive Transformations

Reductive organocatalytic electrosynthesis dates back to the late 1960s, when Grimshaw and co-workers described the use of alkaloids in catalytic amounts for the reduction of 4-methylcoumarins to optically active 3,4-dihydro-4-methylcoumarins in a divided cell at constant potential [79,80]. The transformation resulted in only modest chiral induction and moderate yields, the latter due to competing dimerization of the intermediate coumarin radical. Nevertheless, the organocatalytic strategy inspired the work of other groups working in this field [81–83] and chiral amines have been used as additives in catalytic amounts for asymmetric reduction of various ketones, carboxylic acids and oximes as well as for reductive dehalogenation [84–97]. Electrocatalytic reductions have also been carried out using metal catalysts, including a Rh^{III} polypyridyl complex for the hydrogenation of acetophenone with modest chiral induction [98]. Other early electroreductive metal-catalyzed enantioselective processes include cobalt catalysis in the form of vitamin B_{12} [99–101], as well as nickel catalysis [102] for dehalogenations, isomerizations, reductive cleavages, etc.

3.1. Carboxylation Using CO_2

The electroreductive use of carbon dioxide (CO_2) as a C1 synthon is a strategy that receives continued interest for direct reduction to fuels and bulk chemicals, as well as for incorporation in more complex organic molecules [103–107]. In the context of asymmetric electrosynthesis, both organocatalytic and metal-catalyzed examples can be found. In similarity with the early asymmetric electroreductive protocols, alkaloids have found their use in asymmetric electroreductive carboxylation of ketones with carbon dioxide. Lu and co-workers utilized catalytic amounts of cinchonidine and cinchonine to form (*R*)-or (*S*)-atrolactic acid with *ee*'s up to 30%, respectively, using a stainless steel cathode and Mg sacrificial anode in an undivided cell and constant current conditions [108]. Furthermore, cinchonine was used in catalytic amounts for the electrocarboxylation of 4-methylpropiophenone to form the (*S*)-configured product (up to 33% *ee*), using a similar experimental setup [109]. A few years later, the same group reported moderate yields and *ee*'s (up to 49% *ee*) in the electrocarboxylation of 2-acetonaphthone under constant current conditions, using 2.5 mol% of cinchonidine with a sacrificial Mg anode and stainless steel cathode in an undivided cell under atmospheric CO_2 pressure at 0 °C [110]. In successive work, the group utilized a metal-catalyzed strategy for the electrocarboxylation of 1-phenylethyl chloride, employing a Co^{II}-(R,R)(salen) complex (15 mol%) [111]. Optically active 2-phenylpropionic acid could be obtained in 37% yield and 83% *ee* under potentiostatic conditions in an undivided cell with glassy carbon cathode and Mg anode at 50 °C (Scheme 13), while no reaction was observed in absence of a Co source. With the support of CV analysis, the transformation was proposed to proceed via one-electron transfer to give an anionic [Co^{I}-(R,R)(salen)] complex, which could react with the substrate to form a Co^{III} organocobalt intermediate. One-electron reduction of this species to a Co^{II} anionic complex followed by direct nucleophilic attack to CO_2 ensures the asymmetric induction. However, if homolytic cleavage of the Co-R bond occurs, the generated [PhCH(CH)$_3$] radical can prompt a competing background reaction (upon reduction to the corresponding anion and attack to CO_2) to give the racemic acid. Detection of styrene by GC-MS and the observation of dimer as major side-product supported the proposed formation of radical intermediates.

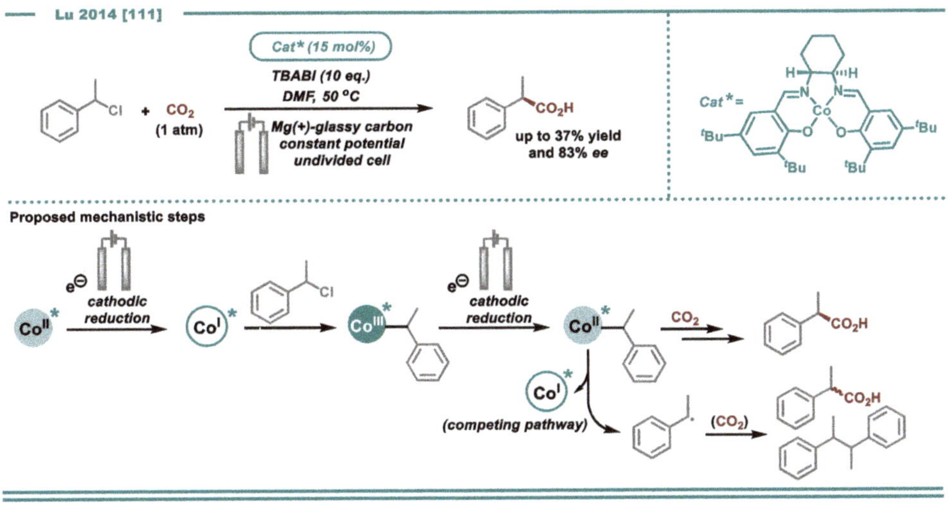

Scheme 13. Cobalt-catalyzed reductive carboxylation of benzylic chlorides.

In 2018, Mei and co-workers developed a Pd-catalyzed electrocarboxylation of allyl esters, providing α-aryl carboxylic acids in high yields and regioselectivity [112]. In the report, the enantioselective variant of the method was developed using $Pd(OAc)_2$ (7.5 mol%) in combination with a chiral bidentate triarylphosphine ligand (8 mol%) under constant current in an undivided cell with a Pt cathode and a Mg sacrificial anode, under 1 atmosphere of CO_2. Moderate yield (66%) and enantioselectivity (67% *ee*) could be obtained (Scheme 14). No reaction was observed when the current was replaced with common chemical reductants as Mn^0 or Zn^0, whereas the reaction proceeded even in the absence of the Pd catalyst, albeit with low conversion. The use of EtOH as an additive was crucial to grant higher yields and regioselectivities, although further insight about its role was not disclosed. Based on CV measurements, the authors proposed that initial reduction of Pd^{II} to Pd^0 followed by oxidative addition of the allyl acetate generates a cationic π-allylpalladium^{II} complex. This species equilibrates to the favored terminal η1-allylpalladium^{II} species, which is reduced at the cathode. The resulting nucleophilic Pd^0 complex reacts with CO_2 to form a magnesium carboxylate salt with Mg^{2+} from the anode, while Pd^0 re-enters the catalytic cycle. Direct attack of the η1-allylpalladium^{II} species to CO_2 was ruled out since the transformation did not occur in absence of electric current.

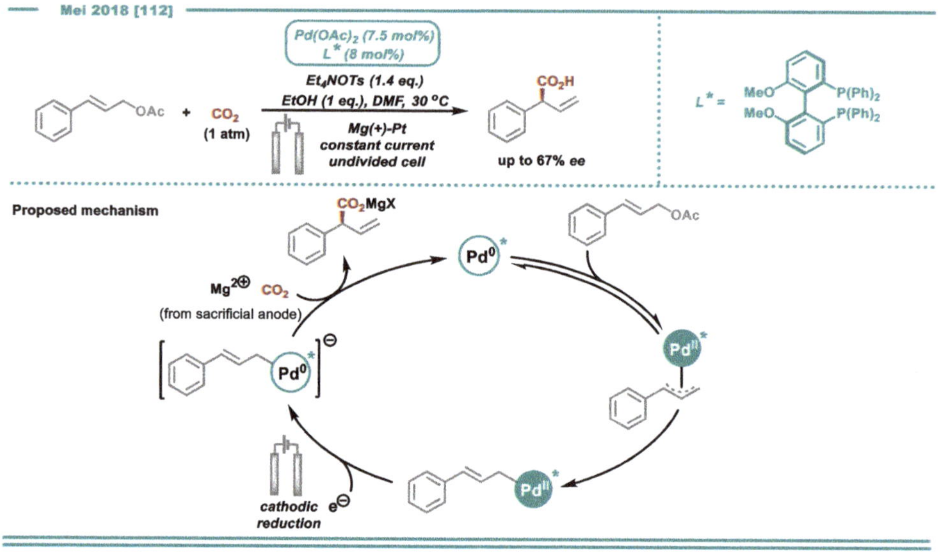

Scheme 14. Cathodically driven Pd-catalyzed carboxylation of cinnamyl acetate.

3.2. Cross-Electrophile Couplings

Reductive dehalogenation is a classic transformation in electrosynthesis [113] that includes stereoselective examples using cobalt catalysis [99,100]. The principle is relevant for transition metal-catalyzed reductive cross-electrophile couplings [114–116], and an enantioselective coupling of alkenyl and benzyl halides using a chiral Ni catalyst was demonstrated by Reisman and co-workers under electrochemical conditions [117]. Electrochemical catalyst activation and turnover granted C(sp^2)–C(sp^3) bond formation without the need for sensitive organometallic reagents or metal powder reductants, and products bearing allyl stereocenters were obtained under mild conditions. $NiCl_2$ dme (10 mol%) was employed as catalyst precursor together with an indanyl-substituted bisoxazoline ligand (20 mol%) in the presence of NaI in an undivided cell with a reticulated vitreous carbon (RVC) cathode and a Zn sacrificial anode at 0 °C, furnishing the products in good yields and high *ee*'s (up to 95%) (Scheme 15). Control experiments confirmed that all components (Ni, ligand, current, additive) were essential for the reaction outcome. Although not specifically discussed in the paper, the cross-coupling has been considered to follow a sequential reduction mechanism [118,119] taking into account results that proved inconsistent with a radical chain mechanism for the non-electrochemically-driven variant of the reaction [120]. In the sequential reduction pathway, a Ni^0 complex would be generated from electroreduction of the Ni^{II} precursor, and the C(sp^2) coupling partner (the alkenyl bromide) would undergo oxidative addition with this Ni^0 species. The resulting Ni^{II} intermediate is reduced at the cathode to a Ni^I intermediate with concomitant loss of halide, while the following oxidative addition (of the racemic benzyl chloride) was defined as the stereoconvergent step. Upon reductive elimination, the Ni^{III} complex is envisioned to liberate the enantioenriched product and forms a Ni^I-Cl species, which is again reduced to Ni^0 to allow for catalyst turnover.

Another asymmetric cross-electrophile coupling was reported by Mei and co-workers to afford enantioselective electrochemical homocoupling of aryl bromides to axially chiral biaryls [121]. By the use of 10 mol% $NiCl_2$·glyme with chiral pyridine-oxazoline ligands in presence of NaI and 4A molecular sieves in an undivided cell equipped with a Ni cathode and a sacrificial Fe anode at 0 °C, yields up to 91% and stereoselectivities up to 96% *ee* were obtained (Scheme 16). Control experiments showed that the reaction does not occur in absence of Ni or current, while Mn^0 or Zn^0 as alternative reductants afforded lower yields and slightly lower *ee*'s. CV analysis indicated that the Ni catalyst is preferentially reduced over the substrate, and that the latter can undergo oxidative addition to Ni^0. The authors suggested a reductive coupling mechanism analogous to that suggested for Reisman's work (Scheme 15), in which cathodic reduction of the initial complex to Ni^0, followed by oxidative addition of the aryl bromide, results in a Ni^{II} complex. Subsequent cathodic reduction to Ni^I and oxidative addition of another molecule of the aryl bromide substrate generates a Ni^{III} intermediate. Reductive elimination releases the biaryl product, while the Ni^I species undergoes further electrochemical reduction to close the catalytic cycle. However, the authors state that other pathways could not be ruled out at this stage.

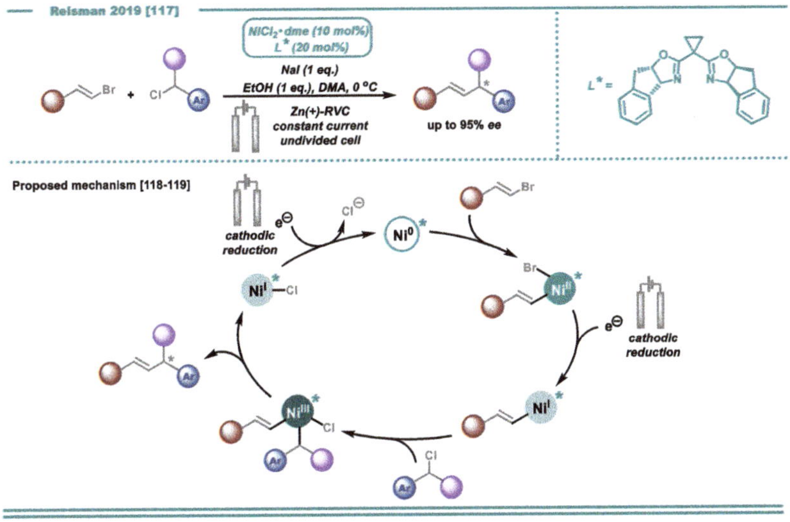

Scheme 15. Electrochemically driven cross-electrophile coupling with Ni-catalysis.

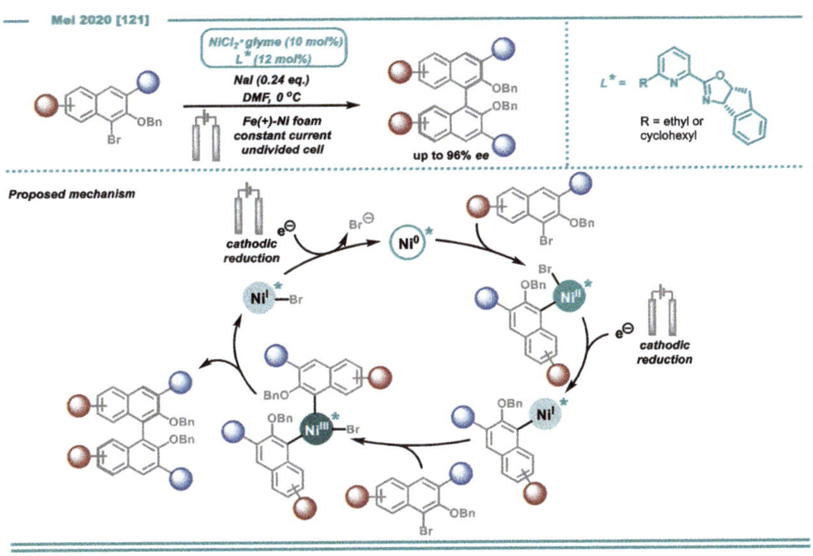

Scheme 16. Reductive cross-electrophile coupling for the formation of axially chiral biaryls using Ni-catalysis under electrochemical conditions.

4. Conclusions and Outlook

Asymmetric catalysis under electrosynthetic conditions is a rapidly expanding research field that has equipped the organic chemist toolbox with new methods for enantioselective synthesis of organic molecules in the last few years, not the least for electrochemically driven metal catalysis. In comparison with the use of pre-functionalized chiral electrodes, chiral electrolytes or solvents or pre-modified substrates with chiral auxiliaries, homogeneous catalysis for the conversion of prochiral substrates appears as an economical, user-friendly and modular approach for asymmetric electrosynthesis [3–6]. It can be noted that the use of electricity to promote enantioselective oxidative processes is currently considerably more explored compared to its use in reductive transformations. As such, further developments on the latter topic can be anticipated.

Electrosynthesis offers the possibility to bypass the use of (super)stoichiometric and commonly hazardous and toxic chemical redox reagents, which results in increased safety and atom efficiency of the processes. On the other hand, large amounts of electrolyte are typically required to provide sufficient conductivity to the organic medium, which clearly hampers atom efficiency. Development of easily recyclable electrolytes, economic ionic liquids or microfluidic systems that make electrolytes superfluous due to the small distance between cathode and anode are interesting possibilities to address this problem that may promote the transition from academic research to industrial processes for the synthesis of fine chemicals [122,123]. Up until now, however, such approaches have not been applied for asymmetric catalysis. Furthermore, the energy efficiency of electrosynthetic processes is not necessarily optimized as more current is typically required to transform starting material into product compared to what theory would suggest. This Faradaic loss may be partly addressed by the use of redox mediators for indirect electrolysis. As redox mediators can allow for redox events to occur at lower potentials (in absolute numbers), this may also enable higher selectivities and functional group tolerance [15]. There is plenty of room for innovation in this field, not the least for reductive transformations, and inspiration for electrosynthetic applications is likely to be found in photoredox catalysis, as well as the proceedings of the organic battery community [124].

It can be noted that chiral amines and derivatives thereof play a key role for asymmetric catalysis in an electrosynthetic setting, either as organocatalysts or as ligands in metal-catalyzed systems. In this light, future development of other chiral inductors will be interesting. For example, chiral hypervalent iodine reagents were recently demonstrated to induce optical activity in organic molecules under electrosynthetic conditions [125]. As electrochemical synthesis of hypervalent halogen compounds has been demonstrated viable [126–129] and the use of halides and halogen compounds as electrochemical redox mediators is already known [40,69,70,130], future developments in this area are anticipated.

Author Contributions: Conceptualization, C.M. and H.L.; writing—original draft preparation, C.M. and H.L.; writing—review and editing, C.M. and H.L.; supervision, H.L.; project administration, H.L.; funding acquisition, H.L. All authors have read and agreed to the published version of the manuscript.

Funding: This research was funded by The Swedish Research Council, grant number 2015-06466, and Stiftelsen Olle Engkvist Byggmästare. Magnus Bergvalls stiftelse C.F. Lundströms stiftelse (The Royal Swedish Academy of Agriculture and Forestry) Stiftelsen Lars Hiertas minne.

Acknowledgments: We gratefully acknowledge Piret Villo for valuable feedback on the manuscript.

Conflicts of Interest: The authors declare no conflict of interest.

References

1. Trost, B.M. Asymmetric catalysis: An enabling science. *Proc. Nat. Acad. Sci. USA* **2004**, *101*, 5348. [CrossRef]
2. Ojima, I. *Catalytic Asymmetric Synthesis*, 3rd ed.; John Wiley & Sons, Inc.: Hoboken, NJ, USA, 2010.
3. Chang, X.; Zhang, Q.; Guo, C. Asymmetric Electrochemical Transformations. *Angew. Chem. Int. Ed.* **2020**, *59*, 12612–12622. [CrossRef] [PubMed]
4. Lin, Q.; Li, L.; Luo, S. Asymmetric Electrochemical Catalysis. *Chem. Eur. J.* **2019**, *25*, 10033–10044. [CrossRef] [PubMed]

5. Ghosh, M.; Shinde, V.S.; Rueping, M. A review of asymmetric synthetic organic electrochemistry and electrocatalysis: Concepts, applications, recent developments and future directions. *Beilstein J. Org. Chem.* **2019**, *15*, 2710–2746. [CrossRef] [PubMed]
6. Ogawa, K.A.; Boydston, A.J. Recent Developments in Organocatalyzed Electroorganic Chemistry. *Chem. Lett.* **2015**, *44*, 10–16. [CrossRef]
7. Wang, K.; Kong, W. Recent Advances in Transition Metal-Catalyzed Asymmetric Radical Reactions. *Chin. J. Chem.* **2018**, *36*, 247–256. [CrossRef]
8. Saha, D. Catalytic Enantioselective Radical Transformations Enabled by Visible Light. *Chem. Asian J.* **2020**, *15*, 2129–2152. [CrossRef]
9. Sibi, M.P.; Manyem, S.; Zimmerman, J. Enantioselective Radical Processes. *Chem. Rev.* **2003**, *103*, 3263–3296. [CrossRef]
10. Jiang, C.; Chen, W.; Zheng, W.-H.; Lu, H. Advances in asymmetric visible-light photocatalysis. *2015–2019, Org. Biomol. Chem.* **2019**, *17*, 8673–8689.
11. Siu, J.C.; Fu, N.; Lin, S. Catalyzing Electrosynthesis: A Homogeneous Electrocatalytic Approach to Reaction Discovery. *Acc. Chem. Res.* **2020**, *53*, 547–560. [CrossRef]
12. Francke, R. Integrating Catalytic Processes and Modern Electrolyte Concepts into Electrosynthesis. *CHIMIA* **2020**, *74*, 49–56. [CrossRef]
13. Kärkäs, M.D. Electrochemical strategies for C–H functionalization and C–N bond formation. *Chem. Soc. Rev.* **2018**, *47*, 5786–5865. [CrossRef] [PubMed]
14. Yan, M.; Kawamata, Y.; Baran, P.S. Synthetic Organic Electrochemical Methods Since 2000: On the Verge of a Renaissance. *Chem. Rev.* **2017**, *117*, 13230–13319. [CrossRef]
15. Francke, R.; Little, R.D. Redox catalysis in organic electrosynthesis: Basic principles and recent developments. *Chem. Soc. Rev.* **2014**, *43*, 2492–2521. [CrossRef] [PubMed]
16. Pollok, D.; Waldvogel, S.R. Electro-organic synthesis—A 21st century technique. *Chem. Sci.* **2020**. [CrossRef]
17. Krištofíková, D.; Modrocká, V.; Mečiarová, M.; Šebesta, R. Green Asymmetric Organocatalysis. *ChemSusChem* **2020**, *13*, 2828–2858. [CrossRef] [PubMed]
18. Atobe, M. Organic electrosynthesis in flow microreactor. *Curr. Opin. Electrochem.* **2017**, *2*, 1–6. [CrossRef]
19. Schäfer, H.J. Contributions of organic electrosynthesis to green chemistry. *Comptes Rendus Chim.* **2011**, *14*, 745–765. [CrossRef]
20. Moeller, K.D. Using Physical Organic Chemistry To Shape the Course of Electrochemical Reactions. *Chem. Rev.* **2018**, *118*, 4817–4833. [CrossRef]
21. Heard, D.; Lennox, A. Electrode Materials in Modern Organic Electrochemistry. *Angew. Chem. Int. Ed.* **2020**, accepted. [CrossRef]
22. Sandford, C.; Edwards, M.A.; Klunder, K.J.; Hickey, D.P.; Li, M.; Barman, K.; Sigman, M.S.; White, H.S.; Minteer, S.D. A synthetic chemist's guide to electroanalytical tools for studying reaction mechanisms. *Chem. Sci.* **2019**, *10*, 6404–6422. [CrossRef] [PubMed]
23. Wu, R.; Ma, C.; Zhu, Z. Enzymatic electrosynthesis as an emerging electrochemical synthesis platform. *Curr. Opin. Electrochem.* **2020**, *19*, 1–7. [CrossRef]
24. Cadoux, C.; Milton, R.D. Recent Enzymatic Electrochemistry for Reductive Reactions. *ChemElectroChem* **2020**, *7*, 1974–1986. [CrossRef]
25. Kohlmann, C.; Märkle, W.; Lütz, S. Electroenzymatic synthesis. *J. Mol. Catal. B Enzym.* **2008**, *51*, 57–72. [CrossRef]
26. Zhu, L.; Wang, D.; Jia, Z.; Lin, Q.; Huang, M.; Luo, S. Catalytic Asymmetric Oxidative Enamine Transformations. *ACS Catal.* **2018**, *8*, 5466–5484. [CrossRef]
27. Bui, N.-N.; Ho, X.-H.; Mho, S.-i.; Jang, H.-Y. Organocatalyzed α-Oxyamination of Aldehydes Using Anodic Oxidation. *Eur. J. Org. Chem.* **2009**, *2009*, 5309–5312. [CrossRef]
28. Ho, X.-H.; Mho, S.-i.; Kang, H.; Jang, H.-Y. Electro-Organocatalysis: Enantioselective α-Alkylation of Aldehydes. *Eur. J. Org. Chem.* **2010**, *2010*, 4436–4441. [CrossRef]
29. Jensen, K.L.; Franke, P.T.; Nielsen, L.T.; Daasbjerg, K.; Jørgensen, K.A. Anodic Oxidation and Organocatalysis: Direct Regio- and Stereoselective Access to meta-Substituted Anilines by α-Arylation of Aldehydes. *Angew. Chem. Int. Ed.* **2010**, *49*, 129–133. [CrossRef]
30. Fu, N.; Li, L.; Yang, Q.; Luo, S. Catalytic Asymmetric Electrochemical Oxidative Coupling of Tertiary Amines with Simple Ketones. *Org. Lett.* **2017**, *19*, 2122–2125. [CrossRef]

31. Osa, T.; Kashiwagi, Y.; Yanagisawa, Y.; Bobbitt, J.M. Enantioselective, electrocatalytic oxidative coupling of naphthol, naphthyl ether and phenanthrol on a TEMPO-modified graphite felt electrode in the presence of (–)-sparteine (TEMPO = 2,2,6,6-tetramethylpiperidin-1-yloxyl). *J. Chem. Soc. Chem. Commun.* **1994**, *21*, 2535–2537. [CrossRef]
32. Kashiwagi, Y.; Yanagisawa, Y.; Kurashima, F.; Anzai, J.-i.; Osa, T.; Bobbitt, J.M. Enantioselective electrocatalytic oxidation of racemic alcohols on a TEMPO-modified graphite felt electrode by use of chiral base (TEMPO = 2,2,6,6-tetramethylpiperidin-1-yloxyl). *Chem. Commun.* **1996**, *24*, 2745–2746. [CrossRef]
33. Yoshinori, Y.; Yoshitomo, K.; Futoshi, K.; Jun-ichi, A.; Tetsuo, O.; Enantioselective, B.J.M. Electrocatalytic Lactonization of Methyl-substituted Diols on a TEMPO-modified Graphite Felt Electrode in the Presence of (−)-Sparteine. *Chem. Lett.* **1996**, *25*, 1043–1044.
34. Kashiwagi, Y.; Kurashima, F.; Kikuchi, C.; Anzai, J.-i.; Osa, T.; Bobbitt, J.M. Enantioselective electrocatalytic oxidation of racemic sec-alcohols using a chiral 1-azaspiro [5.5]undecane-N-oxyl radical. *Tetrahedron Lett.* **1999**, *40*, 6469–6472. [CrossRef]
35. Kashiwagi, Y.; Kurashima, F.; Kikuchi, C.; Anzai, J.-i.; Osa, T.; Bobbitt, J.M. Enantioselective electrocatalytic oxidation of racemic amines using a chiral 1-azaspiro[5.5]undecane N-oxyl radical. *Chem. Commun.* **1999**, 1983–1984. [CrossRef]
36. Kuroboshi, M.; Yoshihisa, H.; Cortona, M.N.; Kawakami, Y.; Gao, Z.; Tanaka, H. Electro-oxidative kinetic resolution of sec-alcohols by using an optically active N-oxyl mediator. *Tetrahedron Lett.* **2000**, *41*, 8131–8135. [CrossRef]
37. Tanaka, H.; Kawakami, Y.; Goto, K.; Kuroboshi, M. An aqueous silica gel disperse electrolysis system. N-Oxyl-mediated electrooxidation of alcohols. *Tetrahedron Lett.* **2001**, *42*, 445–448. [CrossRef]
38. Shiigi, H.; Mori, H.; Tanaka, T.; Demizu, Y.; Onomura, O. Chiral azabicyclo-N-oxyls mediated enantioselective electrooxidation of sec-alcohols. *Tetrahedron Lett.* **2008**, *49*, 5247–5251. [CrossRef]
39. Wang, F.; Stahl, S.S. Electrochemical Oxidation of Organic Molecules at Lower Overpotential: Accessing Broader Functional Group Compatibility with Electron−Proton Transfer Mediators. *Acc. Chem. Res.* **2020**, *53*, 561–574. [CrossRef]
40. Minato, D.; Arimoto, H.; Nagasue, Y.; Demizu, Y.; Onomura, O. Asymmetric electrochemical oxidation of 1,2-diols, aminoalcohols, and aminoaldehydes in the presence of chiral copper catalyst. *Tetrahedron* **2008**, *64*, 6675–6683. [CrossRef]
41. Gao, P.-S.; Weng, X.-J.; Wang, Z.-H.; Zheng, C.; Sun, B.; Chen, Z.-H.; You, S.-L.; Mei, T.-S. Cu^{II}/TEMPO-Catalyzed Enantioselective C(sp^3)–H Alkynylation of Tertiary Cyclic Amines through Shono-Type Oxidation. *Angew. Chem. Int. Ed.* **2020**, *59*, 15254–15259. [CrossRef]
42. Shono, T.; Hamaguchi, H.; Matsumura, Y. Electroorganic chemistry. XX. Anodic oxidation of carbamates. *J. Am. Chem. Soc.* **1975**, *97*, 4264–4268. [CrossRef]
43. Lu, F.-Y.; Chen, Y.-J.; Chen, Y.; Ding, X.; Guan, Z.; He, Y.-H. Highly enantioselective electrosynthesis of C2-quaternary indolin-3-ones. *Chem. Commun.* **2020**, *56*, 623–626. [CrossRef] [PubMed]
44. Mukherjee, S.; Yang, J.W.; Hoffmann, S.; List, B. Asymmetric Enamine Catalysis. *Chem. Rev.* **2007**, *107*, 5471–5569. [CrossRef] [PubMed]
45. Li, L.; Li, Y.; Fu, N.; Zhang, L.; Luo, S. Catalytic Asymmetric Electrochemical α-Arylation of Cyclic β-Ketocarbonyls with Anodic Benzyne Intermediates. *Angew. Chem. Int. Ed.* **2020**, *59*, 14347–14351. [CrossRef] [PubMed]
46. Campbell, C.D.; Rees, C.W. Reactive intermediates. Part I. Synthesis and oxidation of 1- and 2-aminobenzotriazole. *J. Chem. Soc. C* **1969**, 742–747. [CrossRef]
47. Kato, H.; Nakazawa, S.; Kiyosawa, T.; Hirakawa, K. Heterocycles by cycloaddition. Part II. Cycloaddition–extrusion reactions of five-membered mesoionic compounds with benzyne: Prepartion of benz[*c*]azole and benzo[*c*]thiophen derivatives. *J. Chem. Soc. Perkin Trans. 1* **1976**, 672–675. [CrossRef]
48. Rigby, J.H.; Holsworth, D.D.; James, K. Vinyl isocyanates in synthesis. [4 + 2] Cycloaddition reactions with benzyne addends. *J. Org. Chem.* **1989**, *54*, 4019–4020. [CrossRef]
49. Sakurai, H.; Sakaba, H.; Nakadaira, Y. Facile preparation of 2,3-benzo-1,4-diphenyl-7-silanorbornadiene derivatives and the first clear evidence of silylene-to-disilene thermal rearrangement. *J. Am. Chem. Soc.* **1982**, *104*, 6156–6158. [CrossRef]
50. Cresp, T.M.; Wege, D. The addition of benzyne to azulene. *Tetrahedron* **1986**, *42*, 6713–6718. [CrossRef]

51. Huang, X.; Zhang, Q.; Lin, J.; Harms, K.; Meggers, E. Electricity-driven asymmetric Lewis acid catalysis. *Nat. Catal.* **2019**, *2*, 34–40. [CrossRef]
52. Zhang, Q.; Chang, X.; Peng, L.; Guo, C. Asymmetric Lewis Acid Catalyzed Electrochemical Alkylation. *Angew. Chem. Int. Ed.* **2019**, *58*, 6999–7003. [CrossRef] [PubMed]
53. Jeffrey, J.L.; Terrett, J.A.; MacMillan, D.W.C. O–H hydrogen bonding promotes H-atom transfer from α C–H bonds for C-alkylation of alcohols. *Science* **2015**, *349*, 1532–1536. [CrossRef] [PubMed]
54. Shaw, M.H.; Shurtleff, V.W.; Terrett, J.A.; Cuthbertson, J.D.; MacMillan, D.W.C. Native functionality in triple catalytic cross-coupling: sp^3 C–H bonds as latent nucleophiles. *Science* **2016**, *352*, 1304. [CrossRef] [PubMed]
55. Le, C.; Liang, Y.; Evans, R.W.; Li, X.; MacMillan, D.W.C. Selective sp ^{3} C-H alkylation via polarity-match-based cross-coupling. *Nature* **2017**, *547*, 79–83. [CrossRef]
56. Zhang, X.; MacMillan, D.W.C. Direct aldehyde C–H arylation and alkylation via the combination of nickel, hydrogen atom transfer, and photoredox catalysis. *J. Am. Chem. Soc.* **2017**, *139*, 11353–11356. [CrossRef]
57. Twilton, J.; Christensen, M.; DiRocco, D.A.; Ruck, R.T.; Davies, I.W.; MacMillan, D.W.C. Selective Hydrogen Atom Abstraction through Induced Bond Polarization: Direct a-Arylation of Alcohols through Photoredox, HAT, and Nickel Catalysis. *Angew. Chem. Int. Ed.* **2018**, *57*, 5369–5373. [CrossRef] [PubMed]
58. Yang, H.-B.; Feceu, A.; Martin, D.B.C. Catalyst-Controlled C–H Functionalization of Adamantanes Using Selective H-Atom Transfer. *ACS Catal.* **2019**, *9*, 5708–5715. [CrossRef]
59. Chang, X.; Zhang, J.; Zhang, Q.; Guo, C. Merging Electrosynthesis and Bifunctional Squaramide Catalysis in the Asymmetric Detrifluoroacetylative Alkylation Reactions. *Angew. Chem. Int. Ed.* **2020**, accepted. [CrossRef]
60. Dhawa, U.; Tian, C.; Wdowik, T.; Oliveira, J.C.A.; Hao, J.; Ackermann, L. Enantioselective Palladaelectro-Catalyzed C–H Activations by Transient Directing Groups: Expedient Access to Helicenes. *Angew. Chem. Int. Ed.* **2020**, *59*, 13451–13457. [CrossRef]
61. Hentges, S.G.; Sharpless, K.B. Asymmetric induction in the reaction of osmium tetroxide with olefins. *J. Am. Chem. Soc.* **1980**, *102*, 4263–4265. [CrossRef]
62. Sharpless, K.B.; Amberg, W.; Bennani, Y.L.; Crispino, G.A.; Hartung, J.; Jeong, K.S.; Kwong, H.L.; Morikawa, K.; Wang, Z.M. The osmium-catalyzed asymmetric dihydroxylation: A new ligand class and a process improvement. *J. Org. Chem.* **1992**, *57*, 2768–2771. [CrossRef]
63. Xu, D.; Crispino, G.A.; Sharpless, K.B. Selective asymmetric dihydroxylation (AD) of dienes. *J. Am. Chem. Soc.* **1992**, *114*, 7570–7571. [CrossRef]
64. Hoi-Lun, K.; Sorato, C.; Ogino, Y.; Hou, C.; Sharpless, K.B. Preclusion of the "second cycle" in the osmium-catalyzed asymmetric dihydroxylation of olefins leads to a superior process. *Tetrahedron Lett.* **1990**, *31*, 2999–3002. [CrossRef]
65. Minato, M.; Yamamoto, K.; Tsuji, J. Osmium tetraoxide catalyzed vicinal hydroxylation of higher olefins by using hexacyanoferrate(III) ion as a cooxidant. *J. Org. Chem.* **1990**, *55*, 766–768. [CrossRef]
66. Ogino, Y.; Chen, H.; Kwong, H.-L.; Sharpless, K.B. On the timing of hydrolysis/reoxidation in the osmium-catalyzed asymmetric dihydroxylation of olefins using potassium ferricyanide as the reoxidant. *Tetrahedron Lett.* **1991**, *32*, 3965–3968. [CrossRef]
67. Amundsen, A.R.; Balko, E.N. Preparation of chiral diols by the osmium-catalysed, indirect anodic oxidation of olefins. *J. Appl. Electrochem.* **1992**, *22*, 810–816. [CrossRef]
68. Torii, S.; Liu, P.; Tanaka, H. Electrochemical Os-Catalyzed Asymmetric Dihydroxylation of Olefins with Sharpless' Ligand. *Chem. Lett.* **1995**, *24*, 319–320. [CrossRef]
69. Torii, S.; Liu, P.; Bhuvaneswari, N.; Amatore, C.; Jutand, A. Chemical and Electrochemical Asymmetric Dihydroxylation of Olefins in $I_2-K_2CO_3-K_2OsO_2(OH)_4$ and $I_2-K_3PO_4/K_2HPO_4-K_2OsO_2(OH)_4$ Systems with Sharpless' Ligand. *J. Org. Chem.* **1996**, *61*, 3055–3060. [CrossRef]
70. Tanaka, H.; Kuroboshi, M.; Takeda, H.; Kanda, H.; Torii, S. Electrochemical asymmetric epoxidation of olefins by using an optically active Mn-salen complex. *J. Electroanal. Chem.* **2001**, *507*, 75–81. [CrossRef]
71. Irie, R.; Noda, K.; Ito, Y.; Matsumoto, N.; Katsuki, T. Catalytic asymmetric epoxidation of unfunctionalized olefins. *Tetrahedron Lett.* **1990**, *31*, 7345–7348. [CrossRef]
72. Zhang, W.; Loebach, J.L.; Wilson, S.R.; Jacobsen, E.N. Enantioselective epoxidation of unfunctionalized olefins catalyzed by salen manganese complexes. *J. Am. Chem. Soc.* **1990**, *112*, 2801–2803. [CrossRef]
73. Katsuki, T. Mn-salen catalyst, competitor of enzymes, for asymmetric epoxidation. *J. Mol. Catal. A* **1996**, *113*, 87–107. [CrossRef]

74. Page, P.C.B.; Marken, F.; Williamson, C.; Chan, Y.; Buckley, B.R.; Bethell, D. Enantioselective Organocatalytic Epoxidation Driven by Electrochemically Generated Percarbonate and Persulfate. *Adv. Synth. Catal.* **2008**, *350*, 1149–1154. [CrossRef]
75. Nguyen, B.H.; Perkins, R.J.; Smith, J.A.; Moeller, K.D. Photovoltaic-driven organic electrosynthesis and efforts toward more sustainable oxidation reactions. *Beilstein J. Org. Chem.* **2015**, *11*, 280–287. [CrossRef]
76. Nguyen, B.H.; Redden, A.; Moeller, K.D. Sunlight, electrochemistry, and sustainable oxidation reactions. *Green Chem.* **2014**, *16*, 69–72. [CrossRef]
77. Fu, N.; Song, L.; Liu, J.; Shen, Y.; Siu, J.C.; Lin, S. New Bisoxazoline Ligands Enable Enantioselective Electrocatalytic Cyanofunctionalization of Vinylarenes. *J. Am. Chem. Soc.* **2019**, *141*, 14480–14485. [CrossRef]
78. Song, L.; Fu, N.; Ernst, B.G.; Lee, W.H.; Frederick, M.O.; DiStasio, R.A.; Lin, S. Dual Electrocatalysis Enables Enantioselective Hydrocyanation of Conjugated Alkenes. *Nat. Chem.* **2020**, *12*, 747–754. [CrossRef]
79. Gourley, R.N.; Grimshaw, J.; Millar, P.G. Electrochemical reduction in the presence of tertiary amines: An asymmetric synthesis of 3, 4-dihydro-4-methylcoumarin. *Chem. Commun.* **1967**, 1278–1279. [CrossRef]
80. Gourley, R.N.; Grimshaw, J.; Millar, P.G. Electrochemical reactions. Part VIII. Asymmetric induction during the reduction of coumarins modified by the presence of tertiary amines. *J. Chem. Soc. C* **1970**, 2318–2323. [CrossRef]
81. Höweler, U.; Schoo, N.; Schäfer, H.-J. Enantioselective cathodic reduction of 4-substituted coumarins with alkaloids as catalysts, 2. AM1 and force-field study of the transition-state model. *Liebigs Ann. Chem.* **1993**, *1993*, 609–614. [CrossRef]
82. Schoo, N.; Schäfer, H.-J. Electroorganic Synthesis, 54. Enantioselective Cathodic Reduction of 4-Substituted Coumarins with Alkaloids as Catalysts, 1. *Liebigs Ann. Chem.* **1993**, *1993*, 601–607. [CrossRef]
83. Nielsen, M.F.; Batanero, B.; Löhl, T.; Schäfer, H.J.; Würthwein, E.-U.; Fröhlich, R. Enantioselective Cathodic Reduction of 4-Methylcoumarin: Dependence of Selectivity on Reaction Conditions and Investigation of the Mechanism. *Chem. Eur. J.* **1997**, *3*, 2011–2024. [CrossRef]
84. Kariv, E.; Terni, H.A.; Gileadi, E. The role of quinidine in induction of asymmetric synthesis at mercury cathodes. *Electrochim. Acta* **1973**, *18*, 433–441. [CrossRef]
85. Chen, B.-L.; Xiao, Y.; Xu, X.-M.; Yang, H.-P.; Wang, H.; Lu, J.-X. Alkaloid induced enantioselective electroreduction of acetophenone. *Electrochim. Acta* **2013**, *107*, 320–326. [CrossRef]
86. Kariv, E.; Terni, H.A.; Gileadi, E. Asymmetric Induction by Alkaloids in Electrolytic Reductions. *J. Electrochem. Soc.* **1973**, *120*, 639. [CrossRef]
87. Hermolin, J.; Kopilov, J.; Gileadi, E. Asymmetric reduction of 2- and 4-acetylpyridine in the presence of strychnine. *J. Electroanal. Chem.* **1976**, *71*, 245–248. [CrossRef]
88. Kopilov, J.; Shatzmiller, S.; Kariv, E. Asymmetric induction in reduction of acetyl pyridines at a mercury cathode. *Electrochim. Acta* **1976**, *21*, 535–536. [CrossRef]
89. Kopilov, J.; Kariv, E.; Miller, L.L. Asymmetric, cathodic reduction of acetylpyridines. *J. Am. Chem. Soc.* **1977**, *99*, 3450–3454. [CrossRef]
90. Yadav, A.K.; Singh, A. Enantioselective Cathodic Reduction of Some Prochiral Ketones in the Presence of (1R,2S)-(-)-N,N-Dimethylephedrinium Tetrafluoroborate at a Mercury Pool Cathode. *Bull. Chem. Soc. Jpn.* **2002**, *75*, 587–588. [CrossRef]
91. Yadav, A.K.; Manju, M.P.; Chhinpa, R. Enantioselective cathodic reduction of some prochiral ketones in the presence of (−)-N,N′-dimethylquininium tetrafluoroborate at mercury cathode. *Tetrahedron Asymmetry* **2003**, *14*, 1079–1081. [CrossRef]
92. Vago, M.; Williams, F.J.; Calvo, E.J. Enantioselective electrocatalytic hydrogenation of ethyl pyruvate on carbon supported Pd electrodes. *Electrochem. Commun.* **2007**, *9*, 2725–2728. [CrossRef]
93. Jubault, M.; Raoult, E.; Armand, J.; Boulares, L. Effect of cathodic potential on the electrochemical synthesis of optically active amino-acids. *J. Chem. Soc. Chem. Commun.* **1977**, 250–251. [CrossRef]
94. Jubault, M.; Raoult, E.; Peltier, D. Preprotonation of the substrate by protonated alkaloid in asymmetric electrosynthesis. *J. Chem. Soc. Chem. Commun.* **1979**, 232–233. [CrossRef]
95. Jubault, M. Effect of alkaloid concentration in asymmetric electrosynthesis. *J. Chem. Soc. Chem. Commun.* **1980**, 953–954. [CrossRef]
96. Park, J.W.; Choi, M.H.; Park, K.K. Electrocatalytic reduction of benzoylformic acid mediated by methyl viologen. *Tetrahedron Lett.* **1995**, *36*, 2637–2638. [CrossRef]

97. Hazard, R.; Jaouannet, S.; Tallec, A. Stereochemistry of electroreductions of bromocyclopropanes: 1-asymmetric electrochemical synthesis by reduction at a mercury cathode in the presence of adsorbed alkaloids. *Tetrahedron* **1982**, *38*, 93–102. [CrossRef]
98. Moutet, J.-C.; Cho, L.Y.; Duboc-Toia, C.; Ménage, S.; Riesgo, E.C.; Thummel, R.P. Heterogeneous and homogeneous asymmetric electrocatalytic hydrogenation with rhodium(III) complexes containing chiral polypyridyl ligands. *New J. Chem.* **1999**, *23*, 939–944. [CrossRef]
99. Ohno, T.; Nishioka, T.; Hisaeda, Y.; Murakami, Y. Hydrophobic vitamin B12: Part 13. Asymmetric reaction of hydrophobic vitamin B12 under electrochemical conditions and rationalization of enantioselectivity based on conformational analysis. *J. Mol. Struct.* **1994**, *308*, 207–218. [CrossRef]
100. Hisaeda, Y.; Nishioka, T.; Inoue, Y.; Asada, K.; Hayashi, T. Electrochemical reactions mediated by vitamin B12 derivatives in organic solvents. *Coord. Chem. Rev.* **2000**, *198*, 21–37. [CrossRef]
101. Su, H.; Walder, L.; Zhang, Z.-d.; Scheffold, R. Asymmetric Catalysis by Vitamin B_{12}. The isomerization of achiral epoxides to optically active allylic alcohols. *Helv. Chim. Acta* **1988**, *71*, 1073–1078. [CrossRef]
102. Franco, D.; Riahi, A.; Hénin, F.; Muzart, J.; Duñach, E. Electrochemical Reduction of a Racemic Allyl β-Keto Ester Catalyzed by Nickel Complexes: Asymmetric Induction. *Eur. J. Org. Chem.* **2002**, *2002*, 2257–2259. [CrossRef]
103. Hori, Y. *Modern Aspects of Electrochemistry*; Vayenas, C.G., White, R.E., Gamboa-Aldeco, M.E., Eds.; Springer: New York, NY, USA, 2008; pp. 89–189.
104. Liu, Y.; Li, F.; Zhang, X.; Ji, X. Recent progress on electrochemical reduction of CO_2 to methanol. *Curr. Opin. Green Sustain. Chem.* **2020**, *23*, 10–17. [CrossRef]
105. Nitopi, S.; Bertheussen, E.; Scott, S.B.; Liu, X.; Engstfeld, A.K.; Horch, S.; Seger, B.; Stephens, I.E.L.; Chan, K.; Hahn, C.; et al. Progress and Perspectives of Electrochemical CO2 Reduction on Copper in Aqueous Electrolyte. *Chem. Rev.* **2019**, *119*, 7610–7672. [CrossRef] [PubMed]
106. Lv, J.-J.; Jouny, M.; Luc, W.; Zhu, W.; Zhu, J.-J.; Jiao, F. A Highly Porous Copper Electrocatalyst for Carbon Dioxide Reduction. *Adv. Mater.* **2018**, *30*, 1803111. [CrossRef] [PubMed]
107. Zhu, W.; Kattel, S.; Jiao, F.; Chen, J.G. Shape-Controlled CO_2 Electrochemical Reduction on Nanosized Pd Hydride Cubes and Octahedra. *Adv. Energy Mater.* **2019**, *9*, 1802840. [CrossRef]
108. Zhang, K.; Wang, H.; Zhao, S.-F.; Niu, D.-F.; Lu, J.-X. Asymmetric electrochemical carboxylation of prochiral acetophenone: An efficient route to optically active atrolactic acid via selective fixation of carbon dioxid. *J. Electroanal. Chem.* **2009**, *630*, 35–41. [CrossRef]
109. Zhao, S.-F.; Zhu, M.-X.; Zhang, K.; Wang, H.; Lu, J.-X. Alkaloid induced asymmetric electrocarboxylation of 4-methylpropiophenone. *Tetrahedron Lett.* **2011**, *52*, 2702–2705. [CrossRef]
110. Chen, B.-L.; Tu, Z.-Y.; Zhu, H.-W.; Sun, W.-W.; Wang, H.; Lu, J.-X. CO_2 as a C1-organic building block: Enantioselective electrocarboxylation of aromatic ketones with CO_2 catalyzed by cinchona alkaloids under mild conditions. *Electrochim. Acta* **2014**, *116*, 475–483. [CrossRef]
111. Chen, B.-L.; Zhu, H.-W.; Xiao, Y.; Sun, Q.-L.; Wang, H.; Lu, J.-X. Asymmetric electrocarboxylation of 1-phenylethyl chloride catalyzed by electrogenerated chiral [CoI(salen)]−complex. *Electrochem. Commun.* **2014**, *42*, 55–59. [CrossRef]
112. Jiao, K.-J.; Li, Z.-M.; Xu, X.-T.; Zhang, L.-P.; Li, Y.-Q.; Zhang, K.; Mei, T.-S. Palladium-catalyzed reductive electrocarboxylation of allyl esters with carbon dioxide. *Org. Chem. Front.* **2018**, *5*, 2244–2248. [CrossRef]
113. Shatskiy, A.; Lundberg, H.; Kärkäs, M.D. Organic Electrosynthesis: Applications in Complex Molecule Synthesis. *ChemElectroChem* **2019**, *6*, 4067–4092. [CrossRef]
114. Everson, D.A.; Weix, D.J. Cross-Electrophile Coupling: Principles of Reactivity and Selectivity. *J. Org. Chem.* **2014**, *79*, 4793–4798. [CrossRef] [PubMed]
115. Goldfogel, M.J.; Huang, L.; Weix, D.J. Cross-Electrophile Coupling: Principles and New Reactions. In *Nickel Catalysis in Synthesis: Methods and Reactions*; Ogoshi, S., Ed.; Wiley-VCH: Weinheim, Germany, 2020; Volume 352.
116. Truesdell, B.L.; Hamby, T.B.; Sevov, C.S. General C(sp2)–C(sp3) Cross-Electrophile Coupling Reactions Enabled by Overcharge Protection of Homogeneous Electrocatalysts. *J. Am. Chem. Soc.* **2020**, *142*, 5884–5893. [CrossRef] [PubMed]
117. DeLano, T.J.; Reisman, S.E. Enantioselective Electroreductive Coupling of Alkenyl and Benzyl Halides via Nickel Catalysis. *ACS Catal.* **2019**, *9*, 6751–6754. [CrossRef]

118. Zhang, S.-K.; Samanta, R.C.; del Vecchio, A.; Ackermann, L. Evolution of High-Valent Nickela-Electrocatalyzed C−H Activation: From Cross(-Electrophile)-Couplings to Electrooxidative C−H Transformations. *Chem. Eur. J.* **2020**, *26*, 10936–10947. [CrossRef]
119. Poremba, K.E.; Dibrell, S.E.; Reisman, S.E. Nickel-Catalyzed Enantioselective Reductive Cross-Coupling Reactions. *ACS Catal.* **2020**, *10*, 8237–8246. [CrossRef]
120. Cherney, A.H.; Reisman, S.E. Nickel-Catalyzed Asymmetric Reductive Cross-Coupling Between Vinyl and Benzyl Electrophiles. *J. Am. Chem. Soc.* **2014**, *136*, 14365–14368. [CrossRef]
121. Qiu, H.; Shuai, B.; Wang, Y.-Z.; Liu, D.; Chen, Y.-G.; Gao, P.-S.; Ma, H.-X.; Chen, S.; Mei, T.-S. Enantioselective Ni-Catalyzed Electrochemical Synthesis of Biaryl Atropisomers. *J. Am. Chem. Soc.* **2020**, *142*, 9872–9878. [CrossRef]
122. Mo, Y.; Lu, Z.; Rughoobur, G.; Patil, P.; Gershenfeld, N.; Akinwande, A.I.; Buchwald, S.L.; Jensen, K.F. Microfluidic electrochemistry for single-electron transfer redox-neutral reactions. *Science* **2020**, *368*, 1352–1357. [CrossRef]
123. Leech, M.C.; Garcia, A.D.; Petti, A.; Dobbs, A.P.; Lam, K. Organic electrosynthesis: From academia to industry. *React. Chem. Eng.* **2020**, *5*, 977–990. [CrossRef]
124. Anson, C.W.; Stahl, S.S. Mediated Fuel Cells: Soluble Redox Mediators and Their Applications to Electrochemical Reduction of O2 and Oxidation of H2, Alcohols, Biomass, and Complex Fuels. *Chem. Rev.* **2020**, *120*, 3749–3786. [CrossRef] [PubMed]
125. Gao, W.-C.; Xiong, Z.-Y.; Pirhaghani, S.; Wirth, T. Enantioselective Electrochemical Lactonization Using Chiral Iodoarenes as Mediators. *Synthesis* **2019**, *51*, 276–284. [CrossRef]
126. Broese, T.; Francke, R. Electrosynthesis Using a Recyclable Mediator–Electrolyte System Based on Ionically Tagged Phenyl Iodide and 1,1,1,3,3,3-Hexafluoroisopropanol. *Org. Lett.* **2016**, *18*, 5896–5899. [CrossRef] [PubMed]
127. Doobary, S.; Sedikides, A.T.; Caldora, H.P.; Poole, D.L.; Lennox, A.J.J. Electrochemical Vicinal Difluorination of Alkenes: Scalable and Amenable to Electron-Rich Substrates. *Angew. Chem. Int. Ed.* **2020**, *59*, 1155–1160. [CrossRef] [PubMed]
128. Francke, R. Electrogenerated hypervalent iodine compounds as mediators in organic synthesis. *Curr. Opin. Electrochem.* **2019**, *15*, 83–88. [CrossRef]
129. Massignan, L.; Tan, X.; Meyer, T.H.; Kuniyil, R.; Messinis, A.M.; Ackermann, L. C−H Oxygenation Reactions Enabled by Dual Catalysis with Electrogenerated Hypervalent Iodine Species and Ruthenium Complexes. *Angew. Chem. Int. Ed.* **2020**, *59*, 3184–3189. [CrossRef]
130. Liu, K.; Song, C.; Lei, A. Recent advances in iodine mediated electrochemical oxidative cross-coupling. *Org. Biomol. Chem.* **2018**, *16*, 2375–2387. [CrossRef]

catalysts

MDPI

Review

Asymmetric Ring-Opening of Epoxides Catalyzed by Metal–Salen Complexes

Anna Lidskog †, Yutang Li † and Kenneth Wärnmark *

Centre for Analysis and Synthesis, Lund University, P.O Box 124, SE-22100 Lund, Sweden; Anna.Lidskog@chem.lu.se (A.L.); Yutang.Li@chem.lu.se (Y.L.)
* Correspondence: kenneth.warnmark@chem.lu.se; Tel.: +46-46-222-8217
† These authors equally contributed to this work.

Received: 4 June 2020; Accepted: 18 June 2020; Published: 23 June 2020

Abstract: The asymmetric ring-opening of epoxides is an important reaction in organic synthesis, since it allows for the enantioselective installation of two vicinal functional groups with specific stereochemistry within one step from a highly available starting material. An effective class of catalysts for the asymmetric ring-opening of epoxides is metal–salen complexes. This review summarizes the development of metal–salen catalyzed enantioselective desymmetrization of *meso*-epoxides and kinetic resolution of epoxides with various nucleophiles, including the design and application of both homogeneous- and heterogeneous epoxide-opening catalysts as well as multi-metallic covalent and supramolecular catalytic systems.

Keywords: asymmetric catalysis; salen complexes; ring-opening; epoxide; kinetic resolution

1. Introduction

The availability of enantiomerically pure or enriched compounds is crucial for several different fields of chemistry including pharmaceutical, biological, agricultural, and materials chemistry [1,2]. The high demand for chiral compounds has inspired extensive research into the development of practical and efficient methods for their preparation. One of the main tools for obtaining enantiomerically pure compounds is asymmetric catalysis. In the field of asymmetric catalysis, enantiomerically pure catalysts are employed to transform prochiral or racemic substrates into valuable enantioenriched compounds.

Chiral metal–salen complexes are privileged catalysts, meaning that they are demonstrating enantioselectivity over a wide range of reactions and substrates [3,4]. Salen ligands have attracted much attention due to the fact of their tunable steric and electronic properties as well as their ability to coordinate a large number of different metal ions and stabilize them in various oxidation states [5,6]. In addition, the commercial availability of starting materials, such as enantiomerically pure vicinal diamines, and well-established synthetic procedures allows for the facile preparation of a wide variety of enantiomerically pure salen ligands and their corresponding metal complexes in high yields (Scheme 1).

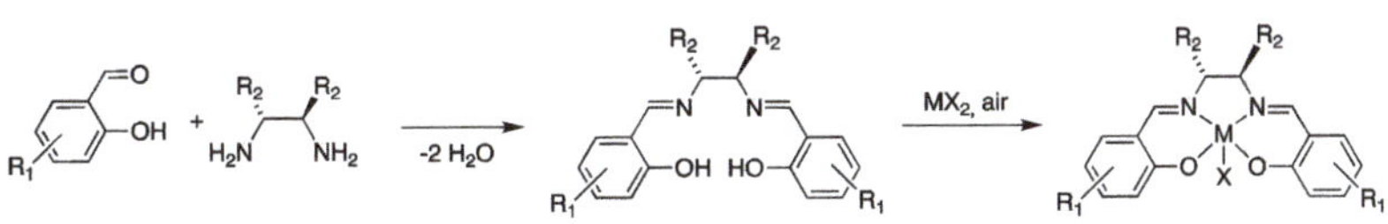

Scheme 1. The synthesis of enantiomerically pure salen ligands and corresponding metal–salen complexes. The ligand synthesis is a Schiff base reaction. The formation of the metal–salen complex is illustrated as the insertion of the metal in oxidation state +II followed by oxidation in air to oxidation state +III.

Epoxides represent an important class of compounds in organic synthesis due to the fact of their high availability and facile, stereoselective, and often regioselective nucleophilic ring-opening, leading to multifunctional organic compounds in very few steps [7]. Achiral and racemic epoxides can be easily prepared by the oxidation of simple alkene precursors, and much effort has also been focused on the development of enantioselective epoxidations [8].

The asymmetric ring-opening (ARO) of epoxides affords enantiopure vicinal difunctionalized organic compounds with two adjacent stereocenters [9,10]. The relative stereochemistry of two stereocenters in the ring-opened product depends on the configuration of the epoxide (Scheme 2). The use of chiral Lewis acids, such as metal–salen complexes, has been shown to significantly increase the reactivity and enantioselectivity of ARO reactions. An alternative way of obtaining the same product would be to first perform an enantioselective epoxidation, followed by selective ring-opening of the enantiomerically pure epoxide.

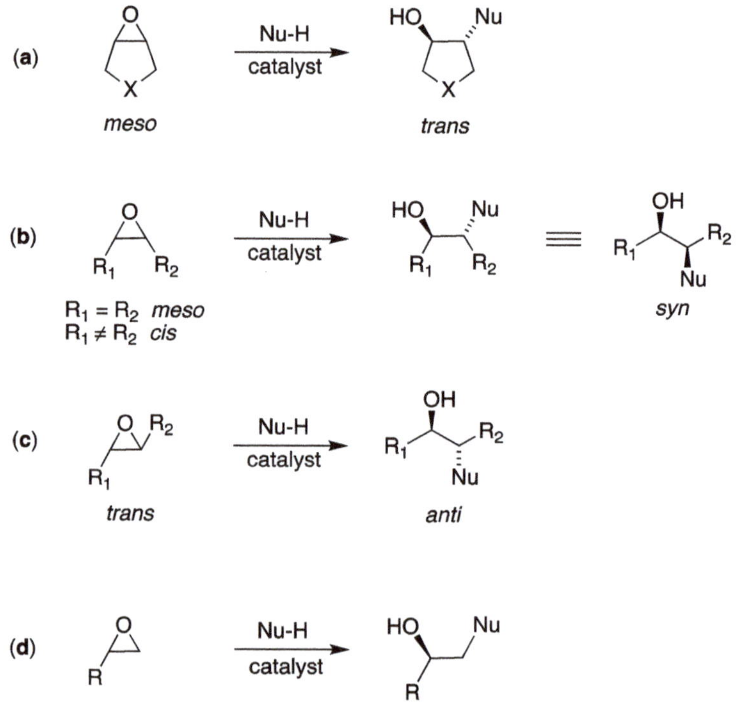

Scheme 2. The asymmetric ring-opening of epoxides and the resulting relative stereochemistry of the ring-opened products. Ring-opening of (**a**) cyclic *meso*-epoxides to yield a ring-opened product with *trans*-stereochemistry; (**b**) acyclic *meso*-epoxides and *cis*-epoxides to yield ring-opened product with *syn*-stereochemistry; (**c**) *trans*-epoxides to yield ring-opened product with *anti*-stereochemistry; (**d**) terminal epoxides.

There are two main types of epoxides that have been used in ARO reactions: achiral (*meso*) epoxides and racemic epoxides. In the former case, the desymmetrization of *meso*-epoxides with a suitable optically pure catalyst leads to the formation of chiral vicinally substituted alcohols with *trans*- (for cyclic *meso*-epoxides) and *syn* stereochemistry (for acyclic *meso*-epoxides) in up to 100% yield and enantiomeric excess. In the latter case, the catalyst has to be able to differentiate between the two enantiomers and preferentially transform one of them into the ring-opened product with high regioselectivity and enantioselectivity. This process is a kinetic resolution (KR). While the maximum yield of the ring-opened product is only 50%, this method also allows for the simultaneous enantioenrichment of the unreacted epoxide, enabling, for example, the preparation of optically enriched terminal epoxides.

In this review, we present an overview of the development and application of metal–salen complexes as catalysts for the ARO of epoxides. Previous reviews in this area have either been focused on the ARO of epoxides in general and covered a number of different salen and non-salen catalysts [9,10] or focused on the use of metal–salen complexes as catalysts for several different reactions [4]. There are also a number of reviews focused on multi-metallic salen complexes [11–13] and heterogeneous salen complexes [14,15] but, again, covering several different types of reactions. As several of the mentioned reviews were published 5–15 years ago, we here provide an up-to-date and comprehensive overview of the field, covering the literature until the beginning of 2020.

2. Desymmetrization of *meso*-Epoxides

2.1. With Azides

The enantioselective ring-opening of epoxides with azides is a very important reaction due to the facile conversion of the products into valuable vicinal amino alcohols of very high optical purity (Scheme 3) [16–19]. The first report of a metal–salen catalyst used for an ARO reaction came from Jacobsen and coworkers [20] in 1995 with the use of Cr(III)salen complex **1** for the ARO of epoxides with trimethylsilylazide ($TMSN_3$) as nucleophile (Scheme 4). Since then, considerable efforts have been invested into improving this catalytic system, including developing multi-metallic catalysts and employing different heterogenization strategies. The Cr(III)salen complexes remain the most investigated catalysts for the ARO of epoxides with azides as nucleophile, and only a few less efficient salen complexes based on other metals can be found in the literature [21–23].

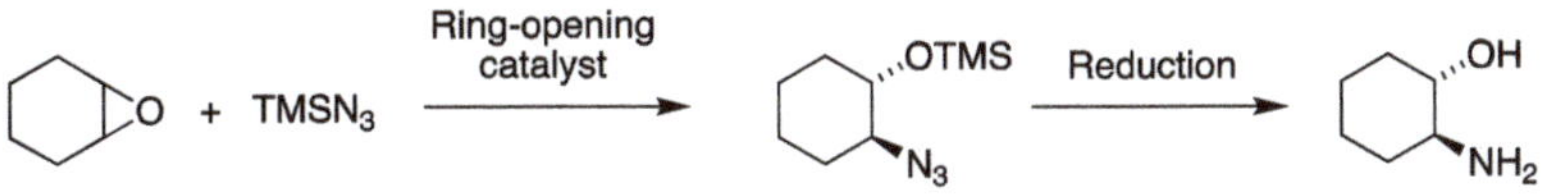

Scheme 3. The asymmetric ring-opening of cyclohexene oxide with $TMSN_3$ followed by reduction to synthesize the corresponding *trans*-1,2-amino alcohol with high enantioselectivity.

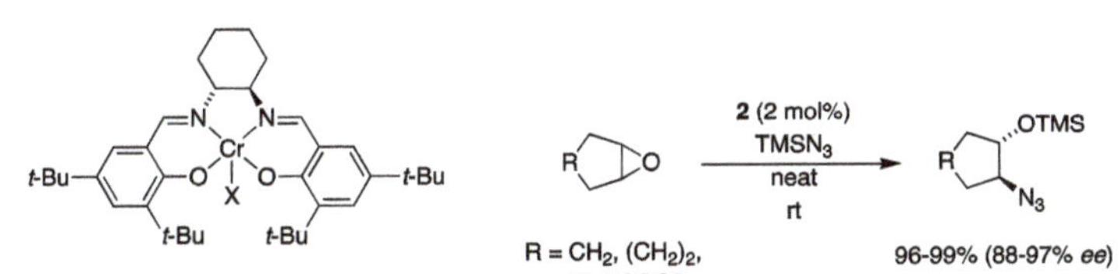

Scheme 4. The asymmetric azidolysis of *meso*-epoxides catalyzed by Cr–salen complexes.

The seminal work of Jacobsen and coworkers on chiral Cr–salen complexes for the asymmetric catalysis of the azidolysis of *meso*-epoxides have yielded invaluable insight in the development of catalysts for the ARO of epoxides. Drawing on knowledge gained in their previous work on the asymmetric epoxidation of alkenes [24–26], they found that by using the same salen ligand and exchanging the manganese(III) for chromium(III) (complex **1**), the ARO of cyclic *meso*-epoxides with $TMSN_3$ as nucleophile could be catalyzed in high yields and with high enantioselectivities (up to 99% yield and 97% *ee*, Scheme 4). The reactions could be conducted either in ethereal solution or under solvent-free conditions with similar yields and enantioselectivities [20,27].

The stability of the catalyst under the reaction conditions facilitated the mechanistic studies, which showed that the active catalyst was the Cr–salen azide complex **2** (Scheme 4). This complex could be isolated from the reaction mixture and recycled up to 10 times at 1 mol% loading with maintained reactivity and enantioselectivity. Kinetic studies revealed a second-order dependence of the reaction rate on the catalyst, suggesting that the catalyst played a dual role in the mechanism: activating both the electrophile and the nucleophile in a bimetallic rate-determining step (Scheme 5) [28]. The proposed mechanism was also supported by the observation of significant non-linear effects of the enantiomeric composition of the catalyst on the enantioselectivity of the reaction [29]. Studies also showed that $TMSN_3$ was not directly involved in the catalytic cycle but rather served as a source of HN_3 in the presence of trace amounts of water [28].

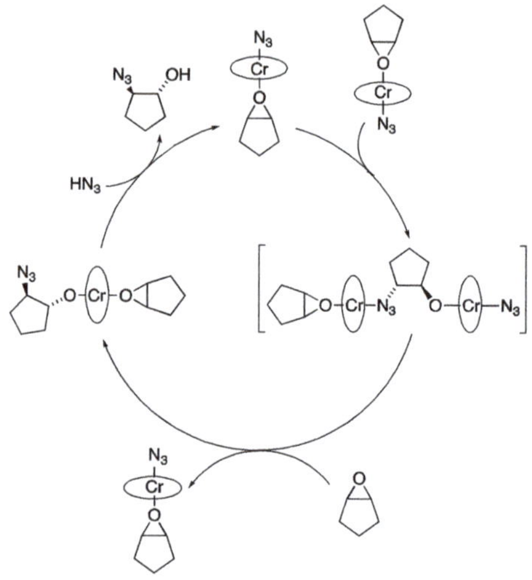

Scheme 5. The mechanism for the ARO of cyclopentene oxide with $TMSN_3$ catalyzed by Cr–salen complexes as proposed by Jacobsen et al. [28].

The proposal of a bimetallic rate-determining step led to the synthesis and application of dimeric catalysts, both in order to gain further mechanistic insights and in the hope that enforcing the cooperativity would lead to improved catalytic activity and enantioselectivity. Jacobsen and coworkers [30] designed and synthesized a number of covalently linked dimeric complexes with different position and length of the linker (Figure 1), which were evaluated as catalysts for the solvent-free ARO of cyclopentene oxide with $TMSN_3$. The positioning of the linker was based on two limiting geometries envisioned for the transition state of the ARO of *meso*-epoxides catalyzed by monomeric complex **2**: "head-to-head" and "head-to-tail" (Figure 2). Complex **3** was designed to evaluate the "head-to-head" geometry and complexes **4–10** were designed to evaluate the "head-to-tail" geometry.

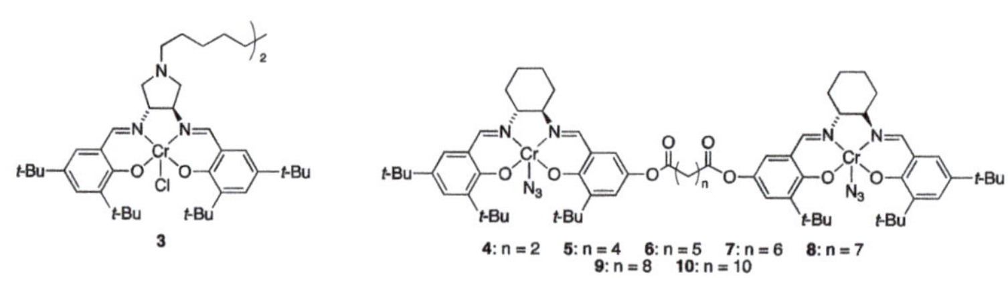

Figure 1. Bimetallic catalysts designed to evaluate the "head-to-head" geometry (catalyst **3**) and the "head-to-tail" geometry (catalyst **4–10**) of the bimolecular catalysis.

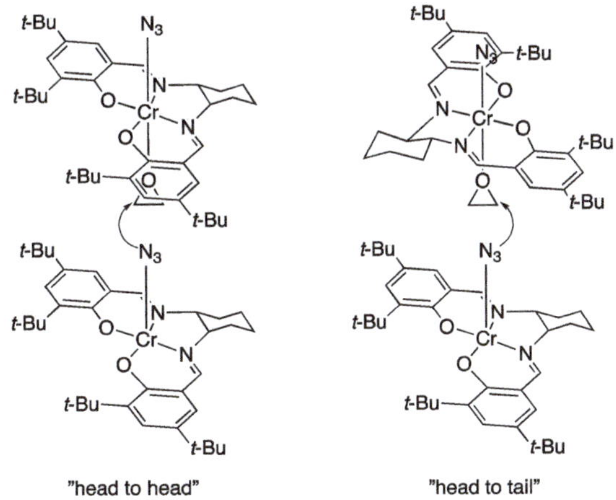

Figure 2. The two limiting geometries for the enantioselectivity-determining transition state of the asymmetric ring-opening of *meso*-epoxides catalyzed by Cr–salen complex **2** as proposed by Jacobsen et al. [30].

All of the investigated dimeric complexes could be used under solvent-free conditions and gave increased reaction rates in the ARO of cyclopentene oxide with $TMSN_3$ compared to monomeric analogues. While the complex based on the "head-to-head" alignment (complex **3**) gave very low enantioselectivity (8% *ee*), the "head-to-tail" aligned complexes **4–10** gave enantioselectivities similar to those obtained with monomeric analogues (90–94% *ee*). Complexes **4–10** also performed efficient catalysis at concentrations one order of magnitude below the lower limit of reactivity for the monomeric analogues. Kinetic studies showed the participation of both inter- and intramolecular pathways in the ARO, with complex **6** ($n = 5$) demonstrating the highest rate constants for both the inter- and intramolecular pathways out of the investigated dimeric catalysts (**4–10**). The intramolecular cooperative catalysis could also be observed by the decreasing non-linear effects in the ARO of cyclopentene oxide with $TMSN_3$ with decreasing concentration of the dimeric catalyst [29,30].

Inspired by Jacobsen's dimeric Cr–salen complexes (Figure 1), Wärnmark and coworkers designed and synthesized several heterobimetallic dimeric complexes, based on the idea that some metal–salen complexes might be more efficient at activating the epoxide and others might be better at activating the nucleophile. This was supported by the initial screening of different 1:1 combinations of monometallic salen complexes with different metal ions as catalysts in the ARO of cyclohexene oxide with $TMSN_3$, where synergistic effects in terms of reactivity and enantioselectivity were observed when Cr–monosalen complexes were combined with Mn– or Co–monosalen complexes. The design of the dimeric complexes included a novel design principle, where C_2- and C_s-symmetric ligands are desymmetrized by the insertion of two different metal ions, giving catalysts of pseudo-C_2 [31] and pseudo-C_s [32] symmetry, respectively (Figure 3). The catalysts were applied in the ARO of a number of *meso*-epoxides with $TMSN_3$. The pseudo-C_2 catalysts, Cr(III)-Co(III) catalyst **12** gave the best results; the reaction could be performed under solvent-free conditions and with very low catalyst loading (0.01 mol%). The ring-opened products were obtained in excellent yields and enantioselectivities (up to 99% yield and 94% *ee*). The Cr(III)–Co(III) catalyst **12** (Figure 3) exhibited higher enantioselectivity than monomeric analogue **1** (Scheme 4) and dimeric homobimetallic complex **6** (Figure 1) in the ARO of cyclohexene oxide under the same reaction conditions. Cr(III)–Mn(III) catalyst **11** (Figure 3) displayed the highest reactivity, with a turnover frequency (TOF) fifty times higher than monomeric **1** and five times higher than homometallic complex **6** (Figure 1) under the same reaction conditions. The study also included a number of dimeric catalysts with different diamine backbones, although the best results were obtained for the catalysts shown in Figure 3 with *trans*-1,2-diaminocyclohexyl backbones. For the pseudo-C_s

catalysts **13** and **14** (Figure 3), the ligand itself is achiral, and the chirality is induced only by the coordination of two different metal ions. Even so, this weak source of chirality was enough to induce enantioselectivity and the complexes were found to catalyze the ARO of cyclopentene oxide with $TMSN_3$ under solvent-free conditions, giving the ring-opened product in >99% yield and 63–76% *ee*.

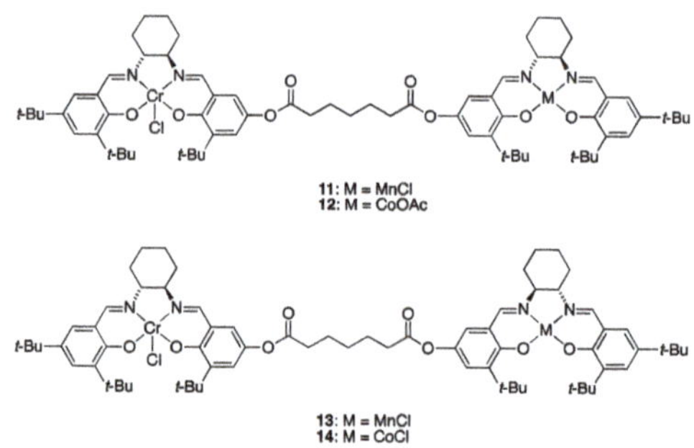

Figure 3. Pseudo-C_2 and pseudo-C_s-symmetric heterobimetallic bissalen complexes.

Another way of exploiting the cooperative mechanism is by designing catalysts where metal–salen complexes are brought into close proximity by non-covalent interactions. Hence, Mirkin and coworkers [33] developed an allosteric supramolecular catalyst by synthesizing a Cr–salen complex with a 2-diphenylphosphanylethylsulfanyl linker at each end of the salen ligand. The dimeric allosteric catalyst was formed by connecting two Cr–salen complexes through coordination to Rh(I) at each end (Figure 4). The distance between the two Cr–salen units (changing between the closed and open forms) could be allosterically controlled by the reversible binding of a CO molecule and a chloride ion to each Rh(I) center (Figure 4). The catalytic properties were investigated in the ARO of cyclohexene oxide with $TMSN_3$. Using the closed form of the catalyst (complex **15**) gave 68% *ee* and a 20-fold rate enhancement compared to the monomeric catalyst **1** under the same conditions. The open form of the catalyst (complex **16**) generated a further doubling of the reaction rate. However, no *ee* values of the product were reported from the catalysis performed with the open form of the catalyst. Due to the poor solubility of the catalyst, all reactions were performed in benzonitrile.

The same research group also developed catalytically active molecular tweezers based on the same design principle with a single Rh(I) center acting as the hinge [34]. This complex allowed for reactions to be run in THF instead of benzonitrile. Using the catalyst in the ARO of cyclohexene oxide with $TMSN_3$ yielded products with up to 80% *ee*, compared to 26% *ee* obtained with complex **1** under the same reaction conditions. The closed form of the catalyst maintained a high enantioselectivity over a range of concentrations, while the open (linear) form exhibited significantly decreased enantioselectivity at lower concentrations.

Wärnmark and coworkers [35,36] explored hydrogen-bonded supramolecular catalysts, designing Cr–salen complexes capable of forming heterodimers through complementary hydrogen-bonding motifs (complexes **17** and **18** in Figure 5). In addition, complexes containing an alkyl strap were also synthesized (complexes **19** and **20** in Figure 5) in order to force the hydrogen-bonding moieties of the monomers into a favorable conformation to assist the aggregation of the desired cyclic heterodimer. The ARO of cyclohexene oxide and cyclopentene oxide with $TMSN_3$ in toluene was used to evaluate the supramolecular systems. Kinetic studies showed that the supramolecular systems **17** + **18** and **19** + **20**, respectively, gave higher reaction rates than monosalen catalyst **1** (Scheme 4) with the strapped systems showing the highest catalytic efficacy. The effect was most pronounced at lower catalyst concentrations. As an example, at 5 mM catalyst concentration in toluene, the strapped system **19** + **20** gave eight times higher initial rate than monosalen catalyst **1** for the ARO of cyclohexene oxide. Studies also indicated that more than one reaction mechanism may be involved in the catalysis involving these systems. The induced enantioselectivity was, however, significantly lower (<10% *ee*) than what was obtained with catalyst **1**. Equilibrium studies on similar studies by the Wärnmark group [37] showed that only a fraction of the supramolecular assemblies contained the cyclic dimeric structure in solution. Hence, the catalysis was, instead, most likely conducted by a mixture of open, linearly aggregated species.

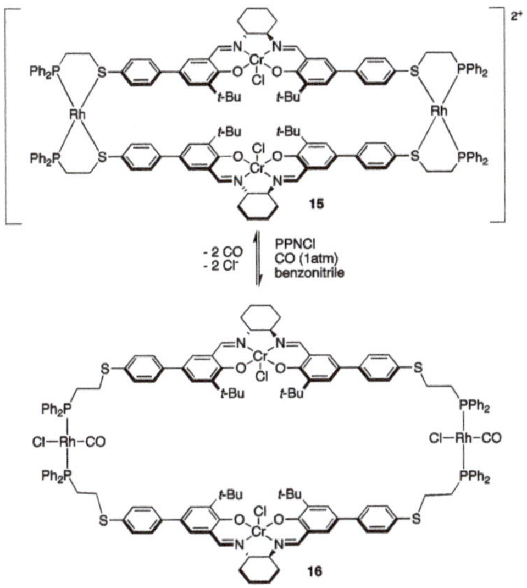

Figure 4. An allosteric supramolecular bissalen catalyst; closed form (top, **15**) and open form (bottom, **16**). PPNCl = bis(triphenylphosphoranylidene)ammonium chloride.

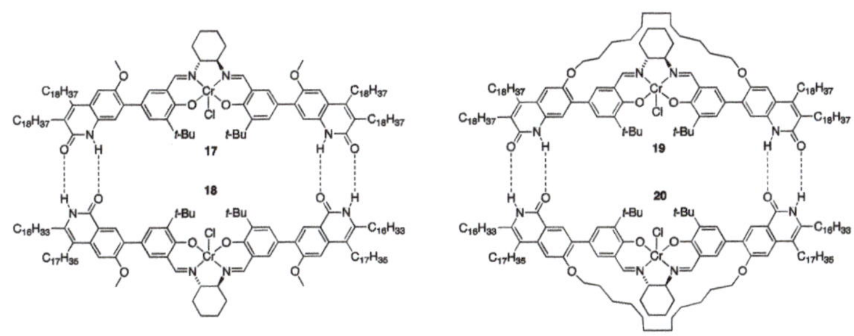

Figure 5. Monomeric catalysts **17**–**20**, drawn as hydrogen-bonded cyclic dimeric aggregates.

One area of interest has been the development of heterogeneous metal–salen catalysts, as this potentially allows for an easier product separation and catalyst recovery [14,15,38]. Several different strategies have been explored for the immobilization of metal–salen complexes on different solid supporting materials [39]. In addition to the reactivity and enantioselectivity, the heterogeneous catalysts also need to be evaluated in terms of stability, leaching, and recyclability.

Early examples include polymer-supported chiral Cr–salen complexes [40] and cationic chiral Cr–salen complexes incorporated into the cavities of zeolites and the interlamellar region of montmorillonite [41]. Both of these strategies gave low enantioselectivities in the ARO of cyclic *meso*-epoxides with $TMSN_3$. The decreased enantioselectivities compared to homogeneous catalysts were attributed to a changed steric environment around the complex [40,41] and loss of the cooperative catalytic effect [41].

Garcia and co-workers [42] developed heterogeneous catalysts by binding chiral Cr–salen complexes to solid silicates. The complexes were anchored on functionalized silicates through aminopropyl tethers, either by complexation with the metal (complex **21**, Figure 6) or by covalent linkage to the ligand (complex **22**, Figure 6). The complexes anchored to the solid through coordination to the chromium catalyzed the ARO of cyclohexene oxide with $TMSN_3$ with high yield (93–99%) and up to 70% *ee* but were found to undergo extensive leaching of the complex into the diethyl ether solution. In contrast, the covalently linked complexes showed no leaching but generated only moderate yields and low enantiomeric excesses in the ring-opening of cyclohexene oxide (43–66% yield and 8–18% *ee*). The decreased enantioselectivity compared to homogeneous catalysts was attributed to a likely change in the reaction mechanism.

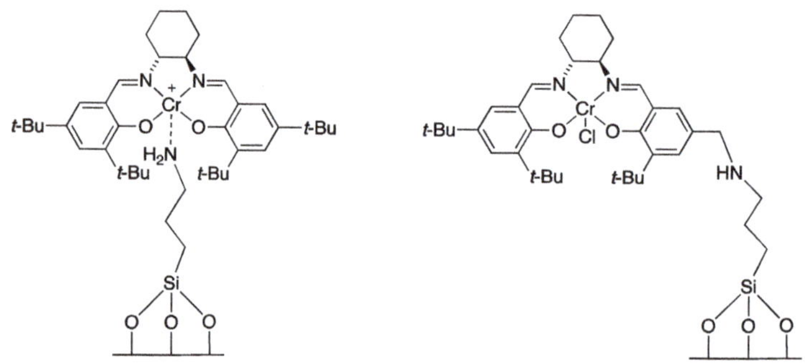

Figure 6. Cr–salen complexes anchored to functionalized silicates SiO_2, ITQ-2 and MCM-41.

Jacobs and coworkers [43] conducted further investigations of heterogeneous catalyst **21** (Figure 6) in the ARO of *meso*-epoxides. They found that the level of Cr–salen complex leaching is largely dependent on the choice of solvent. While homogeneously catalyzed ARO reactions of epoxides are usually performed in ethereal solvents, in heterogeneous catalysis, these solvents greatly enhanced the leaching by facilitating ligand exchange processes. Using apolar non-coordinating solvents like hexane gave leaching of <1% in the ARO of cyclohexene oxide and cyclopentene oxide with $TMSN_3$ as nucleophile, while maintaining excellent yields and good enantioselectivities (up to 77% *ee*). Recycling experiments showed that the catalyst could be recycled up to 10 times without loss of reactivity and enantioselectivity, although the reaction time had to be increased after each run.

Jacobs and coworkers also explored monomeric complex **1** (Scheme 4) and dimeric complex **23** (Figure 7) impregnated on unfunctionalized silica [44,45]. The heterogeneous catalysts were used in the ARO of cyclopentene oxide and cyclohexene oxide with $TMSN_3$. All reactions were performed in hexane. The catalyst based on monomeric complexes exhibited moderate reactivity and enantioselectivity (47–98% yield and 30–45% *ee*), and while leaching was limited, the silica support suffered some deterioration after repeated experiment [45]. The catalyst based on dimeric complex **23** (Figure 7) gave higher yields (74–98%) and improved, although still moderate, enantioselectivities (50–68% *ee*). Using the catalyst in a continuous-flow reactor reduced the deterioration of the support, although these experiments were only performed for the KR of 1,2-epoxyhexane with $TMSN_3$ [44]. In later studies, the dimeric Cr–salen complex was also immobilized in a silica-supported ionic liquid, using ionic liquid [bmim][PF_6] (bmim = 1-butyl-3-methylimidazolium) [46]. This type of heterogenization resulted in both increased reactivity and enantioselectivity compared to the dimeric catalyst impregnated on silica (93% yield and 75% *ee* in the ARO of cyclohexene oxide with $TMSN_3$), as the catalyst in the supported ionic liquid phase was more accessible for reaction than when adsorbed on a silica surface. The catalyst and the ionic liquid could be recovered by Soxhlet extraction with acetone and recycled without loss of enantioselectivity. As reported for the catalysts impregnated on silica, the support shows some deterioration over time, but this could again be improved by using a continuous-flow reactor.

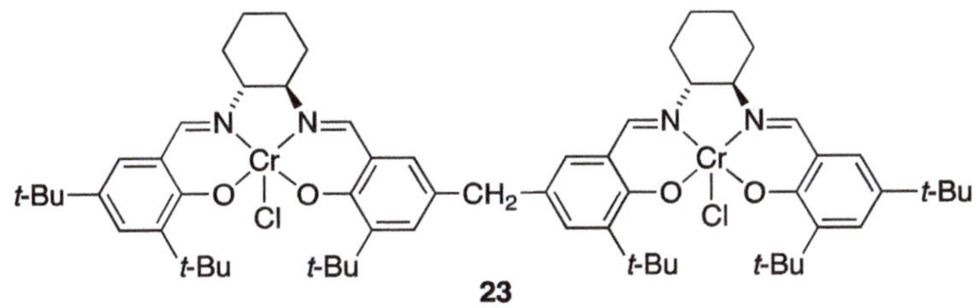

23

Figure 7. Dimeric Cr–salen complex **23**.

Another approach to the heterogenization of Cr–salen complexes is to use dendritic structures as a soluble support. Hence, Keilitz and Haag [47] developed catalysts based on Cr–salen complexes immobilized on hyperbranched polyglycerol (complexes **24–27**, Figure 8). The catalysts consist of Cr–salen analogues with a pyrrolidine backbone, which are linked to the support by linkers of different lengths. The catalysts were used in the ARO of cyclopentene oxide and cyclohexene oxide with $TMSN_3$, using diethyl ether as solvent. The dendrimeric catalysts displayed significant rate enhancements compared to a monomeric analogue, but decreased enantioselectivity. By increasing the length of the linker, the enantioselectivity was improved (from 16% to 64% *ee* for the ARO of cyclopentene oxide). The best results were obtained with a C10 linker (catalyst **26**) which gave the ring-opened products with *ee* values of 48% (for the product from cyclohexene oxide) and 64% (for the product from cyclopentene oxide). Full conversions were reported for all reactions.

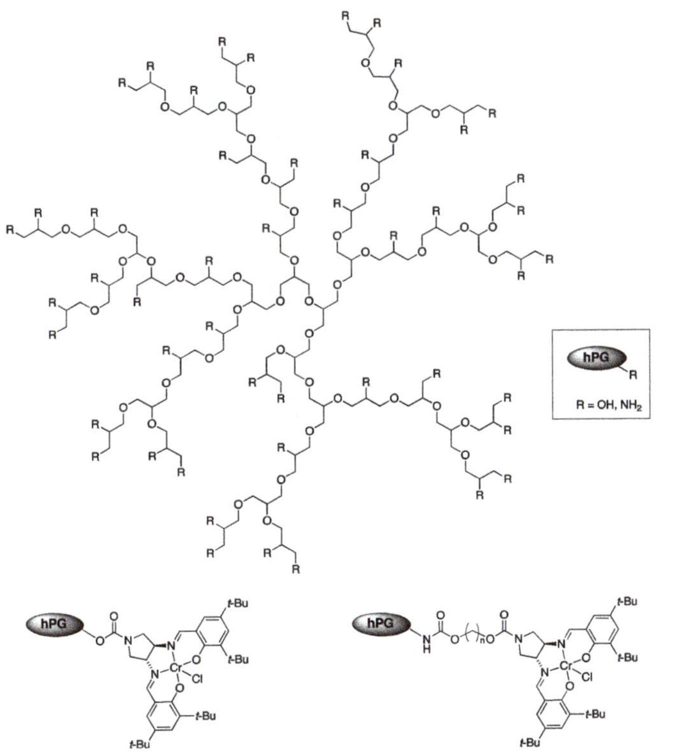

24

25: n = 6
26: n = 10
27: n = 18

Figure 8. Cr–salen complexes immobilized on hyperbranched polyglycerol (hPG).

Schulz and coworkers [38] synthesized a chiral calixsalen-type chromium complex that was employed as a heterogeneous catalyst in the ARO of cyclohexene oxide and 3,4-epoxytetrahydrofuran with $TMSN_3$. The macrocyclic catalyst **28** contains 2–5 repeating thiophene-salen units and was used as an oligomeric mixture (Figure 9). The reaction was performed under heterogeneous conditions in *tert*-butyl methyl ether (TBME), and the catalyst could be recovered by filtration and recycled several times. The products were obtained in good yields (62–90%) and moderate enantioselectivities (42–62% *ee*).

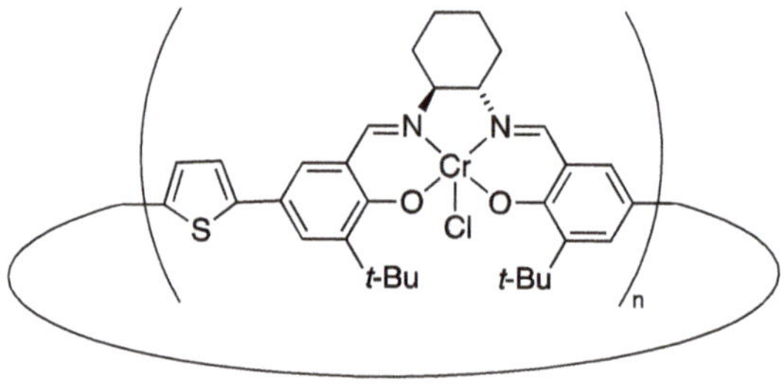

28: n = 2-5

Figure 9. Macrocyclic calixsalen catalyst **28**.

Inspired by Weck's [48] work on hydrolytic kinetic resolution (HKR) of terminal epoxides, Liu and coworkers [49] studied the catalytic properties of macrocyclic oligomer-supported Cr–salen catalysts in the ARO of cyclohexene oxide with $TMSN_3$. Catalysts with different linker lengths between the salen complex and the oligomers were synthesized and compared (catalysts **29–31**, Figure 10). The catalysts were obtained as mixtures of different ring sizes (dimers to decamers) and used as such in the catalytic studies. Reactions carried out with 0.2 mol% catalyst loading in diethyl ether gave significantly higher catalytic activity, enantioselectivity and reaction rates than for the monomeric catalyst **1** (Scheme 4) under the same conditions. The catalyst with the shortest linker (catalyst **29**) showed the highest reaction rate (17 times higher initial TOF than monomeric catalyst **1**) and the catalyst with the longest linker (catalyst **31**) gave the highest enantioselectivity (82% *ee*), demonstrating the importance of distance and relative orientation of the Cr–salen complexes in multi-metallic catalysts. The catalysts could be recovered by precipitation with acetone and recycled up to five times with maintained reactivity and enantioselectivity.

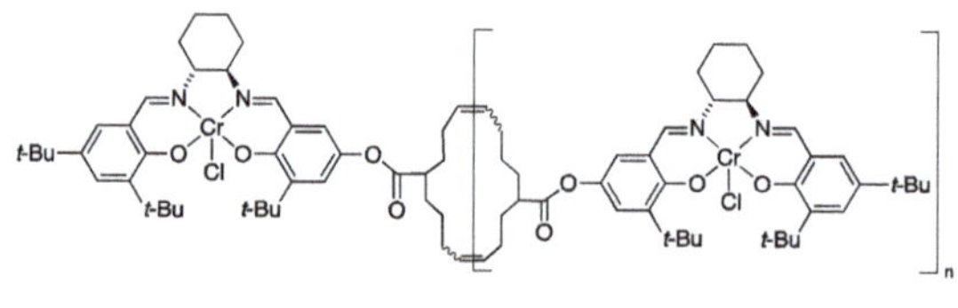

29: n = 1-9

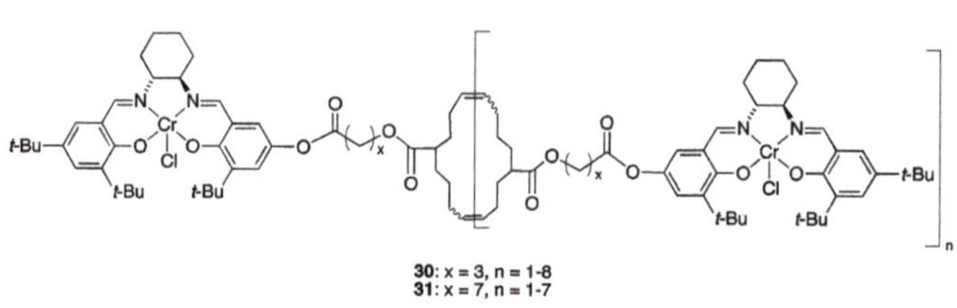

30: x = 3, n = 1-8
31: x = 7, n = 1-7

Figure 10. Macrocyclic oligomer-supported Cr–salen catalysts.

In an effort to improve catalyst recycling, Song and coworkers [50] investigated the use of ionic liquids based on 1-butyl-3-methylimidazolium salts ([bmim][X], Figure 11) as reaction medium in the ARO of cyclic *meso*-epoxides with $TMSN_3$ catalyzed by Cr–salen complex **1** (Scheme 4). By performing the reactions in ionic liquids, the products could be extracted with hexane, while the catalyst remained in the ionic liquid phase. The recovered catalyst could be recycled several times without loss of reactivity or enantioselectivity. The study also found that the nature of the anionic counter ion has a large influence on the reaction, where reactions performed in [bmim][PF_6] gave high yields and enantioselectivies (Scheme 6), comparable to those obtained under homogeneous conditions [20], while reactions carried out in more hydrophilic ionic liquids like [bmim][OTf] hardly gave any product.

Figure 11. 1-Butyl-3-methylimidazolium salts.

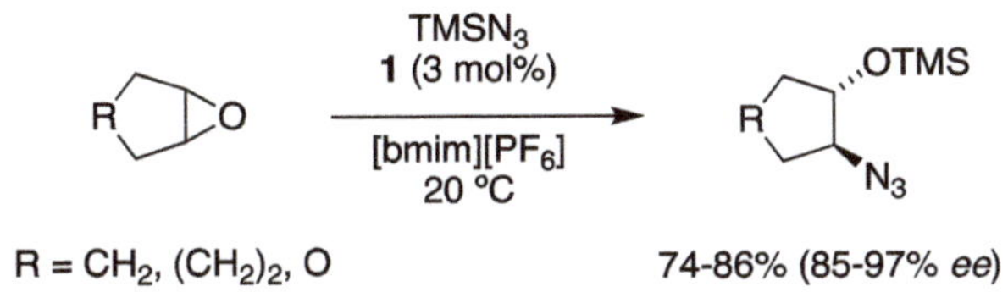

Scheme 6. The asymmetric azidolysis of cyclopentene oxide catalyzed by Cr–salen complex **1** (Scheme 4) in [bmim][PF_6].

Cui and coworkers [51] prepared chiral coordination cages with metal–salen linkers. The so formed supramolecular nanoreactors consisted of six dicarboxylate linkers based on metal–salen complexes (Mn(III)-, Cr(III)-, and Fe(III)salen complexes) and four Cp_3Zr_3 cluster vertices, giving a tetrahedral cage with a hydrophobic cavity (Figure 12). Remarkably, the mixed-linker cage containing both Mn–salen and Cr–salen linkers was able to catalyze the sequential asymmetric epoxidation/ring-opening of 2,2-dimethyl-2*H*-chromene, giving the product with high yield and enantioselectivity (Scheme 7). The cage catalyst also allowed for very low catalyst loadings and remained active at 0.005–0.01 mol%. The reactions were performed under homogeneous conditions, but the catalyst could be recovered by precipitation by the addition of diethyl ether and recycled up to five times with only a slight decrease of enantioselectivity. The same group also explored metal-organic frameworks (MOFs) containing metal–salen linkers prepared by post-synthetic exchange of ligands. The mixed Cr–Mn MOFs catalyzed the sequential alkene epoxidation/ring-opening of 2,2-dimethyl-2*H*-chromene in good yield and enantioselectivity [52].

Yang and coworkers [53] used Cr–salen complex **1** (Scheme 4) to catalyze the ARO of *meso*-epoxides with $TMSN_3$ as an example reaction to investigate the efficiency of their liquid-solid hybrid catalysts. The hybrid catalysts were designed to bridge homogeneous and heterogeneous catalysis and consisted of a catalyst-containing ionic liquid hosted in a porous solid outer crust (Figure 13). The ionic liquid phase consisted of [bmim][PF_6] mixed with 20% [bmim][BF_4] and a small amount of water, and the silica porous crust was prepared from tetramethoxysilane (TMOS). The catalytic particles were successfully packed into fixed-bed reactors for continuous flow reactions. During the reaction, the reactants could pass through the porous crust to the liquid pool where the "homogeneous" reaction occurred. When used in a continuous flow system with *n*-octane as the mobile phase, the hybrid catalyst demonstrated high catalytic efficiency for the ARO of cyclopentene oxide, cyclohexene oxide and *cis*-2,3-epoxybutane. The products were obtained with excellent yields (>99%) and up to 93% *ee* (for the product from cyclopentene oxide). The reactivity and enantioselectivity could be maintained over 600 h.

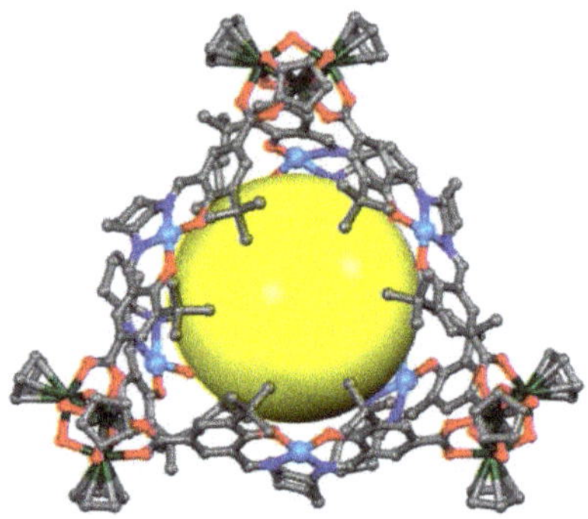

Figure 12. Single crystal X-ray structure of Mn–salen linked coordination cage. The yellow sphere highlights the hydrophobic cavity. Reprinted with permission from Reference [51]. Copyright 2018 American Chemical Society.

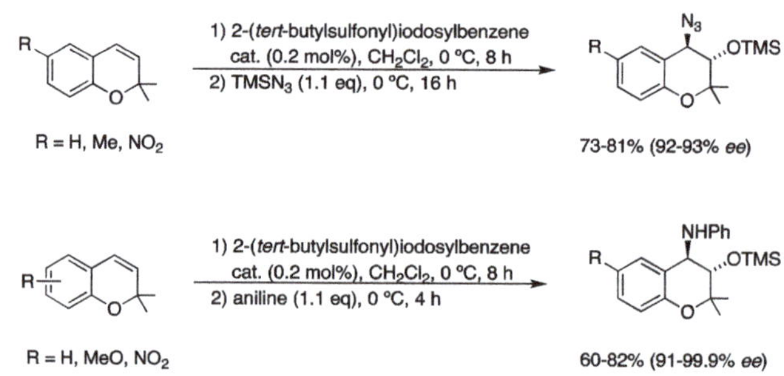

Scheme 7. The sequential asymmetric epoxidation/ring-opening of 2,2-dimethyl-2*H*-chromenes catalyzed by the mixed Mn–Cr coordination cage shown in Figure 12.

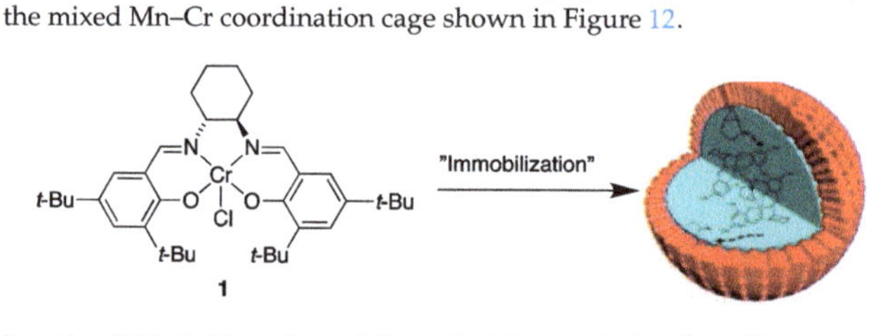

Figure 13. A liquid-solid hybrid catalyst. Adapted with permission from Reference [53]. Copyright 2019 American Chemical Society.

2.2. *With Anilines, Amines, and Carbamates*

An alternative, seemingly more straightforward method for obtaining enantiopure vicinal amino alcohols than first forming the azido alcohol and then reducing it, is to use amines as nucleophiles in the ARO of *meso*-epoxides. However, this approach suffers from an inherent compatibility problem, as the Lewis acidic metal–salen catalyst can be deactivated by complexation with the Lewis basic amine. Nevertheless, there are several examples of successful strategies for overcoming these issues and achieving highly active and enantioselective catalysis of the aminolysis of *meso*-epoxides.

Bartoli and Melchiorre [54] employed Cr–salen complex **1** (Scheme 4) in the ARO of *meso*-stilbene oxide with anilines as nucleophiles. Reactions carried out in dichloromethane with 10 mol% catalyst loading afforded the corresponding amino alcohols with high yields, complete diastereocontrol (only the *syn*-isomer was observed) and high enantioselectivity (Scheme 8). The use of a catalytic amount of Et_3N as an additive was found to give a significantly enhancement of the enantioselectivity (from 76% to 90% *ee* for reaction with aniline) but caused a moderate loss of reactivity.

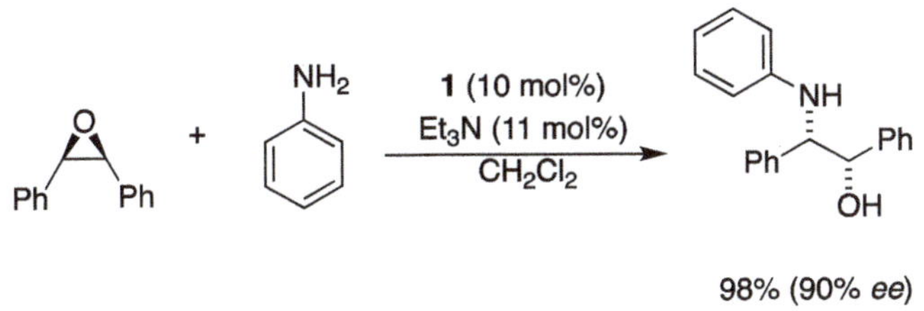

Scheme 8. The asymmetric aminolysis of *meso*-stilbene catalyzed by Cr–salen complex **1** (Scheme 4).

Kureshy and coworkers [55,56] investigated several different approaches to achieve the enantioselective synthesis of vicinal amino alcohols by the ARO of *meso*-epoxides. One such approach is the use of in situ generated monomeric and polymeric Ti(IV)salen complexes from ligands **32** and **33** (Scheme 9) [55]. In order to achieve an efficient asymmetric catalysis, high catalyst loading (20 mol%) and the addition of additives was required. For the ARO of *meso*-stilbene oxide with aniline, the best results were obtained with the enantiopure imine additive **A** (Scheme 9). Under optimized conditions, the ring-opened product was obtained in excellent yield and enantioselectivity (99% yield and >99% *ee*). Noticeably, the use of the opposite enantiomer of the additive resulted in significantly lower enantioselectivity (75% *ee*). The authors suggested that this could indicate synergistic effects between the catalyst and the additive but did not investigate it further. For the ARO of cyclic epoxides, triphenylphosphine was found to be the best additive, although yields and *ee* values were significantly lower than those found for *meso*-stilbene oxide. The monomeric and polymeric catalyst gave comparable yields and enantioselectivities, but the polymeric catalyst generated from ligand **33** showed higher reactivity. In addition, the polymeric catalyst could be precipitated out from the reaction mixture by the addition of *n*-hexane and recycled several times without loss of enantioselectivity.

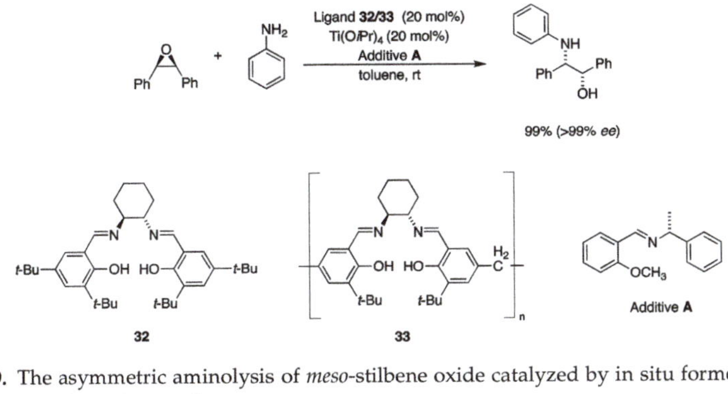

Scheme 9. The asymmetric aminolysis of *meso*-stilbene oxide catalyzed by in situ formed Ti–salen complexes using ligands **32** and **33**.

The same group also synthesized and evaluated a series of enantiopure macrocyclic Cr–salen complexes with different counter ions [56]. After the screening of different reaction conditions, optimal results were obtained for reactions performed in CH_2Cl_2/MeOH (9:1 *v*/*v*) with 0.5 mol% catalyst loading using catalyst **34** (Figure 14). These conditions were used for the ARO of a limited number of cyclic and acyclic *meso*-epoxides with aniline, giving the products in excellent yields (98–99%) and high enantioselectivities (up to 91% *ee*). The catalyst could be recycled up to four times without loss of reactivity or enantioselectivity.

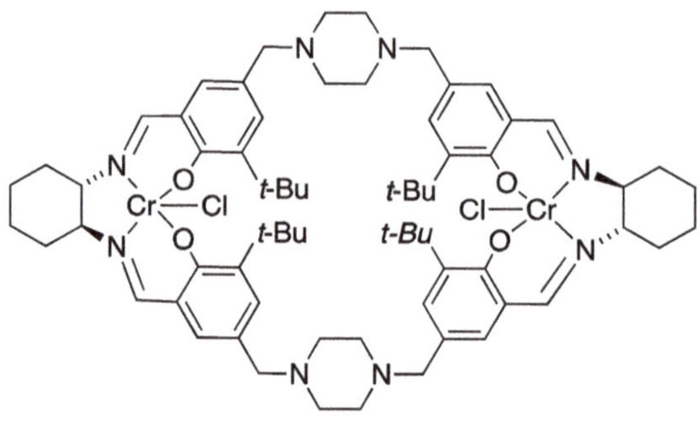

34

Figure 14. Macrocyclic Cr–salen complex **34**.

Jacobsen and coworkers [57,58] used cyclic Co–salen catalyst **35** as a mixture of oligomers in the ARO of cyclic *meso*-epoxides with phenyl carbamate, resulting in the asymmetric synthesis of *N*-protected *trans*-1,2-amino alcohols in high yields and with excellent enantioselectivities (Scheme 10). For six-membered ring epoxides, the addition of the nucleophile was followed by intramolecular cyclization, resulting in *trans*-4,5-disubstituted oxazolidinone products. The ARO of five-membered ring epoxides also proceeded smoothly, although the products did not undergo cyclization, most likely due to the unfavorable strain in *trans*-fused 5-5 ring systems. The products could be deprotected by hydrolysis under basic conditions, and the method could be scaled-up to prepare multigram quantities of the products (Scheme 11).

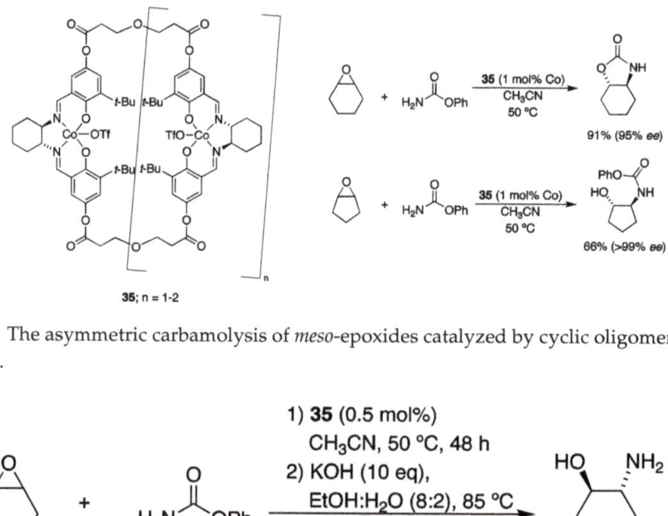

Scheme 10. The asymmetric carbamolysis of *meso*-epoxides catalyzed by cyclic oligomeric Co–salen complex **35**.

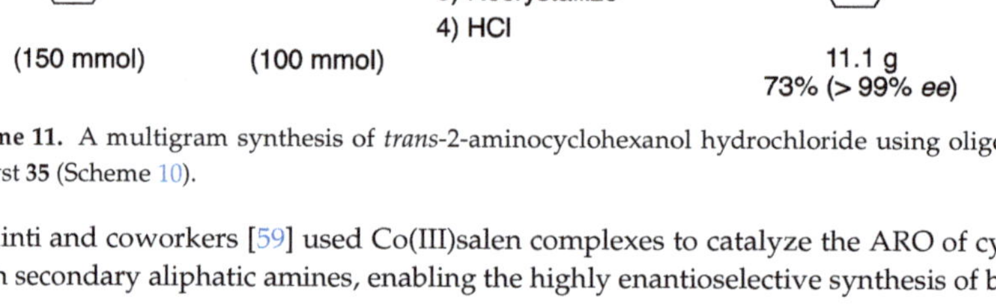

Scheme 11. A multigram synthesis of *trans*-2-aminocyclohexanol hydrochloride using oligomeric catalyst **35** (Scheme 10).

Peddinti and coworkers [59] used Co(III)salen complexes to catalyze the ARO of cyclohexene oxide with secondary aliphatic amines, enabling the highly enantioselective synthesis of biologically important molecules such as vesamicol. Screening of reaction conditions revealed that the solvent had a large impact on the reaction rate and the enantioselectivity. The counter ion was also shown to affect both the catalytic activity of the catalyst and the enantioselectivity of the reaction. For the ARO of cyclohexene oxide with 4-arylpiperidines, the reaction performed with catalyst **36** in *tert*-butyl methyl ether (TBME) gave products with good yields and good to excellent enantioselectivities (Scheme 12).

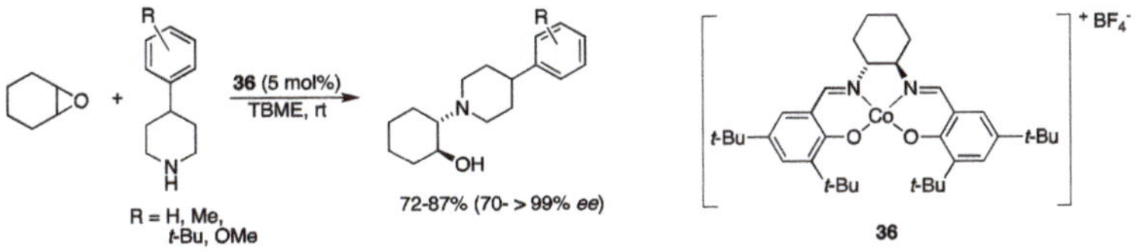

Scheme 12. The asymmetric aminolysis of cyclohexene oxide catalyzed by Co–salen complex **36**.

In the area of heterogeneous catalysis, Islam and coworkers developed catalysts based on a series of metal–salen complexes supported on functionalized mesoporous silica materials. A chiral Fe(III)salen complex was covalently immobilized on mesoporous silica SBA-15 through an aminopropyl linker (Figure 15) [60]. The so-formed heterogeneous catalyst **37** was used in the ARO of cyclohexene oxide with different anilines, giving the products in high yields (85–96%) and with excellent enantioselectivities (96–99% *ee*). Notably, the reactions could be performed under solvent-free conditions and reached complete conversions in just 2–3 h at room temperature. The catalyst could be recycled up to five times with maintained reactivity and enantioselectivity.

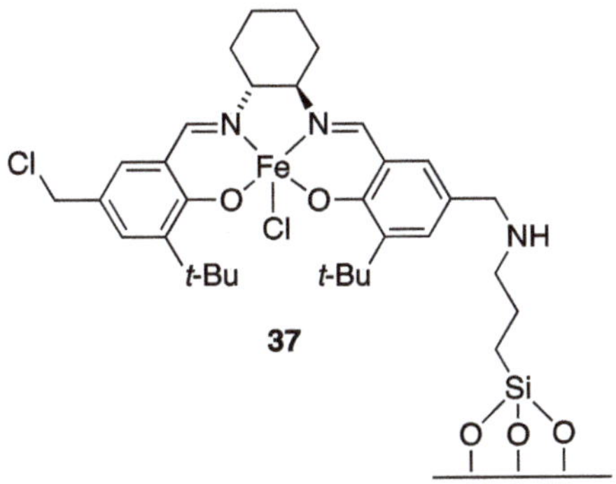

Figure 15. A Fe–salen complex immobilized on mesoporous silica SBA-15.

Islam and coworkers [61] also immobilized Co–salen complexes on mesoporous silica. The Co–salen complexes were grafted on the material through non-covalent interactions between quaternary amine groups on the salen units and carboxylate units on the functionalized silica (catalyst **38**, Figure 16). The material showed excellent catalytic activities for the ARO of cyclohexene oxide with a number of aromatic and cyclic amines under solvent-free conditions, with short reaction times (1–2.5 h), high yields and high enantioselectivities (87–97% yield and 77–99% *ee*). After the reaction, the catalyst could be recovered by precipitation and centrifugation and then recycled without loss of reactivity or enantioselectivity and with no detectable metal leaching.

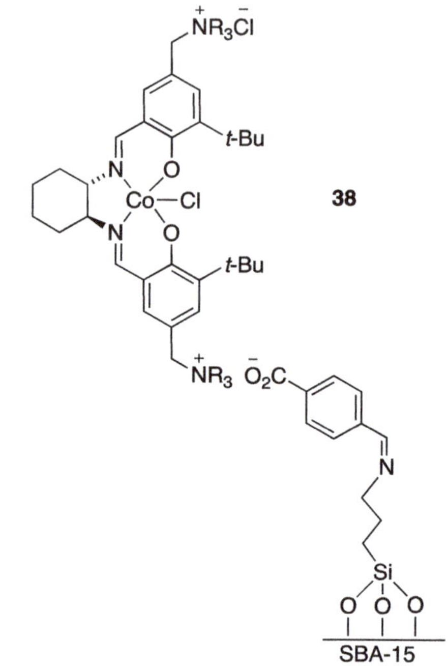

38

Figure 16. A Co–salen complex immobilized on mesoporous silica SBA-15.

Tu and coworkers [62] developed two-dimensional self-supported (i.e., immobilized without the use of an external support) chiral catalysts based on titanium. The catalytic systems consisted of coordination assemblies consisting of heteroditopic ligands containing enantiopure 1,1′-bi-2-naphthol (BINOL) and salen derivatives, where the oxygen bridge in dimeric Ti(IV)salen complexes was used as a crosslinker (Figure 17). Due to the fact of their insolubility in most organic solvents and water, the assemblies were used as heterogeneous catalysts in the ARO of cyclic and acyclic *meso*-epoxides with benzylic and aliphatic amines as nucleophiles. The reactions were performed in toluene and the products were obtained in high yields and enantioselectivities (83–99% yield and 83–98% *ee*). Furthermore, saturated analogues of the metal–salen complexes were also investigated, and the salan-based catalyst **39** was successfully used for the one-pot sequential asymmetric epoxidation/ring-opening of 2,2-dimethyl-4,7-dihydro-1,3-dioxepine, giving the product with high yield and enantioselectivity (Scheme 13). The self-supported catalysts demonstrated very high stability and could be reused up to 20 times without significant loss in yield or enantioselectivity.

Cui and coworkers [63] developed heterogeneous catalysts based on MOFs and coordination cages with metal–salen linkers. The Cr/VO-salen mixed MOF catalysts were fabricated via solvent-assisted linker exchange and used in the ARO of *meso*-stilbene oxide with aniline as nucleophile. The reaction performed in CH_2Cl_2 with 5 mol% catalyst loading gave the product in 87% yield and 76% *ee*. The same group also designed coordination cages with mixed Mn–salen and Cr–salen linkers which were able to catalyze the sequential asymmetric epoxidation/ring-opening of 2,2-dimethyl-2*H*-chromene with anilines, affording the products in good yield (60–82%) and with excellent enantioselectivity (up to 99.9% *ee*) (Scheme 7) [51].

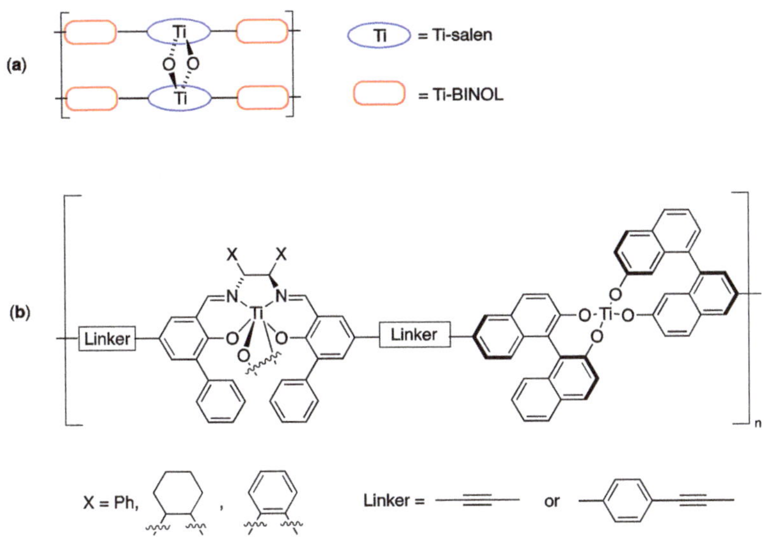

Figure 17. (**a**) A schematic representation of 2D titanium metal–organic coordination assemblies. (**b**) A self-supported chiral Ti–salen catalyst with different linkers and diamine substituents.

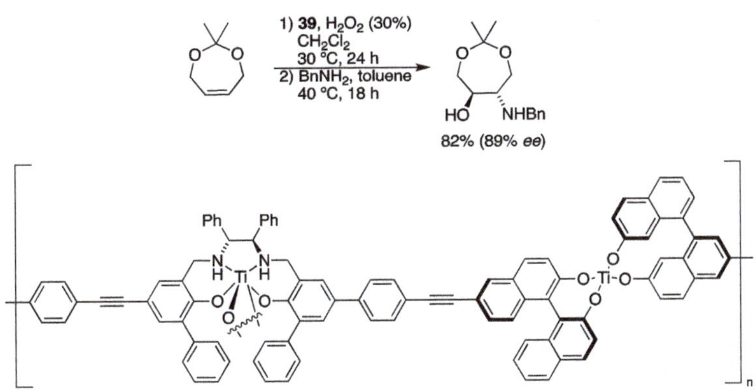

39

Scheme 13. The sequential asymmetric epoxidation/ring-opening of 2,2-dimethyl-4,7-dihydro-1,3-dioxepine catalyzed by self-supported Ti-salan catalyst **39**.

2.3. With Oxygen-Containing Nucleophiles

The hydrolytic asymmetric ring opening of cyclic *meso*-epoxides provides a pathway towards enantioenriched vicinal *trans*-diols and is a valuable complement to other methods such as the catalytic asymmetric dihydroxylation of cyclic alkenes which selectively produce *cis*-diols [64]. Several of the catalysts that have proven effective for the HKR of terminal epoxides have also been investigated in the ARO of *meso*-epoxides [65–68]. While many of the HKR catalysts failed to exhibit high reactivity and enantioselectivity in the ARO of *meso*-epoxides, there are a few examples where the use of Co(III)salen complexes afforded enantioenriched vicinal *trans*-diols in high yields and enantioselectivities.

In 2001, Jacobsen and coworkers [69] reported a mixture of macrocyclic oligosalen complexes (catalyst **40**, Figure 18) as a catalyst for the HKR of terminal epoxides. This catalyst was also proven to be effective in the hydrolytic ARO of cyclohexene oxide. Using this catalyst, *trans*-1,2-cyclohexane diol was synthesized with 98% yield and 94% *ee* (Scheme 14).

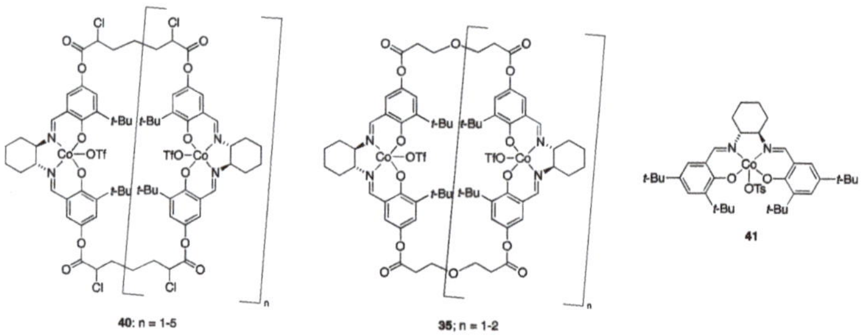

Figure 18. Macrocyclic oligosalen mixtures **40** and **35** and monomeric analogue **41**.

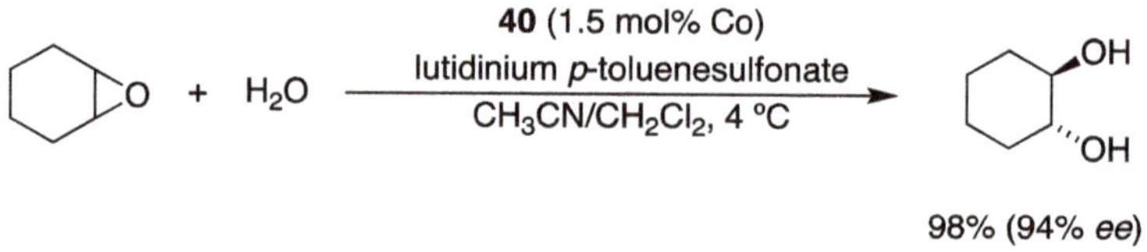

Scheme 14. The asymmetric hydrolysis of cyclohexene oxide catalyzed by cyclic oligomeric Co–salen complex **40** (Figure 18).

Kinetic studies demonstrated that oligosalen catalyst **40** was much more reactive and enantioselective in the hydrolysis of cyclohexene oxide than monomeric analogue **41** (Figure 18). Despite its excellent reactivity, the application of catalyst **40** was limited by its low solubility in the reaction mixture. In addition, reproducibility problems were observed when applying **40** from different batches in the same reaction [69]. To resolve these issues, the same group reported a new generation of oligomeric salen complexes (catalyst **35** in Figure 18) which could be prepared on a multigram scale with 60–66% overall yield [58]. The mixture was found to consist mainly of dimer (n = 1), with a small amount of trimer (n = 2). The catalyst showed improved reactivity and enantioselectivity compared to monosalen complex **41** and oligomeric mixture **40** in the hydrolysis of *meso*-epoxides. The catalyst tolerated a broad scope of substrates and gave the ring-opened products in high yields and enantioselectivities for five- and six-membered cyclic *meso*-epoxides (Scheme 15). The enhanced reactivity of these oligomeric salen complexes compared to monosalen analogues was attributed to the cooperative intramolecular interactions between two salen moieties, which was further supported by the observed first-order kinetic dependence on catalyst concentration.

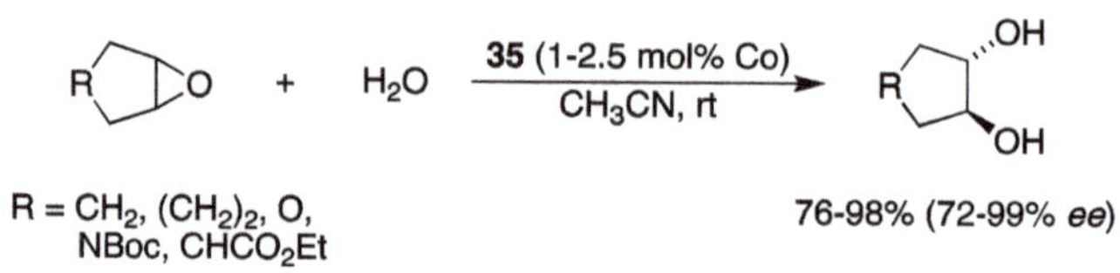

Scheme 15. The asymmetric hydrolysis of cyclic *meso*-epoxides catalyzed by cyclic oligomeric Co–salen complex **35** (Figure 18).

The cooperative mechanism was also exploited by Liu and coworkers in the design of bimetallic catalyst **42** (Figure 19) [70]. Complex **42** is capable of self-assembling into a dimer by aromatic donor-acceptor interactions between naphthalenediimide and pyrene. The catalyst displayed excellent reactivities in HKR reactions of several terminal epoxides as well as in the hydrolytic ARO of cyclohexene oxide, the latter detailed in Scheme 16. In addition, it was found that performing the reactions with a 1:1 mixture of analogues **43** and **44** (Figure 19) gave similar or better results than complex **42**. Using only one of the analogues (either **43** or **44**) required significantly longer reaction times to reach full conversions. This further supported that the aromatic donor-acceptor interactions played an important role for the reactivities of these catalysts.

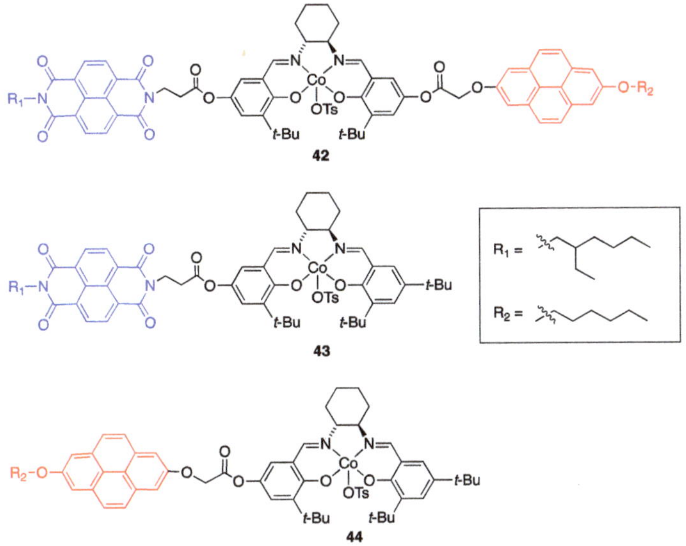

Figure 19. Self-assembling cooperative Co–salen catalysts **42–44** (blue = aromatic acceptor, red = aromatic donor).

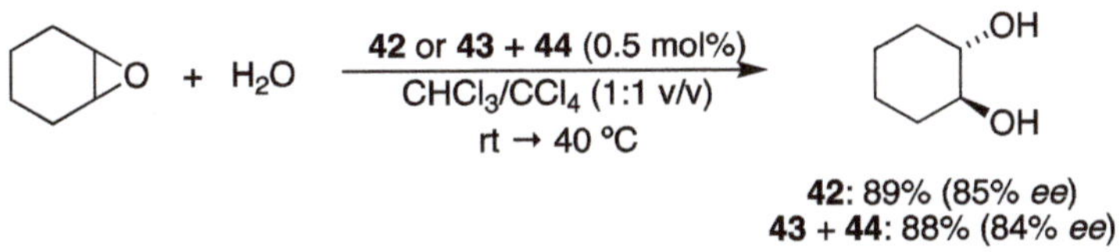

Scheme 16. The asymmetric hydrolysis of cyclohexene oxide catalyzed by cooperative Co–salen complexes **42** or **43** + **44** (Figure 19).

The ARO of *meso*-epoxides with carboxylic acids is of interest owing to its possibility to access monoesters of vicinal diols in high yield and with high enantio- and diastereoselectivity, something which is difficult to achieve with other synthetic methods. Jacobsen's group investigated this type of reaction with various monosalen catalysts, finding that a Co(III)salen complex, generated in situ by stirring Co(II)salen complex **45** (Figure 20) in carboxylic acid in air, was the most effective in catalyzing the reaction [71]. With the addition of diisopropylethylamine, complex **45** catalyzed the reactions between *meso*-epoxides and benzoic acid, affording the corresponding mono-benzoate esters in good to excellent yields and *ee* values (Scheme 17). The substrate scope included cyclic and acyclic *meso*-epoxides with aliphatic and aromatic substituents. The use of one equivalent of weakly coordinating base was necessary to achieve a high rate, enantioselectivity and yield. The authors hypothesized that the beneficial effects might stem from the fact that amines significantly increase the solubility of benzoic acid in TBME. The crystallinity of the mono-benzoate esters also meant that for reactions with moderate *ee* values, the enantiomeric excess could easily be further improved by recrystallization. This methodology was successfully applied to the ARO of *meso*-epoxides with benzoic acid and benzoic acid derivatives as nucleophiles. Using other acids, such as acetic acid or pivalic acid, resulted in lower enantioselectivity and reactivity and was not pursued further.

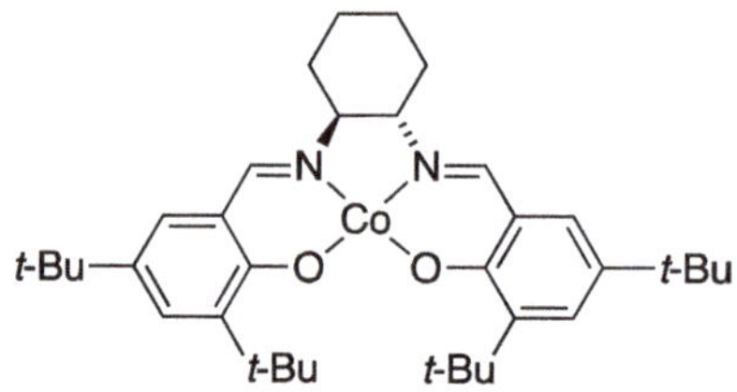

45

Figure 20. Co(II)salen complex **45**.

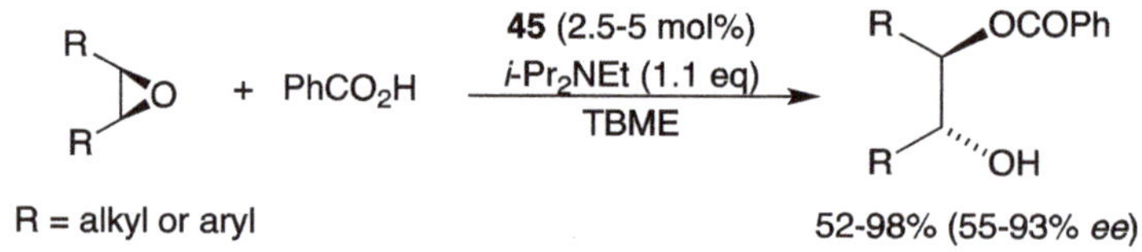

Scheme 17. The asymmetric ring-opening of *meso*-epoxides with benzoic acid catalyzed by Co–salen complex **45** (Figure 20) to generate monoesters.

2.4. With Halogens

Metal–salen complexes have seen limited use as catalysts for the ARO of *meso*-epoxides with halogen nucleophiles. The few examples that can be found in literature are focused on fluorinations, most likely owing to the growing interest in and importance of fluorine substituted organic compounds [72–74].

Haufe and coworkers published a number of studies using Cr–salen complex **46** and different fluoride sources to achieve the enantioselective fluorination of epoxides. The reaction of cyclohexene oxide with KHF_2/18-crown-6 gave (*R*,*R*)-2-fluorocyclohexanol in 64% yield and 55% *ee* when stoichiometric amounts of catalyst **46** was used. Attempts to lower the catalyst loading resulted in a drastic drop in enantioselectivity. The reaction also produced some amount of the corresponding chlorohydrin [75]. Changing the fluoride source to silver fluoride led to improved yields and enantioselectivities, and complete suppression of the formation of the chlorinated by-product, although a high catalyst loading of 50–100 mol% was still required (Scheme 18) [76]. Attempts to lower the necessary quantity of catalyst by using different fluoride sources and reaction conditions proved unsuccessful [77].

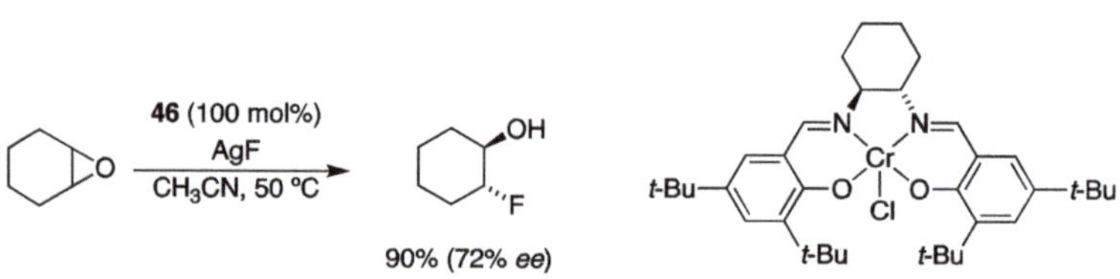

46

Scheme 18. The asymmetric ring-opening of cyclohexene oxide with AgF catalyzed by Cr–salen complex **46**.

Doyle and Kalow reported the enantioselective ring-opening of cyclic *meso*-epoxides with fluoride catalyzed by Co(II)salen complex **47** and an enantiopure amine cocatalyst. The protocol included in situ generation of HF from benzoyl fluoride and 1,1,1,3,3,3-hexafluoro-2-propanol (HFIP). For the ARO of variety cyclic *meso*-epoxides, the use of (-)-tetramisole as the chiral base gave good yields and high enantioselectivities of the resulting fluorohydrins (Scheme 19) [78].

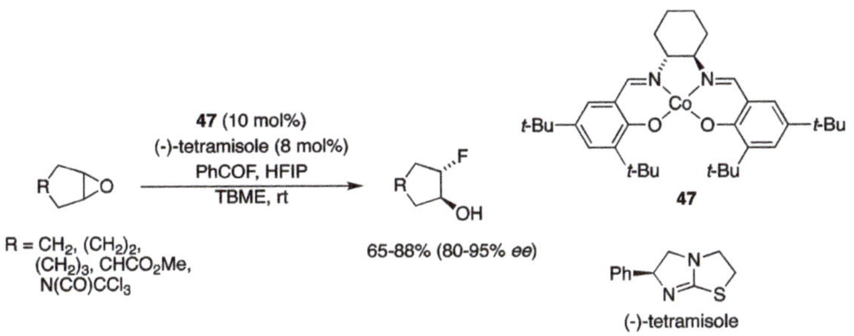

Scheme 19. The asymmetric ring-opening of *meso*-epoxides with HF catalyzed by Co(II)salen complex **47**.

Extensive mechanistic studies of the above reaction resulted in some unexpected and seemingly incompatible data. Kinetic studies demonstrated an apparent first-order dependence on the catalyst while substituent and nonlinear effects supported a bimetallic rate-determining step. Based on this, the authors proposed that the active fluorinating agent was a cobalt-bifluoride complex which formed a fluorine-bridged dimer as a resting state. The amine cocatalyst was proposed to facilitate the dissociation of the dimer (Scheme 20). Further support for the proposed bimetallic mechanism was obtained by the use of the covalently linked Co(II)salen dimer **48** (Figure 21), which resulted in increased reaction rates, extended substrate scope and high enantioselectivity [79].

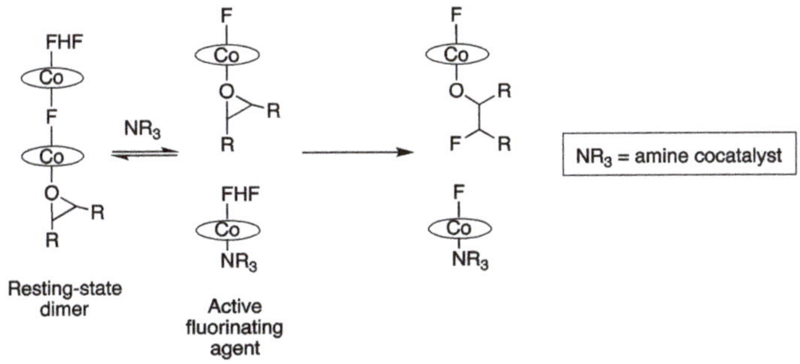

Scheme 20. A part of the proposed mechanism of the asymmetric fluorination of *meso*-epoxides catalyzed by Co–salen **47** (Scheme 19), proposed by Doyle and Kalow [79], showing the dimeric resting-state and its dissociated form.

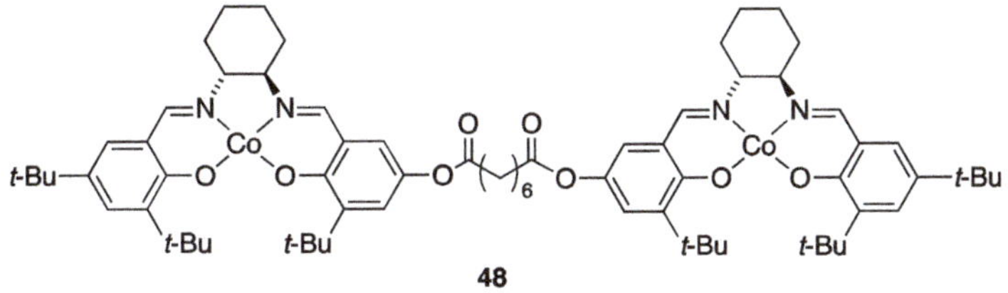

Figure 21. Dimeric Co(II)salen catalyst **48**.

2.5. With Thiols and Selenols

Jacobsen and coworkers applied Cr–salen catalyst **46** (Scheme 18) in the ARO of cyclohexene oxide with benzyl mercaptan as the nucleophile (Scheme 21) [80]. The product was obtained in good yield but with only moderate enantioselectivity. The enantioselectivity could be improved by instead employing 1,4-benzenedimethanethiol as the nucleophile, which resulted in a diastereomeric mixture of C_2-symmetric and *meso* ring-opened products (Scheme 22). The desired C_2-symmetric product could be isolated with high enantioselectivity and transformed into the free thiol by a dissolving metal reduction in a subsequent step.

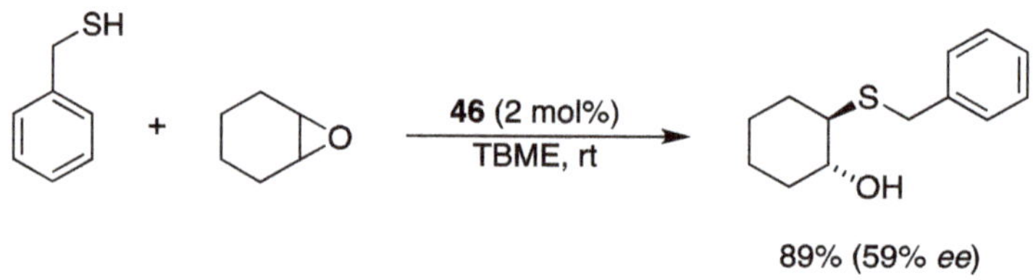

Scheme 21. The asymmetric ring-opening of cyclohexene oxide with benzyl mercaptan catalyzed by Cr–salen complex **46** (Scheme 18).

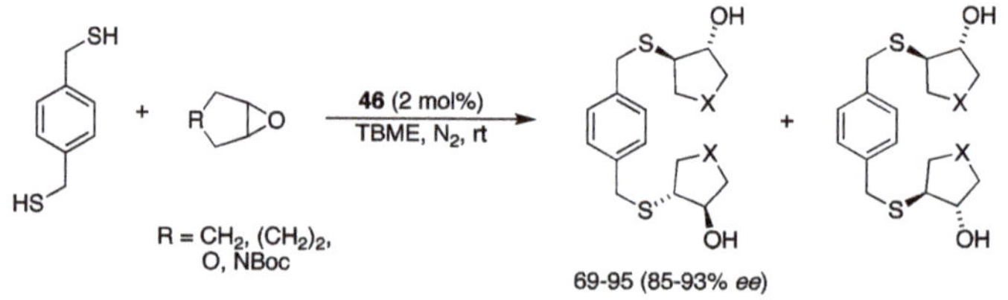

Scheme 22. The asymmetric ring-opening of *meso*-epoxides with 1,4-benzenedimethanethiol catalyzed by Cr–salen complex **46** (Scheme 18).

Many of the reported methods for the ARO of *meso*-epoxides with thiols and selenols employ Ti–salen catalysts. One such method was published by Hou and coworker, who used a chiral Ti(IV)salen complex, formed in situ from Ti(O*i*Pr)$_4$ and salen ligand **32** (Scheme 9). This catalyst afforded vicinal hydroxy sulfides in good yields and moderate enantioselectivities (Scheme 23) [81]. The choice of sulphur nucleophile could also be extended to dithiophosphorus acids, which was explored by Zhou and Tang [82,83]. Their protocol gave the ring-opened product in high yield and good enantioselectivity. Subsequent reduction afforded the corresponding vicinal hydroxy sulfide (Scheme 24).

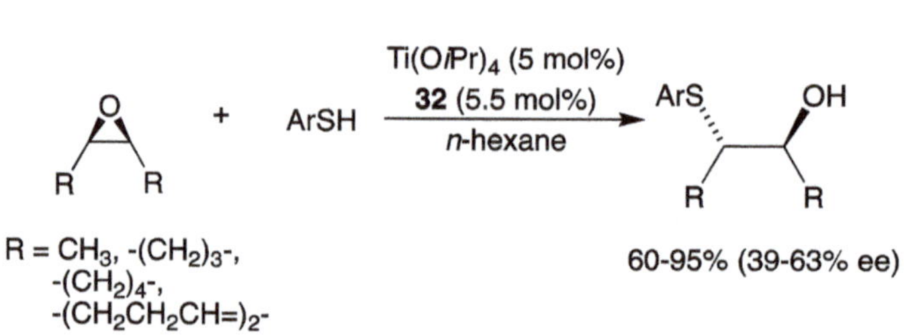

Scheme 23. The asymmetric ring-opening of *meso*-epoxides with thiols catalyzed by a Ti–salen complex generated in situ from Ti(O*i*Pr)$_4$ and salen ligand **32** (Scheme 9).

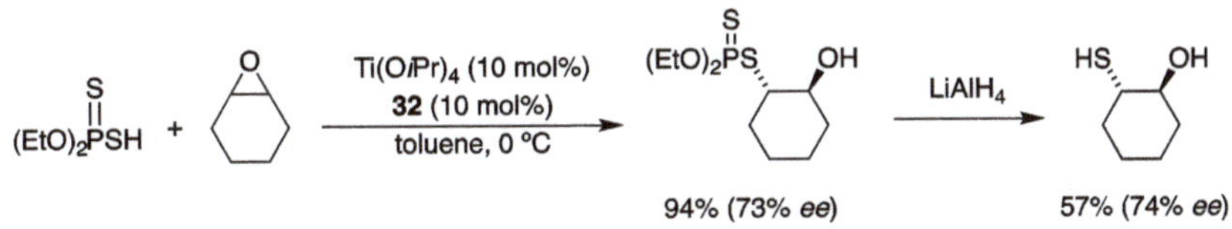

Scheme 24. The asymmetric ring-opening of cyclohexene oxide with dithiophosphorous acid catalyzed by a Ti–salen complex generated in situ from $Ti(OiPr)_4$ and salen ligand **32** (Scheme 9).

For the enantioselective addition of aryl selenols to *meso*-epoxides, Zhu and coworkers reported the use of heterobimetallic titanium-gallium-salen complex **49** (Scheme 25) [84]. The "open" complexes were prepared by first incorporating gallium ($GaMe_3$) and then adding $Ti(OiPr)_4$. Catalyst **49** was used in the ARO of a number of cyclic and acyclic *meso*-epoxides with aryl selenols as nucleophiles, affording the products in high yields and up to 97% *ee*. The authors proposed a strong synergistic effect between the two Lewis acids, with the hard Lewis acid titanium activating the epoxide and the softer gallium coordinating the arylselenol and directing the nucleophilic attack. The same method could also be extended to the ARO of *meso*-epoxides with thiols, resulting in high yields and moderate to high enantioselectivities (up to 92% *ee*) [85,86].

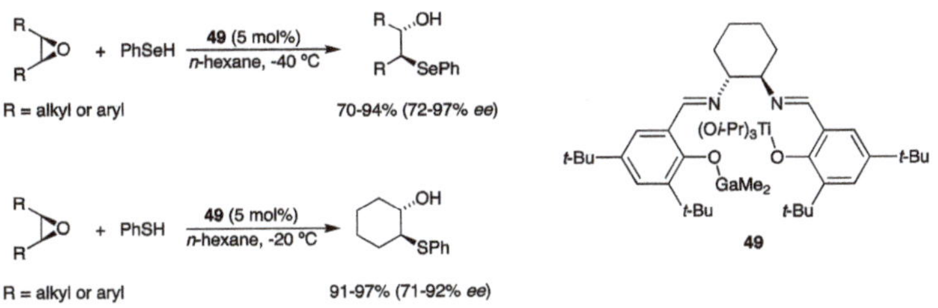

Scheme 25. The asymmetric ring-opening of *meso*-epoxide with aryl selenols and thiols catalyzed by Ti-Ga-salen complex **49**.

Another protocol for the asymmetric synthesis of vicinal hydroxy selenides was reported by Tiecco and Marini, who used Cr–salen complexes to catalyze the ARO of *meso*-epoxides with (phenylseleno)silanes as nucleophiles [87]. Complexes with different counter ions, reactions in different solvents and with different additives were evaluated. The best results were obtained with 5 mol% of complex **50** with BF_4^- as counterion in TBME in the presence of one equivalent of *N,N,N′,N′*-tetramethylethylenediamine (TMEDA) at −10 °C, which gave the products in good to excellent yields and enantioselectivities (Scheme 26). The method gave better results for *meso*-stilbene oxide and its derivatives than for cyclic *meso*-epoxides.

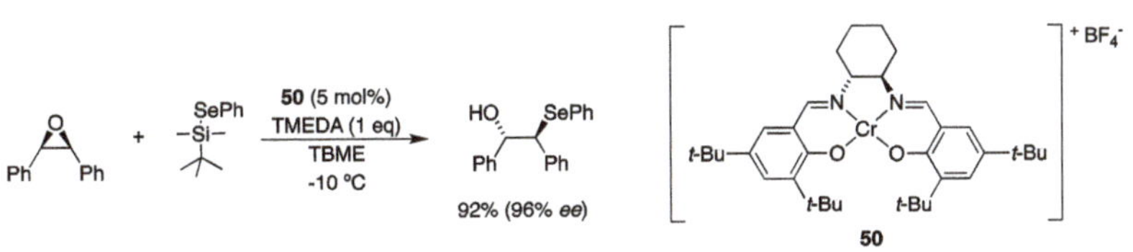

Scheme 26. The asymmetric ring-opening of *meso*-stilbene oxide with (*tert*-butyldimethyl(phenylselanyl)silane as nucleophile catalyzed by Cr–salen complex **50**.

2.6. With Carbon-Containing Nucleophiles

The formation of carbon-carbon bonds in one of the most fundamental and important reactions in organic synthesis. Carbon-carbon bonds form the backbone of essentially all organic molecules, and the asymmetric C-C bond formation is of key importance in the synthesis of optically active and highly functionalized molecules, such as biologically active compounds [88,89]. As such, the catalytic ARO of *meso*-epoxides with carbon-based nucleophiles represents a potentially attractive strategy for achieving this task. However, the use of metal–salen complexes as catalysts for these kinds of reactions has been limited. One example comes from Cozzi and Umani-Ronchi, who used enantiopure Cr–salen catalyst **1** (Scheme 4) for the ARO of *meso*-stilbene oxide with indoles as carbon nucleophiles, giving ring-opened products in excellent yields and enantioselectivities (Scheme 27) [90].

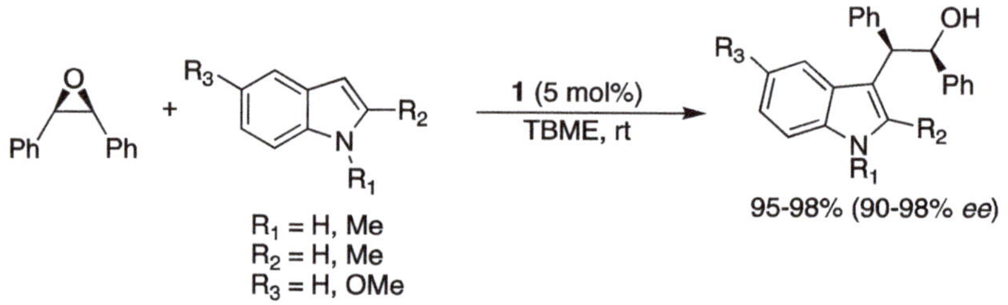

Scheme 27. The asymmetric ring-opening of *meso*-stilbene oxide with indoles catalyzed by Cr–salen complex **1** (Scheme 4).

Pietrusiewicz and coworkers reported the use of Al(III)salen catalyst **51** in the ARO of 3,4-epoxy-1-phenylphospholane-1-oxide with TMSCN (Scheme 28) and $TMSN_3$ as nucleophiles [22]. The products are potential intermediates in the synthesis of phosphasugar derivatives. It was found that the enantioselectivity was greatly influenced by the choice of solvent. For the azide addition, the highest yield was obtained in THF (90%) and best enantioselectivity in dichloromethane (28% *ee*). The use of an ionic liquid as the reaction medium gave high yields but did not increase the enantioselectivity. For the cyanide addition, the ARO of 3,4-epoxy-1-phenylphospholane-1-oxide performed with 10 mol% catalyst in TBME afforded the corresponding cyanohydrin in 56% yield and 72% *ee* (Scheme 28).

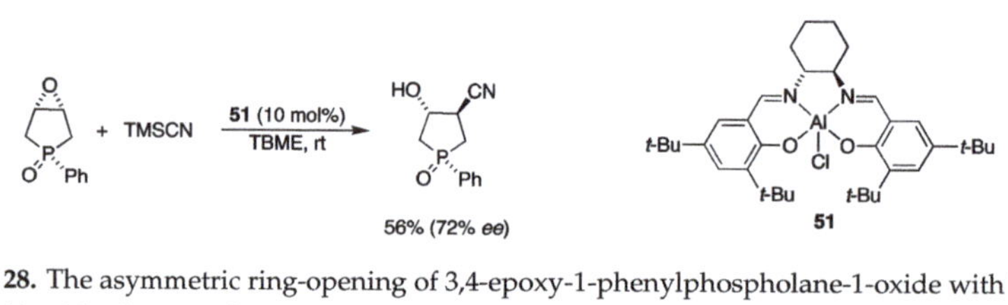

Scheme 28. The asymmetric ring-opening of 3,4-epoxy-1-phenylphospholane-1-oxide with TMSCN catalyzed by Al-salen complex **51**.

Another example is the Y-salen complexes explored by RajanBabu and coworkers [23]. A number of salen complexes with different diamine backbones were employed as catalysts in the ARO of cyclohexene oxide with TMSCN. The best result in terms of enantioselectivity was obtained using binaphthyldiamine ($BINAP(NH_2)_2$) derived salen complex **52**, which gave the ring-opened product in 77% *ee* (Scheme 29). The catalyst loading could be reduced from 2 mol% to 0.1 mol% without significant loss of reactivity or enantioselectivity and the reaction could also be run under solvent-free conditions.

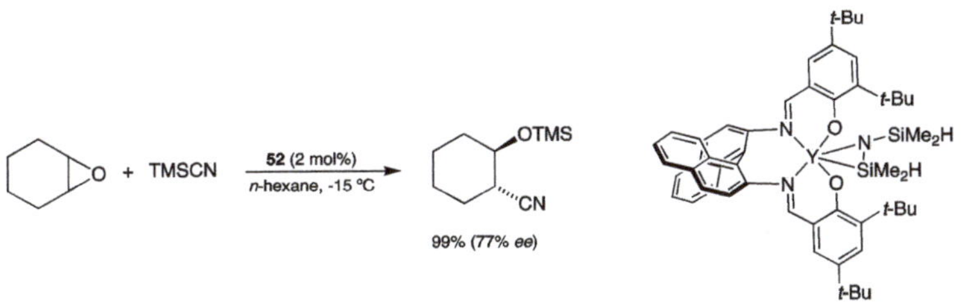

Scheme 29. The asymmetric ring-opening of cyclohexene oxide with TMSCN catalyzed by Y-salen complex **52**.

3. Kinetic Resolution of Epoxides

Some epoxides, typically terminal ones, are difficult to prepare directly with high enantiomeric purity. Since the racemic mixtures are usually readily available at low cost, using an efficient KR becomes an attractive alternative for obtaining enantiomerically pure epoxides. Ideally, this strategy results in both the enantiopure ring-opened product, as well as the enantiomeric enrichment of the unreacted epoxide [91]. The selectivity can often be tuned towards the ring-opened product or the enantioenriched unreacted epoxide by the amount of nucleophile used.

Most of the examples presented herein concern the KR of terminal epoxides, although there are a few examples of the use of 1,2-disubstituted epoxides (mainly with *trans* stereochemistry) and even trisubstituted epoxides. In general, the reactions proceed with almost exclusive regioselectivity for the nucleophilic attack at the least substituted epoxide carbon.

For most of the reported examples of the KR of epoxides, the focus is on the ring-opened product. The most obvious exception is in the HKR, where the use of water as an inexpensive, environmentally friendly and highly available nucleophile makes it an attractive strategy for the enantioenrichment of the unreacted epoxide. As such, many of the references presented herein only report the yields and *ee* values of the product for which the method was optimized. There are also several ways of calculating the yield of the reaction. Although the theoretical maximum yield of the enantiopure ring-opened product is 50% (given that the starting material is used as a racemic mixture), many of the references calculate the yield based on the amount of nucleophile used, which varies a lot between different papers. To facilitate the comparison of different protocols, we have recalculated all yields based on the epoxide, giving a 50% maximum yield for each product. In some references, the results are reported as conversion instead of yield. In those cases, the products are often not isolated, and the conversion is calculated based on the consumption of each of the two enantiomeric starting materials determined by chiral GC separation.

3.1. *With Azides*

Jacobsen and coworkers reported the azidolytic KR of terminal epoxides catalyzed by Cr–salen complex **2** (Scheme 4) [92]. The reactions were run under solvent-free conditions at 0–2 °C. By using 0.5 equivalents of $TMSN_3$, the ring-opened products could be obtained in 40–49% yield and 89–97% *ee*. The reaction exhibited good functional group tolerance (Scheme 30). One of the investigated reactions, the azidolytic KR of epichlorohydrin, was further investigated, as it was discovered that the unreacted epoxide underwent racemization under the reaction conditions. This enabled the dynamic kinetic resolution (DKR) of epichlorohydrin, where, by adding $TMSN_3$ in portions over time, the desired ring-opened product was obtained in 76% yield and 97% *ee* (Scheme 31) [93].

The same protocol was also extended to include 2,2-disubstituted epoxides, using in situ formed HN_3 (formed from $TMSN_3$ and isopropanol) as the nucleophile [94]. Performing the reaction in TBME afforded both ring-opened products and unreacted epoxides in high yields and enantioselectivities

(Scheme 32). This methodology was successfully used for the enantioselective preparation of a key intermediate in a synthesis of the natural product taurospongin A [95].

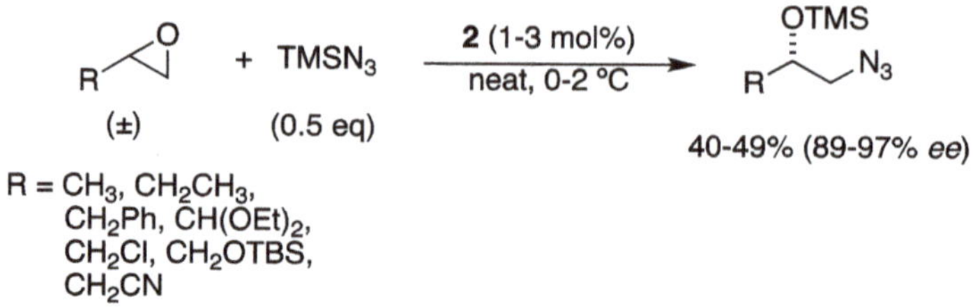

Scheme 30. The azidolytic kinetic resolution terminal epoxides catalyzed by Cr–salen complex **2** (Scheme 4).

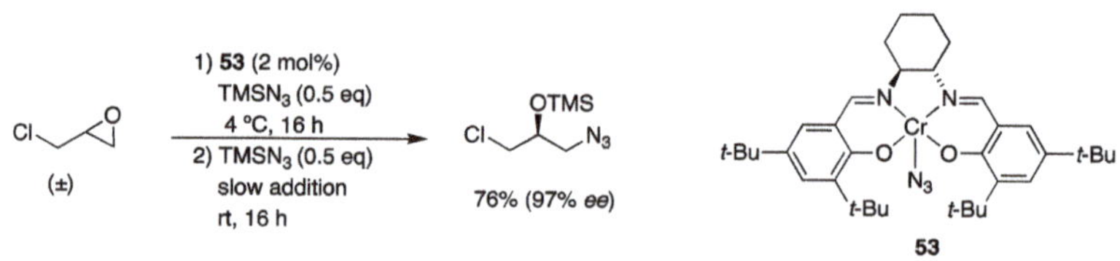

Scheme 31. The dynamic kinetic resolution of epichlorohydrin with $TMSN_3$ catalyzed by Cr–salen complex **53**.

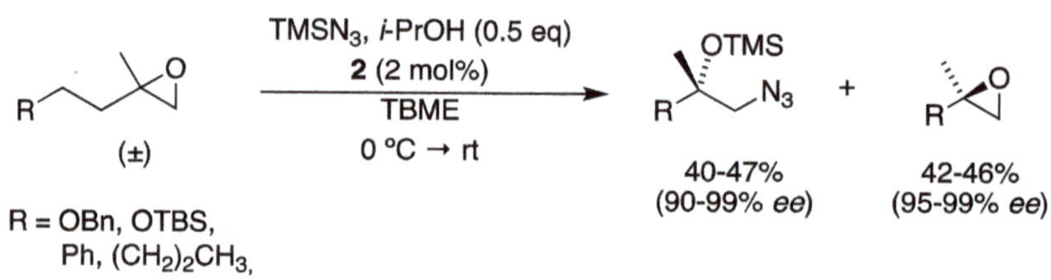

Scheme 32. The azidolytic kinetic resolution of 2,2-disubstituted epoxides catalyzed by Cr–salen complex **2** (Scheme 4).

Jacobs and coworkers used Cr–salen complex **1** (Scheme 4) for the KR of terpene epoxides with $TMSN_3$ [96]. This constitutes a rare example of the kinetic resolution of trisubstituted epoxides. The reaction gave enantioenriched epoxides and ring-opened products in good yields and high diastereomeric excess. Initial experiments revealed that both enantiomers of the Cr–salen complex, (*R*,*R*)-complex **1** (Scheme 4) and (*S*,*S*)-complex **46** (Scheme 18), exhibited similar reactivity and selectivity. It was therefore hypothesized that the observed stereoselectivity of the reaction was induced by the C4 substituent (Scheme 33) and that the Cr–salen complex served only as an activating agent for the azide-transfer. As such, later experiments were performed with the Cr–salen complex **1**(**46**) as a racemic mixture (Scheme 33). For all investigated substrates, the *cis*-diastereomers were selectively transformed and the *trans*-epoxides remained unreacted (*cis* and *trans* refers to the relative stereochemistry of the methyl group and the C4 substituent). Jacobs and coworkers also investigated the effect of micro-wave irradiation on the Cr–salen-catalyzed KR of terminal epoxides, as well as the ARO of *meso*-epoxides. It was found that the reaction rate could be increased by up to three orders of magnitude without any significant loss of enantioselectivity under micro-wave irradiation [97].

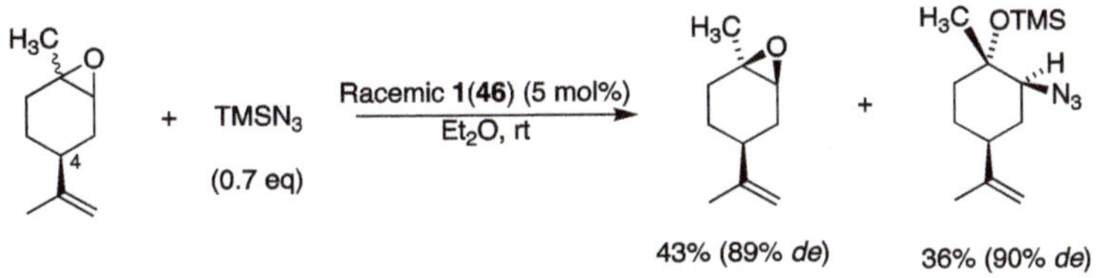

Scheme 33. The kinetic resolution of (-)-limonene-1,2-epoxide with $TMSN_3$ catalyzed by Cr–salen complex **1**(**46**) (Schemes 4 and 18, respectively) as a racemic mixture.

Similar to the ARO of *meso*-epoxides with azides, most examples of metal–salen-catalyzed KRs of epoxides with azides are based on chromium-salen complexes. One example of a different metal–salen complex was published by Kim and coworkers, who used binuclear Co(II)salen complexes bearing Lewis acids of group 13 metals (Scheme 34) [98]. The presence of a group 13 Lewis acid was necessary for catalytic activity, and the dimeric complexes showed enhanced reactivity and enantioselectivity compared to their monomeric analogues. The system allowed for low catalyst loading, where 0.5 mol% of complex **54** catalyzed the azidolytic KR of a number of terminal epoxides, affording ring-opened products in excellent yields and with high enantioselectivity (Scheme 34).

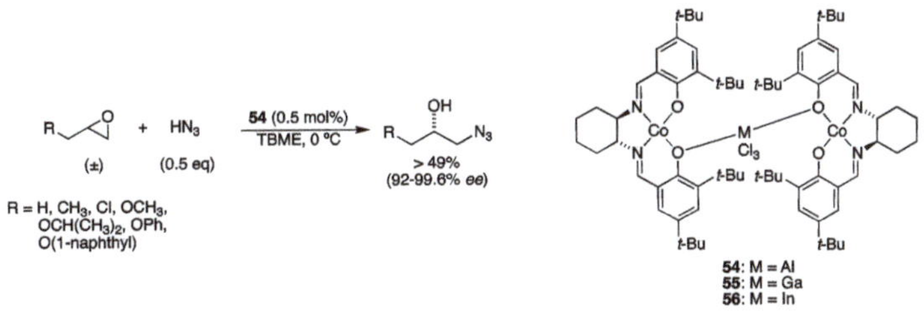

Scheme 34. The azidolytic kinetic resolution of terminal epoxides with catalyzed by dimeric Co(II)salen complexes formed by group 13 metal activation.

For the development of heterogeneous catalysts for the KR of epoxides, the same strategies as for the ARO of *meso*-epoxides have been employed. Early examples of polymer-supported Cr–salen complexes only induced low enantioselectivities in the ring-opened products, but exhibited good stability and recyclability [40]. Jacobs and coworkers have published extensive work on the impregnation of Cr–salen complexes on silica and silica-supported ionic liquids. Both monomeric complex **1** (Scheme 4) and dimeric complex **23** (Figure 7) were separately physiosorbed on silica and evaluated as catalysts in the KR of 1,2-epoxyhexane and 1,2-epoxyoctane, achieving quantitative conversions and high enantioselectivities (up to 98% *ee* for ring-opened products). The dimeric complexes showed improved enantioselectivity compared to the monomeric analogues. Using the catalyst in a continuous flow reactor decreased the deterioration of the solid support and resulted in quantitative yields and high enantioselectivities for both the ring-opened products and unreacted epoxides (up to 99 and 85% *ee*, respectively) [44,45]. Immobilizing the dimeric complex in a silica-supported ionic liquid further improved the reactivity and enantioselectivity, and both the catalyst and the ionic liquid could be recovered by Soxhlet extraction with acetone [46].

Liu and coworkers investigated the macrocyclic oligomer-supported Cr–salen catalysts **29**–**31** (Figure 10) in the KR of 1,2-epoxyhexane and propylene oxide [49]. The same trend as in the ARO of *meso*-epoxides was found, where a shorter linker between the Cr–salen complex and the macrocycle increased the reaction rate and gave the highest yield (45–48%), while a longer linker gave better enantioselectivity (83–84% *ee*) for the ring-opened product.

Yang and Li constructed an efficient solid nanoreactor by encapsulating Cr–salen complex **1** (Scheme 4) and pyridine inside a mesoporous silica nanocage [99]. The addition of pyridine led to a pronounced increase in TOF and enantioselectivity, which was attributed to the increased nucleophilicity of the Cr–salen complex after coordination to pyridine. The system was investigated in the KR of 1,2-epoxyhexane with $TMSN_3$ under solvent-free conditions and at very low catalyst loading (0.002 mol%), affording both ring-opened product and unreacted epoxide in close to quantitative yield and with high enantioselectivity (91% *ee* and 92% *ee* respectively). The nanoreactor showed high stability and could be recovered and recycled nine times with maintained reactivity and enantioselectivity.

3.2. With Anilines, Amines and Carbamates

Bartoli and Melchiorre investigated different metal–salen catalysts for the KR of epoxides with different nitrogen-based nucleophiles. Cr–salen complex **1** (Scheme 4) was used for the KR of a variety of *trans*-disubstituted aromatic epoxides with aniline and anisidine (Scheme 35). Complete regio- and diastereoselectivity was observed, and the vicinal *anti*-amino alcohol products were obtained in reasonable yields and enantioselectivities (77–99% *ee*). The study also included a rare example of the KR of a trisubstituted epoxide, where applying the protocol on *trans*-2-methyl-2,3-diphenyl oxirane afforded the ring-opened products in 18% yield and 81% *ee* [54].

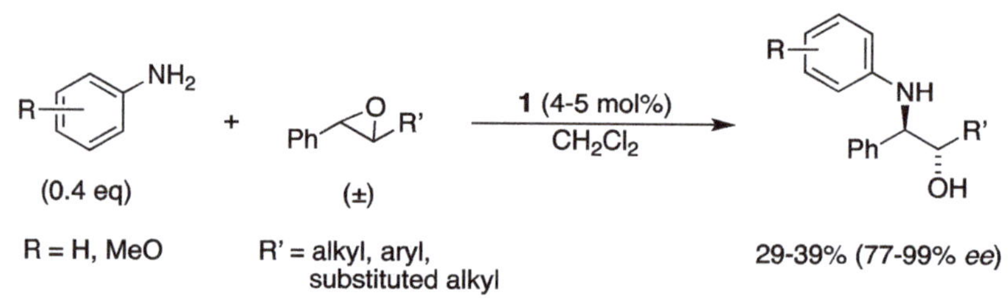

Scheme 35. The aminolytic kinetic resolution of internal *trans*-epoxides catalyzed by Cr–salen complex **1** (Scheme 4).

The same group also investigated the KR of terminal epoxides using *tert*-butyl arbamate as the nucleophile. Co(II)salen complex **45** (Figure 20) was used as a pre-catalyst and oxidized in situ to the corresponding Co(III)salen complex by addition of *p*-nitrobenzoic acid, affording enantiopure (*ee* ≥ 99%) *N*-protected vicinal amino alcohols in high yields and with complete regioselectivity, while the unreacted epoxides could also be isolated with high *ee* values (Scheme 36). The protocol was effective for both linear and relatively hindered aliphatic epoxides, as well as epoxides containing different functional groups [100].

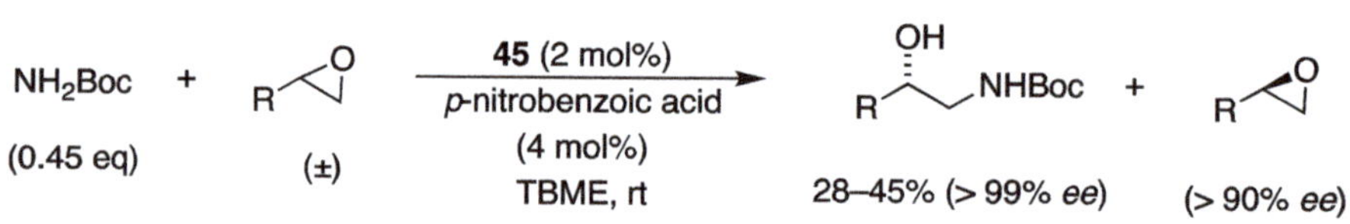

Scheme 36. The carbamolytic kinetic resolution of terminal epoxides employing Co(II)salen complex **45** as a pre-catalyst (Figure 20).

The use of ionic liquids as the reaction medium is considered an environmentally friendly alternative to traditional organic solvents [101]. This was exploited by Kureshy and coworkers in several studies on the KR of epoxides. In one such study, the protocol presented in Scheme 36 was further improved by the use of ionic liquids as reaction medium [102]. Performing the reaction in [bmim][PF_6] afforded both ring-opened products and unreacted epoxides in excellent yields and enantioselectivities for a number of glycidyl ethers and terminal alkyl epoxides (Scheme 37). The reactions were completed in 5–10 h and the catalyst and ionic liquid could be recycled six times with maintained enantioselectivity and reactivity. The choice of nucleophile could be extended to urethane and benzyl carbamate with equally excellent yield and enantioselectivity.

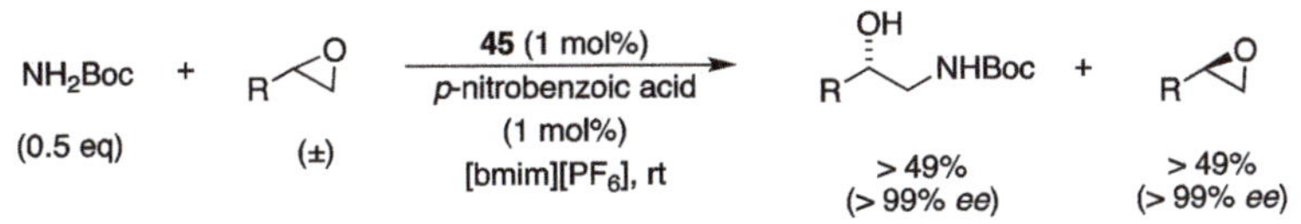

Scheme 37. The carbamolytic kinetic resolution of terminal epoxides in ionic liquid, employing Co(II)salen complex **45** (Figure 20) as the pre-catalyst.

In another study, Kureshy and coworkers investigated the KR of *trans*-stilbene oxide and *trans*-β-methyl styrene oxide with different anilines catalyzed by Cr–salen complex **53** (Scheme 31) [103]. Performing the reaction in [bmim][PF_6] (Figure 11) afforded ring-opened products in 40–49% yield and 63–99% *ee*. The unreacted epoxides could be recovered in quantitative yield and in 60–97% *ee*. The use of ionic liquids had several advantages, including easy product separation (extraction with *n*-hexane), efficient recyclability of the catalyst and the ionic liquid, and significantly reduced reaction times compared to reactions run in conventional organic solvent.

In another effort to improve the recyclability of the catalyst, Kureshy and coworkers employed the enantiomerically pure polymeric Cr–salen complex **57** (Figure 22) in the KR of *trans*-stilbene oxide and *trans*-β-methyl styrene oxide with anilines as nucleophiles [104]. The reactions were performed in dichloromethane. The desired vicinal amino alcohols were obtained in good to excellent yield (35–49%) and high enantiomeric excess (up to 100% *ee* after a single recrystallization). The unreacted epoxide could be recovered in excellent yield (48–51%) and good to high enantiomeric excess (70–98% *ee*). The catalyst could be easily recovered by precipitation with *n*-hexane and recycled up to four times with maintained enantioselectivity. In a separate study, the same reaction was carried out with dimeric and polymeric Cr–salen complexes under microwave irradiation in a CH_2Cl_2/MeOH mixture (1:1), which allowed for very short reaction times (2 min) [105]. The best results were achieved with dimeric complex **23** (Figure 7) in the KR of *trans*-stilbene oxide with different anilines. The ring-opened products were obtained in 45–49% yield and 88–94% *ee* and the unreacted epoxides were isolated in 47–48% yield and 80–92% *ee*. Both dimeric catalyst **23** and polymeric catalyst **57** could be recycled five times without loss of reactivity or enantioselectivity.

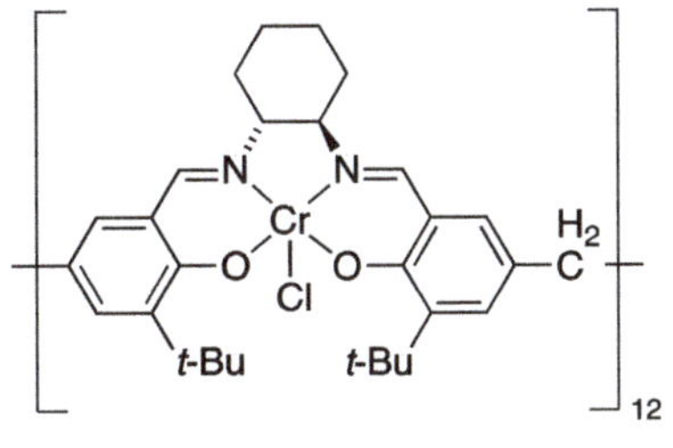

57

Figure 22. Polymeric Cr–salen complex **57**.

The same group also developed a series of Cr–salen complexes with cationic side groups (catalyst **58**–**60**, Figure 23), which were used in the KR of *trans*-epoxides with different anilines [106]. The best results were obtained with catalyst **60** (Figure 23) in the KR of *trans*-stilbene oxide in dichloromethane, where the ring-opened products were obtained in excellent yields (41–49%) and with high enantioselectivities (86–99% *ee*) and the unreacted epoxides were recovered in quantitative yield and 89–99% *ee*. The catalyst could be recycled up to six times without loss of reactivity or enantioselectivity.

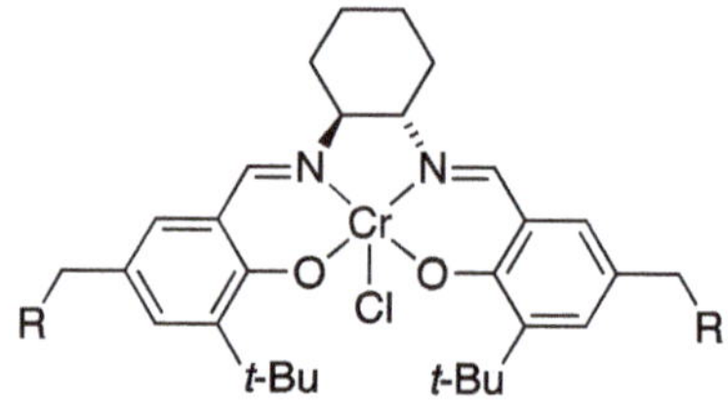

Figure 23. Cr–salen complexes with cationic side chains.

Kureshy and coworkers investigated macrocyclic Cr–salen complex **35** (Figure 14) in the KR of *trans*-epoxides with anilines The ring-opened products were obtained in excellent yields (46–49%) and with high enantioselectivities (up to > 99% *ee*), with concomitant recovery of the unreacted epoxides in quantitative yields and with high enantioselectivities (up to > 99% *ee*) [56]. The reactions were performed in a CH_2Cl_2/MeOH mixture and the catalyst showed excellent recyclability. The same group also developed another set of macrocyclic Cr–salen complexes [107]. Complex **61** (Figure 24) was used as a catalyst for the synthesis of pharmaceutically important β-amino-α-hydroxyl ester derivatives by the KR of aromatic ester epoxides with anilines. The ring-opened products were obtained in high yields and with high diastereoselectivities and enantioselectivities (Scheme 38). The catalyst showed good stability and good recyclability.

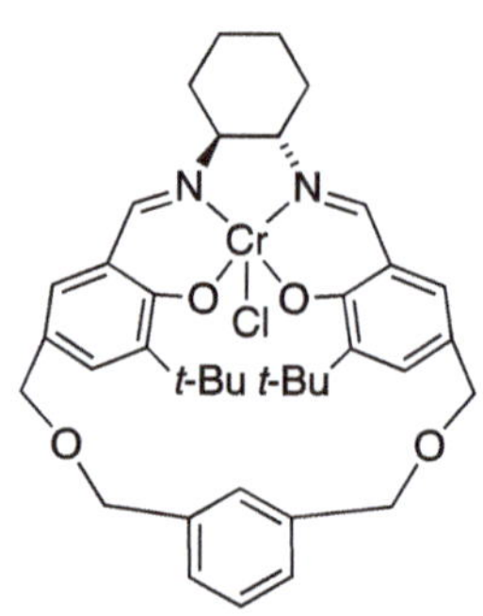

61

Figure 24. Macrocyclic Cr–salen complex **61**.

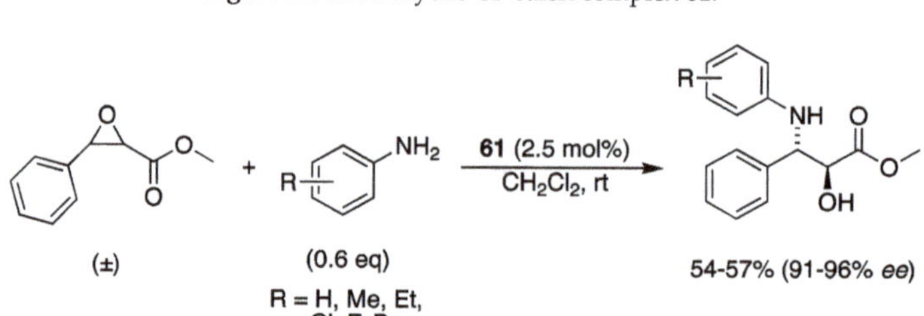

Scheme 38. The aminolytic kinetic resolution of aromatic ester epoxides catalyzed by macrocyclic Cr–salen complex **61** (Figure 24).

For the KR of epoxides with carbamates as nucleophiles, Kureshy and coworkers developed a number of chiral polymeric Co–salen complexes with chiral and achiral linkers. The complexes were evaluated as catalysts in the KR of a number of glycidyl ethers and terminal alkyl epoxides with different carbamates [108]. The reaction with 1 mol% of catalyst **62** (Figure 25) in dichloromethane afforded both epoxides and *N*-protected vicinal amino alcohols in quantitative yield and with excellent enantioselectivity (>99% *ee*) in 16 h. The catalyst could be precipitated with *n*-hexane and reused up to six times with complete retention of enantioselectivity.

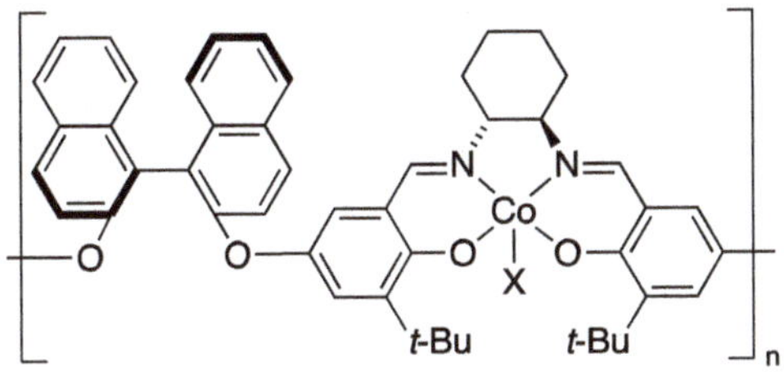

62 (X = *p*-nitrobenzoic acid)

Figure 25. Polymeric Co–salen complex **62**.

Jacobsen and coworkers showed that their oligomeric Co–salen complex **35** (Scheme 10) could be used in the KR of 1,2-epoxyhexane with *tert*-butyl carbamate, where a low catalyst loading (0.2 mol%) was enough to obtain the ring-opened product in high yield and enantioselectivity (Scheme 39) [58].

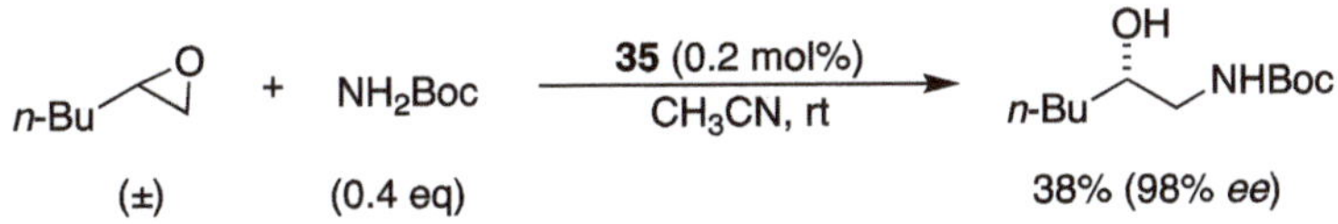

Scheme 39. The carbamolytic kinetic resolution of 1,2-epoxyhexane catalyzed by oligomeric Co–salen complex **35** (Scheme 10).

3.3. *With Water (Hydrolytic Kinetic Resolution)*

Since the first report by Jacobsen in 1997, the hydrolytic kinetic resolution (HKR) of terminal epoxides has been one of the most researched applications of metal–salen catalysts. The use of water as an inexpensive and environmentally friendly nucleophile also makes it an attractive method for the enantioenrichment of epoxides otherwise difficult to obtain. The protocols have mainly used Co(III)salen catalysts and the reactions are characterized by excellent yields and enantioselectivities. Hence, instead of only focusing on obtaining high yields and enantioselectivity, much effort has been focused on increasing cooperativity and reaction kinetics, decreasing catalyst loading, as well as developing heterogeneous systems and new catalytic methodologies in general.

In their original paper, Jacobsen and coworkers found that Co(III)salen complex **63** (Scheme 40), formed in situ from the corresponding Co(II)salen complex, worked very well in the HKR of several terminal alkyl, alkenyl, and aryl epoxides, resulting in unreacted (*S*)-terminal epoxides and (*R*)-1,2-diols with excellent yields (up to 46% and 50%, respectively) and *ee* values (up to 99% and 98%, respectively) [109]. The catalyst could be easily regenerated by treatment with acetic acid in air and reused in two cycles without loss of reactivity or enantioselectivity. Later, the scope was extended to a broad group of epoxides with different steric and electronic environments [110]. In almost all cases, the unreacted epoxides could be obtained in > 99% *ee* (Scheme 40a). Furthermore, by tuning the equivalents of water and catalyst loading, most of the 1,2-diols could be obtained with excellent enantioselectivities (94–99% *ee*) (Scheme 40b).

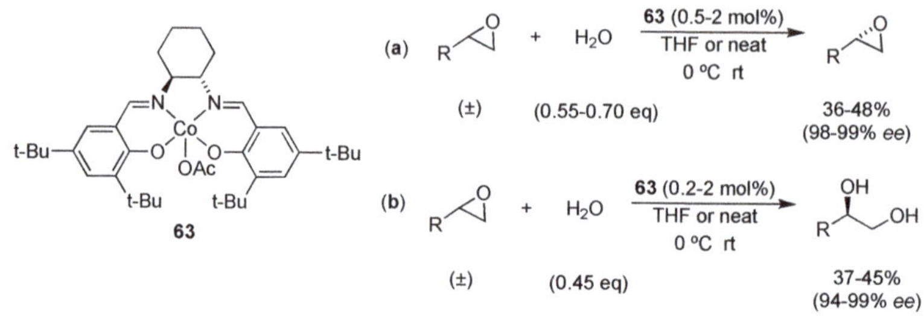

Scheme 40. The hydrolytic kinetic resolution of terminal epoxides catalyzed by Co–salen complex **63**. The displayed reaction conditions are optimized for enantioenrichment of (**a**) (*S*)-epoxides; (**b**) (*R*)-1,2-diols.

For cases where the diol was required in very high enantiomeric excess, a strategy called "double resolution" was suggested and employed in the synthesis of (*S*)-3-chloropropane-1,2-diol (Scheme 41). Epichlorohydrin was first subjected to HKR with (*R*,*R*)-salen complex **64** as the catalyst (Figure 26). The enantioenriched unreacted epoxide was then separated and subjected to another HKR with the opposite enantiomer of the catalyst, (*S*,*S*)-salen complex **63** (Scheme 40), which resulted in enantiopure (*S*)-3-chloropropane-1,2-diol in 41% overall yield.

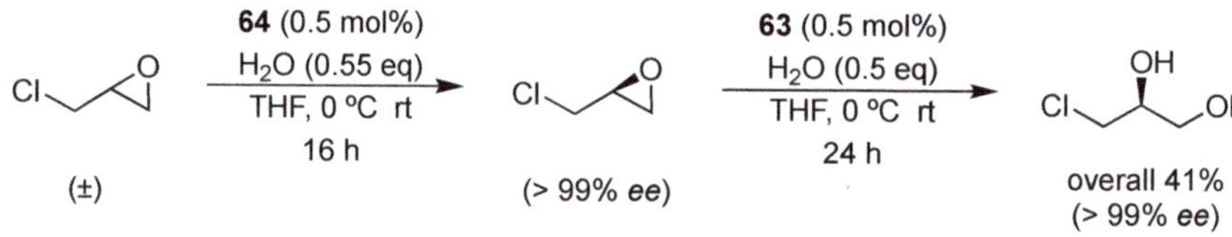

Scheme 41. Preparation of enantiopure (*S*)-3-chloropropane-1,2-diol using a "double resolution" strategy employing catalyst **63** (Scheme 40) and **64** (Figure 26).

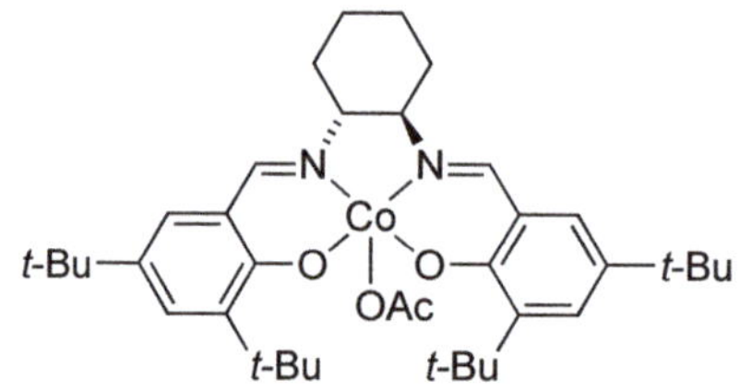

64

Figure 26. Co–salen complex **64**.

The high reactivity and excellent enantioselectivity of complex **63** and **64** in the HKR of a broad range of terminal epoxides set a high standard for catalysts for these reactions. Subsequent studies have largely focused on improving other aspects of the catalysis, including reducing reaction times (average reaction time for catalyst **63** was 12–18 h, sometimes up to 72 h), reducing catalyst loading and enabling solvent-free reactions. Another important issue is catalyst recycling. For instance, homogenous catalyst **63** was recycled as a solid residue by distilling off the diols and unreacted epoxides during work up, a method which is time-consuming and only applicable for sufficiently volatile products. In addition, catalyst **63** needed to be regenerated with acetic acid in air before it could be recycled.

3.3.1. Multi-Metallic Catalysts

Similar to what was observed for the azidolytic ARO of *meso*-epoxides catalyzed by Cr–salen complexes, preliminary kinetic studies indicated a second-order dependence on catalyst **63** in the HKR of terminal epoxides [109]. Consequently, a similar cooperative bimetallic pathway was proposed as a possible mechanism for the HKR of epoxides (Figure 1). Since then, numerous studies have reported multi-metallic catalysts where several salen moieties are linked together in order to facilitate this bimetallic pathway. Larger systems also have the advantage of often being less soluble and therefore more easily recyclable.

Kureshy and coworkers [111,112] synthesized dimeric Co-complexes **65** and **66** (Figure 27). Two Co(III)salen moieties were covalently linked by a methylene or propane-2,2-diyl group, with the expectation that the increased molecular weight would result in an easier catalyst recovery compared to the monosalen complexes. This very straight-forward way of linking two salen moieties together was also employed by Jacobs and coworkers [44] in their design of dimeric Cr–salen complex **23** (Figure 7), which was impregnated on silica and used as a heterogeneous catalyst for the azidolytic ring-opening of *meso*-epoxides. Complex **65** was employed in the HKR of epichlorohydrin, propylene oxide and styrene oxide under solvent-free conditions. Complex **66** was investigated in the HKR of epichlorohydrin and a variety of alkyl terminal epoxides. For all substrates, the ring-opened diols and the unreacted epoxides were obtained with excellent yields (up to 50% for the diols and up to 52% for the epoxides) and enantiomeric excesses (up to 99% for both the diols and the epoxides). In addition, the dimeric complexes could be recycled and reused for up to four cycles without significant loss of reactivity or enantioselectivity [111,112].

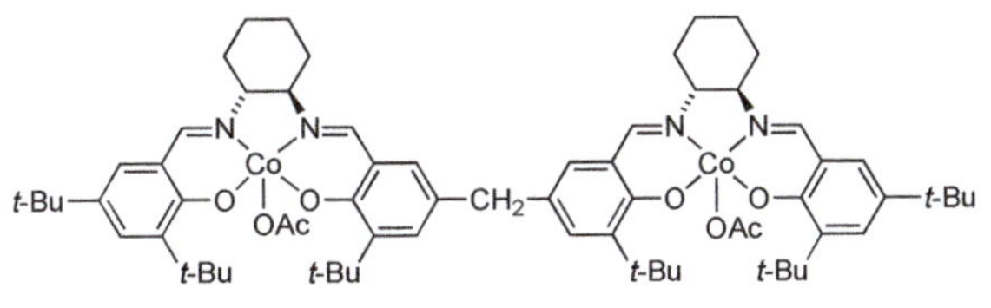

65

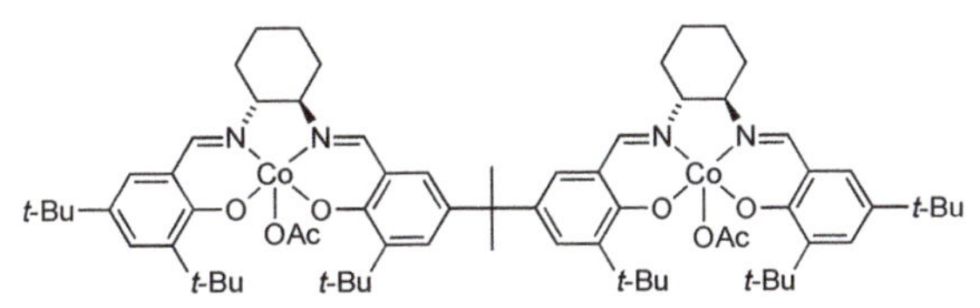

66

Figure 27. Dimeric Co(III)salen complexes **65** and **66**.

In an effort to design versatile and easily recyclable Co–salen complexes capable of cooperative catalysis, Jones and coworkers synthesized styryl-substituted bissalen complex **67**, which could be polymerized to give access to homogeneous catalysts **68** and **69** and heterogeneous catalyst **70** (Figure 28) [113]. All the complexes were investigated as catalysts for the HKR of epichlorohydrin under solvent-free conditions, with reaction conditions optimized for the enantioenrichment of the unreacted epoxide. For homogenous catalysts **67–69**, full conversion of one of the enantiomers (50%) was achieved in 7 h with only 0.02 mol% Co loading, affording highly enantioenriched unreacted epoxide (99% *ee*). As a comparison, the corresponding monosalen catalyst **64** (Figure 26) afforded the unreacted epoxide in 13% *ee* under the same reaction conditions and reaction time. The increased enantioselectivity and reactivity was attributed to the facilitated cooperative pathway with the dimeric catalyst. For the heterogenous catalyst **70**, a slightly higher catalyst loading (0.04 mol% Co) was needed. At this catalyst loading, catalyst **70** achieved full conversion of one of the enantiomers of the epoxide and 99% *ee* for the unreacted epoxide of the opposite absolute configuration in 5 h. Furthermore, the insoluble catalyst **70** could be recovered by simple filtration, then regenerated and reused. The enantioselectivity of catalyst **70** was retained over three cycles, although each cycle required longer reaction time to reach full conversion.

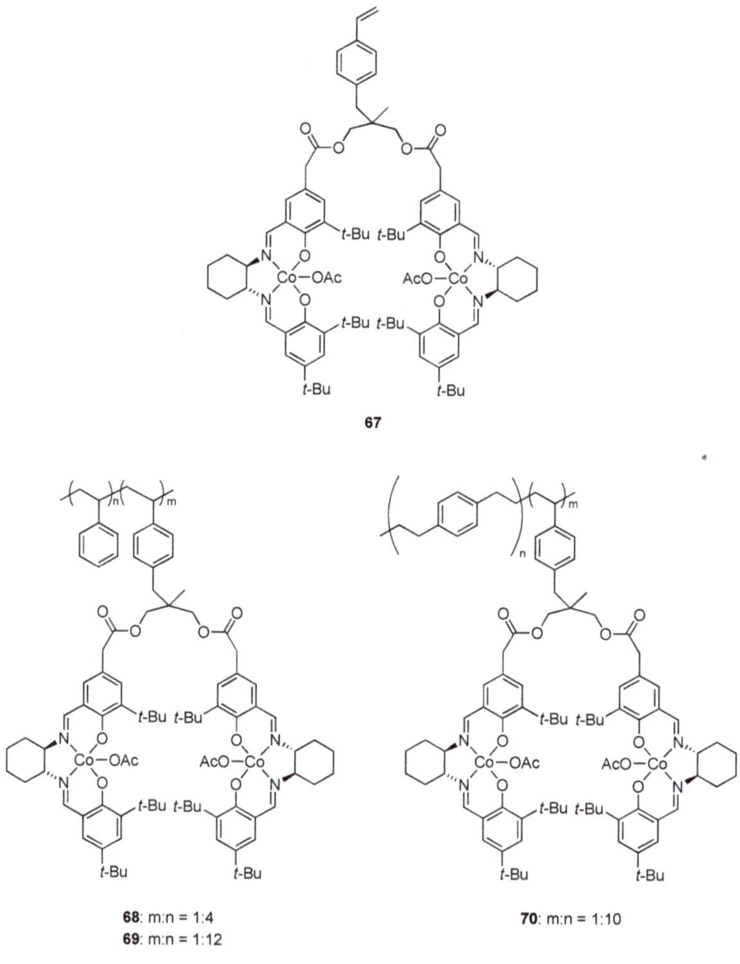

Figure 28. Dimeric Co–salen complex **67** and polymeric complexes **68–70**.

Calix[4]arene-based dimeric Co(III)salen complex **71** (Figure 29) was reported by Wezenberg and Kleij [114]. Two salen moieties were installed on the upper rim of a calix[4]arene, with the expectation of facilitating the cooperative pathway in the HKR of terminal epoxides. Catalyst **71** and monosalen analogues **72** and **73** (Figure 29) were separately employed in the HKR of 1,2-epoxyhexane, epichlorohydrin and styrene oxide in acetonitrile, with conditions optimized for the production of the enantioenriched epoxide. All catalysts gave comparable conversions (32–52%), while catalyst **71** gave slightly lower *ee* values of both the ring-opened product and the unreacted epoxide (83–91% and 83–97%, respectively) than monosalen analogues **72** and **73** (88–95% and 97–99%, respectively). Kinetic studies showed that the HKR reactions employing catalyst **71** predominantly followed an intramolecular cooperative pathway. Comparison of the intra- and intermolecular rate constants for **71** and **72** revealed that although **71** had a significant intramolecular rate constant, as expected, the intermolecular rate constant was lower than for monosalen analogue **72**. This is remarkable since many other bimetallic salen catalysts show both increased intra- and intermolecular rate constants in the HKR reaction compared to the corresponding monosalen catalysts [21]. For catalyst **71**, the lower intermolecular rate constant meant that at higher catalysts loadings, the overall initial rate was lower than for the monosalen analogue **72**, while at lower catalyst loadings the intermolecular pathway was suppressed and catalyst **71** showed higher reaction rates. The authors suggest that the decreased intermolecular reaction rate could be caused by the way the salen moieties were immobilized on the calixarene [114].

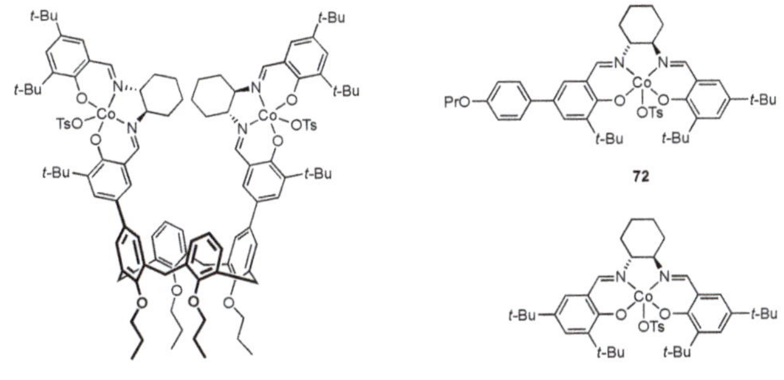

Figure 29. Calix[4]arene-based dimeric Co–salen complex **71** and monomeric analogues **72** and **73**.

Jacobsen and coworkers [58,69,115] developed three generations of oligomeric Co(III)salen complexes. Apart from catalyzing the ARO of *meso*-epoxides with water and carbamate as nucleophiles, the first generation (catalyst **40**, Figure 18) also catalyzed the HKR of terminal epoxides [69]. With enantioselectivity similar to monosalen complex **64** (Figure 26), catalyst **40** allowed a 10- to 50-fold decrease in catalyst loading (mol% Co) and up to 16-fold decrease in reaction time compared to catalyst **64** (Table 1). For the second generation of catalysts (complex **74** and **75** in Figure 30), the major difference was the linker. The use of a pimelate tether gave an oligomeric complex with a more predictable and reliable conformation [115]. Both complexes worked very well in the HKR of terminal epoxides, and the catalyst loading and reaction time were further decreased compared to the first generation (Table 1). The third generation (catalyst **35**, Scheme 10) exhibited better solubility and could be synthesized with high overall yield on a large scale [58]. The catalyst loading in the HKR of terminal epoxides could be reduced down to 0.003 mol% Co while retaining excellent enantioselectivity. Furthermore, its lipophilicity made catalyst **35** a good catalyst for solvent-free reactions (Table 1).

Table 1. Comparison of catalytic properties of oligosalen Co(III) complexes in the hydrolytic kinetic resolution (HKR) of terminal epoxides.

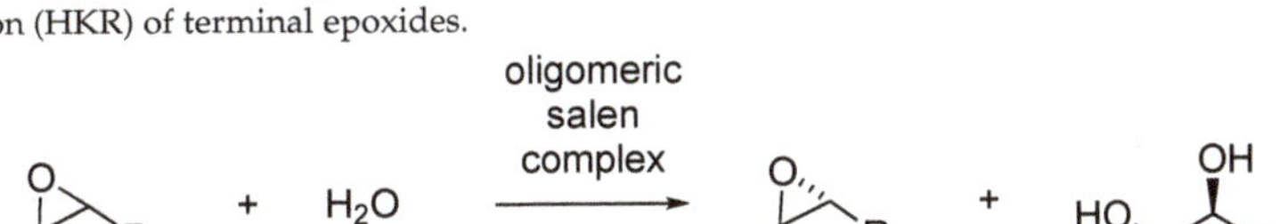

Catalyst	Lowest mol% Co	Epoxide	Diol	Solvent
63 or **64**	0.5	36–48% (98–99% *ee*)	37–45% (94–99% *ee*)	solvent-free or THF
40	0.01	36–45% (99% *ee*)	40–50% (94–98% *ee*)	solvent-free
74 or **75**	0.0004	44–45% (>99% *ee*)	44–51 (97% *ee*)	MeCN or MeCN/CH_2Cl_2
35	0.0003	35–44% (>99% *ee*)	-	solvent-free

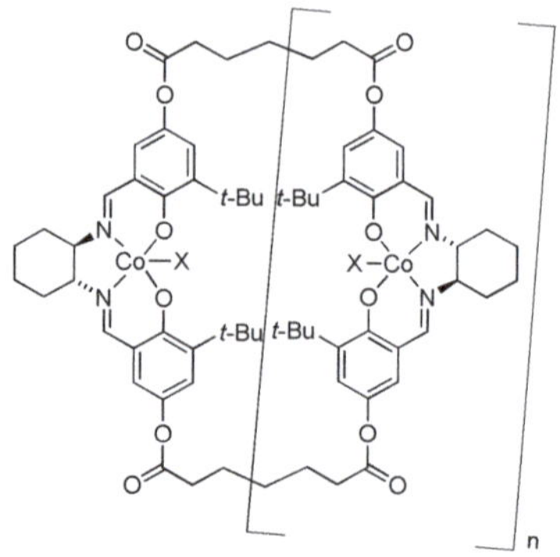

74: X = (1*S*)-(+)-10-camphorsulfonate, n = 1-3
75: X = 3-NO_2-$C_6H_4SO_3$, n = 1-3

Figure 30. Jacobsen and coworkers' [115] second generation of oligosalen catalysts **74** and **75**.

Besides Jacobsen, Weck, and coworkers [48] also developed macrocyclic oligomers. They reported macrocyclic Co(III)salen oligomers **76** and **77** (Figure 31). Unlike other macrocyclic oligosalen complexes where the salen was part of the macrocycle, oligomers **76** and **77** consisted of a central macrocyclic backbone onto which the salen moieties were connected in a pendant-like fashion. This added flexibility afforded excellent reactivity and enantioselectivity in the HKR of terminal epoxides with very low catalyst loadings (down to 0.01 mol% Co). For a number of terminal alkyl epoxides and glycidyl ethers, the unreacted epoxides could be obtained in up to 48% yield and >99% *ee* under solvent-free conditions. More sterically hindered epoxides such as styrene oxide and *tert*-butyloxirane could also be resolved with similar reactivities and enantioselectivities, but higher catalyst loading (0.1–0.25 mol% Co) and longer reaction times were required. Interestingly, a linear polymeric analogue **78** (Figure 31) gave poorer results than macrocyclic catalysts **76** and **77** in terms of both reactivity and enantioselectivity, which reinforced the important role of the macrocyclic structure for this class of catalysts. A later study from the same group described the copolymerized cross-linked macrocyclic Co-oligosalen mixture **79** (Figure 32). With similar catalytic properties to **76** and **77**, oligomeric mixture **79** was more synthetically available and, therefore, more practical for larger scale catalysis [116].

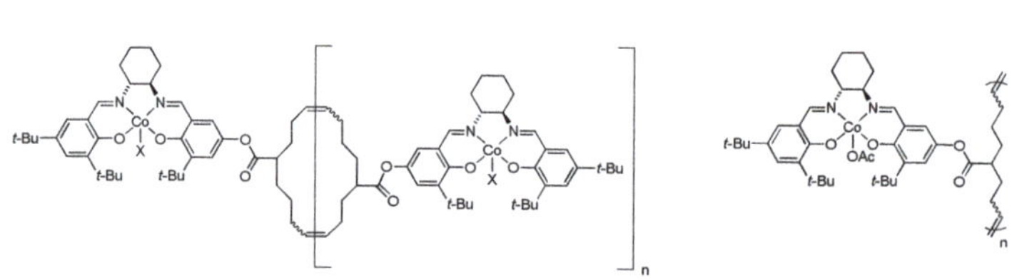

76: X = OAc, n = 1-9
77: X = OTs, n = 1-9

78: n = 48

Figure 31. Macrocyclic oligosalen catalysts **76** and **77**, and linear polymeric analogue **78**.

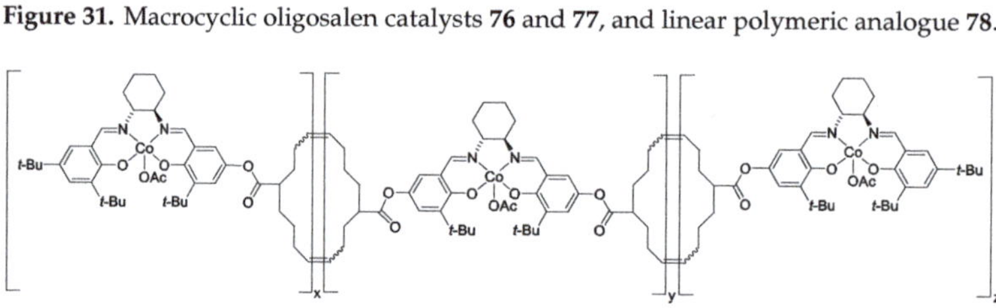

79

Figure 32. Cross-linked macrocyclic oligosalen catalyst **79**.

Khan and coworkers [117] designed a group of cyclic bissalen complexes where two Co–salen moieties were tethered by ethylene glycol chains (**80–84** in Figure 33). Co-complexes **80, 81**, and **84** catalyzed the HKR of a number of terminal epoxides and glycidyl ethers with excellent yields (46–53% for the so formed 1,2-diols and 43–47% for the unreacted epoxides) and enantioselectivities (92–96% *ee* for the 1,2-diols and 96–99% *ee* for the unreacted epoxides) under solvent-free conditions and at low catalyst loadings (down to 0.016 mol% Co). Catalyst **84** could be recycled up to three times with maintained reactivity and enantioselectivity and could be used directly without catalyst regeneration between cycles. For the HKR of epichlorohydrin, catalyst **84** maintained its performance on a multigram scale, illustrating the scalability of the protocol. (*S*)-Propylene oxide obtained from this methodology was applied in a short synthesis of (*R*)-mexiletine with high overall yield (80%) and *ee* (98%).

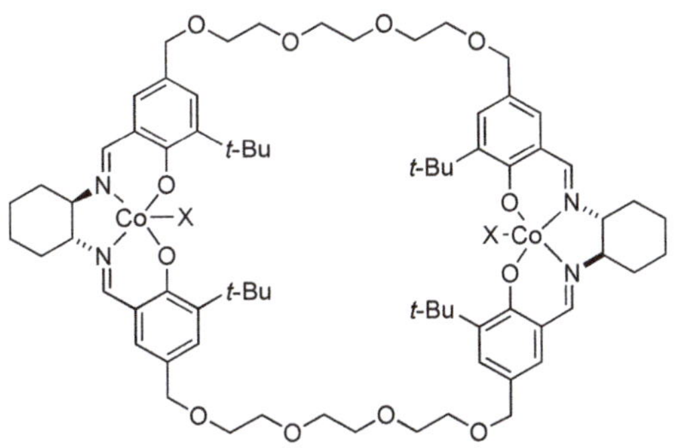

80: X = OAc **81**: X = OTs **82**: X = BF_4
83: X = SbF_6 **84**: X = PF_6

Figure 33. Cyclic bissalen complexes **80–84** containing ethylene glycol chains. The counterions (X) are formally illustrated as bound to the metal.

Kureshy and Ganguly [118] reported a series of polymeric catalysts where the Co–salen moieties were tethered by different chiral and achiral linkers. Preliminary studies of the HKR of 1,2-epoxyhexane revealed that the highest enantioselectivity was obtained with polymeric Co–salen complexes with chiral BINOL linkers where the absolute configuration of the BINOL linker and the salen ligand were opposite (i.e., (*S,R,R*) in catalyst **85** and (*R,S,S*) in catalyst **86**, Figure 34). The preferred enantiomeric pairing of the two components was supported by DFT calculations comparing (*R,S,S*) catalyst with (*S,S,S*) catalyst. When employed in the HKR of a broad range of terminal alkyl and aryl epoxides and glycidyl ethers, catalyst **86** enabled excellent yields and enantioselectivities of both the ring-opened products and the unreacted epoxides (Scheme 42). Catalyst **86** could be recycled six times in the HKR of 1,2-epoxyhexane without loss of enantioselectivity, and only required one catalyst regeneration (between cycle 4 and 5). To show the scalability of the protocol, catalyst **86** was successfully employed in gram-scale syntheses of chiral β-blockers (*S*)-metoprolol, (*S*)-toliprolol and (*S*)-alprenol, where the final products were obtained in moderate overall yields (up to 44%) and excellent enantioselectivities (>99%).

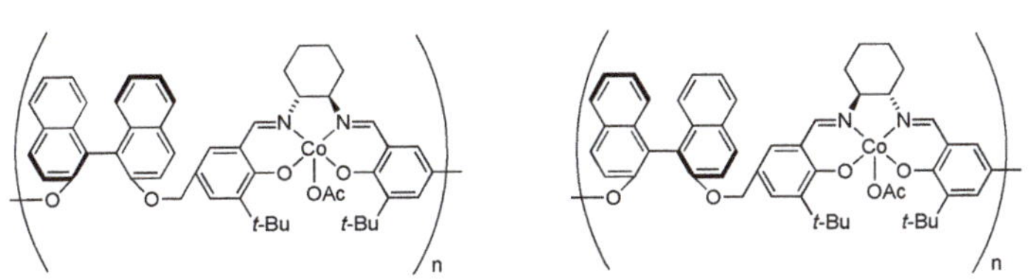

85 **86**

Figure 34. Polymeric Co–salen catalysts with 1,1′-bi-2-naphthol (BINOL) linkers.

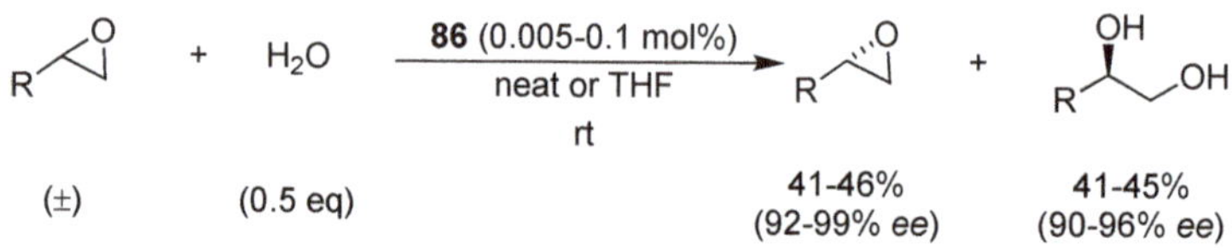

Scheme 42. The hydrolytic kinetic resolution of terminal epoxides catalyzed by catalyst **86** (Figure 34).

Schulz and coworkers [119] reported the polymeric bithiophene-linked Co–salen catalyst **87** (Figure 35). The insoluble catalyst was employed in the dynamic HKR of epibromohydrin in THF. The catalyst was successfully reused in 11 subsequent catalytic runs, giving ring-opened products in good to excellent yield (60–99%) and with consistent enantioselectivity (around 84% *ee*). In a following study, the same group synthesized cyclic calixsalen catalyst **88**, which contained a phenyl linkers instead of the previously used thiophene linkers (Figure 35) [120]. Catalyst **88** could be used either as a pure trimer, a pure tetramer or as a mixture. The best results were obtained with the tetramer, which afforded the ring-opened product in full conversion (>99%) and with high enantioselectivity (92% *ee*) when used in the dynamic HKR of epibromohydrin in THF. The trimer exhibited significantly lower reactivities, both in terms of conversion and reaction rate. All the catalysts could be easily recovered from the reaction mixture by filtration and reused with maintained enantioselectivity, although the yield decreased after each run.

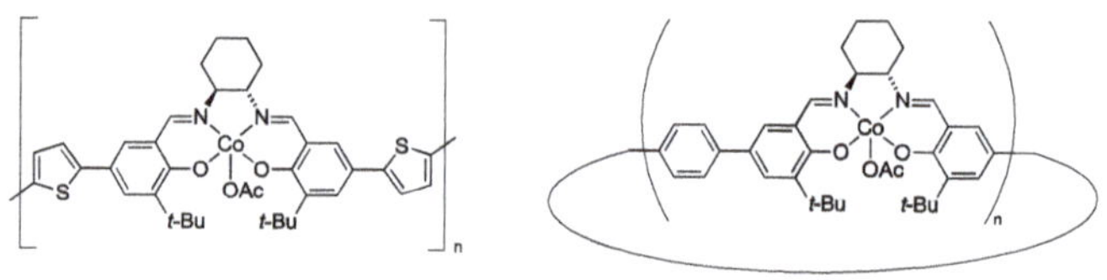

87

88: n = 3-4

Figure 35. Polymeric Co–salen catalyst **87** and calixsalen Co catalyst **88**.

In addition to dimers and macrocyclic oligomers, there are also some examples of other types of multi-salen structures. Jacobsen and coworkers synthesized dendrimeric Co-complexes based on polyamidoamine (PAMAM) central core containing 4, 8, and 16 salen moieties (represented by 8-dendrimer **89** in Figure 36), expecting that the dendritic framework would enforce the bimetallic cooperative pathway in the HKR of 1,2-epoxyhexane in THF [121]. The dendrimeric catalysts exhibited higher reactivities than monomeric and dimeric analogues **90** and **91** (40–42% yield compared to 29–37% yield for the ring-opened product), as well as significantly higher rate constants. All catalysts (including **90** and **91**) afforded the ring-opened product with excellent enantioselectivity (>98% *ee*). For the enantioenrichment of the unreacted epoxide, it was reported that catalyst **89** effected the HKR of 2-cyclohexyloxirane with 50% conversion and 98% *ee*.

Zheng and coworkers developed linear and cross-linked polymeric Co–salen complexes. They reported the linear polymeric Co–salen catalysts **92** and **93** (Figure 37), which catalyzed the HKR of propylene oxide, epichlorohydrin and phenyl glycidyl ether with excellent conversions (up to > 49%) and *ee* values (up to 98% for the unreacted epoxides and the 1,2-diols) in organic solvents (THF and CH_2Cl_2) or under solvent-free conditions [122]. No obvious differences in catalytic performances were observed between the two polymeric catalysts.

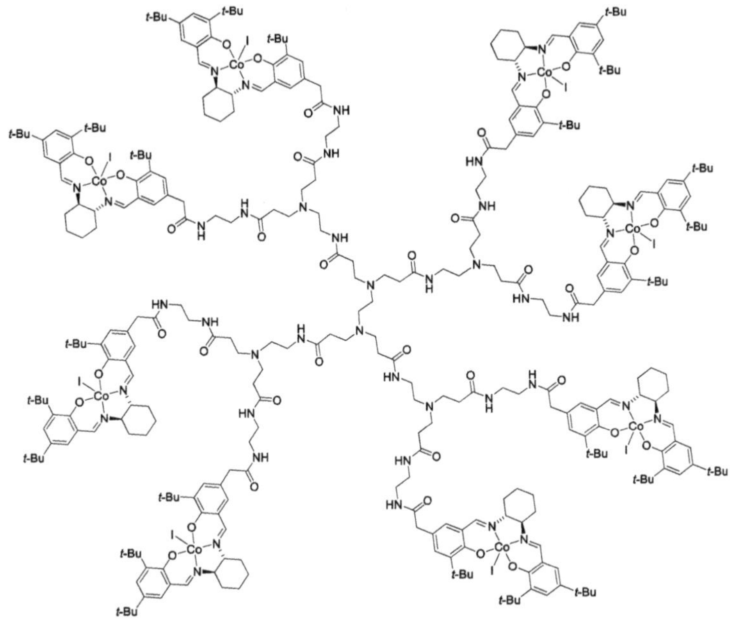

89

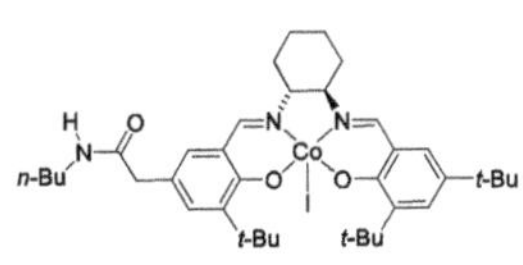

90

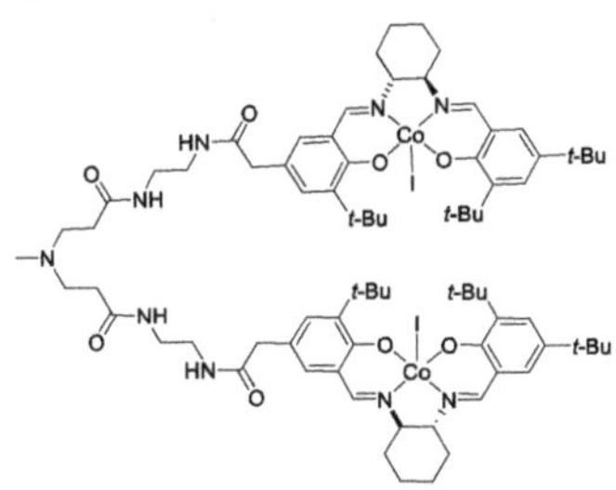

91

Figure 36. Polyamidoamine-based 8-dendrimer **89** and its monomeric (**90**) and dimeric (**91**) analogues.

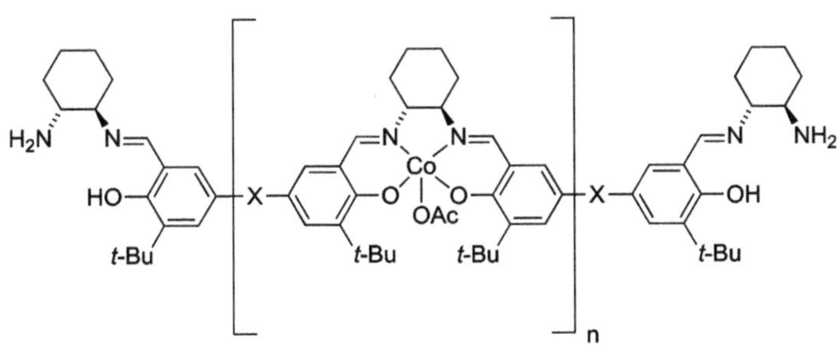

92: X = $C(CH_3)_2$
93: X = CH_2

Figure 37. Linear polymeric Co–salen catalysts **92** and **93**.

The same group also reported the synthesis and catalytic performance of a set of cross-linked copolymers [123]. Catalyst **94** was synthesized by the condensation of enantiopure *trans*-1,2-diaminocyclohexane and a mixture of dialdehyde **95** and trialdehyde **96** (Scheme 43). A series of catalysts, oligomers **94**, were thus synthesized with different ratios of the two aldehydes (ranging from 100% dialdehyde **95** to 100% trialdehyde **96**). The catalysts were applied in the HKR of epichlorohydrin, styrene oxide, and phenyl glycidyl ether under solvent-free conditions. Conversions of 43–53% were reported, with high enantioselectivities for both the ring-opened products and the unreacted epoxides (up to > 99% *ee* and up to 97% *ee*, respectively). The mixed cross-linked polymeric catalysts exhibited slightly higher reactivity than the catalysts based on either pure dialdehyde or pure trialdehyde. Recycling of the catalysts was unsuccessful due to the decomposition of the catalyst during the reaction. The authors attributed this to the hydrolysis of the ester groups under the reaction conditions.

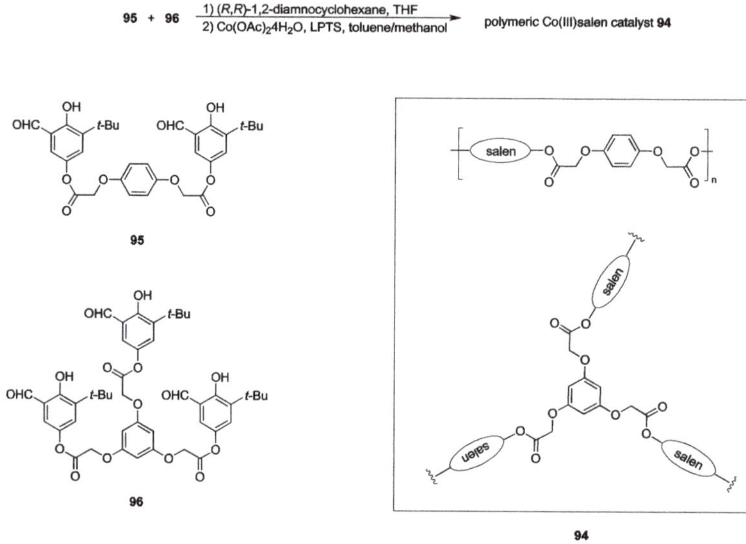

Scheme 43. The synthesis of cross-linked polymeric Co–salen catalyst **94** from dialdehyde **95** and trialdehyde **96**. LPTS = lutidinium *p*-toluenesulfonate.

Weck and Jones [124] reported a set of homopolymerized and copolymerized Co(III)salen complexes, where the salen moieties were installed in a pendant-like fashion on the polystyrene chain (**97**–**100** in Figure 38). Catalysts with different ratios of salen moieties and styrene moieties were investigated in the HKR of epichlorohydrin in CH_2Cl_2. All the catalysts gave 49–55% conversion and afforded the unreacted epoxide with excellent enantioselectivities (>99% *ee*). Copolymeric catalysts **99** and **100** exhibited improved reactivity compared to homopolymeric catalyst **97** (1 h reaction time compared to 2 h). The authors hypothesized that the lower ratio of the salen moieties on the polystyrene chain could make the catalytic sites more accessible to the substrate. The same group also evaluated the optimal flexibility of polymeric Co–salen catalysts by synthesizing a series of homopolymeric catalysts with different lengths of the linker connecting the Co–salen complexes to the polymeric backbone (catalyst **101**–**104** in Figure 38) [125]. The catalysts were employed in the HKR of epichlorohydrin. Kinetic studies revealed that a 6-atom distance (corresponding to m = 1, catalyst **102**) between the polymer backbone and salen unit gave the highest reactivity and enantioselectivity. Catalyst **102** executed the successful HKR of a number of terminal epoxides under solvent-free conditions and with low catalyst loading (0.01–0.1 mol%), affording unreacted epoxides in 43–46% yield and ≥98% *ee*. These papers illustrate the importance of the flexibility and composition of polymer-supported Co–salen catalysts for the HKR of terminal epoxides.

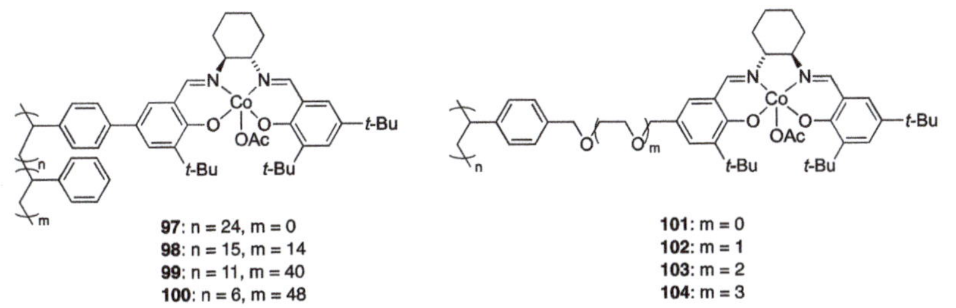

Figure 38. Homo- and copolymerized Co–salen complexes **97–100** and homopolymerized Co–salen complexes **101–104**.

Weberskirch and coworkers [126] synthesized the first water-soluble amphiphilic block copolymeric Co–salen catalyst for the employment in HKR reactions (Figure 39) epoxides. The polymeric complex **105** could aggregate into micellar assemblies in aqueous environment, forming a hydrophobic core with a high local concentration of the active catalyst. The so-formed catalyst **105** enabled the efficient and enantioselective HKR of a number of terminal aryl epoxides and glycidyl ethers. The reactions proceeded with 50–54% conversion and with high-to-excellent enantioselectivities (up to 95% *ee* for the 1,2-diols and >98% *ee* for the unreacted epoxides). The catalyst could be recycled and reused in four successive cycles with consistent *ee* values for the unreacted epoxides, although the reaction time needed to be increased after each cycle to maintain the yield.

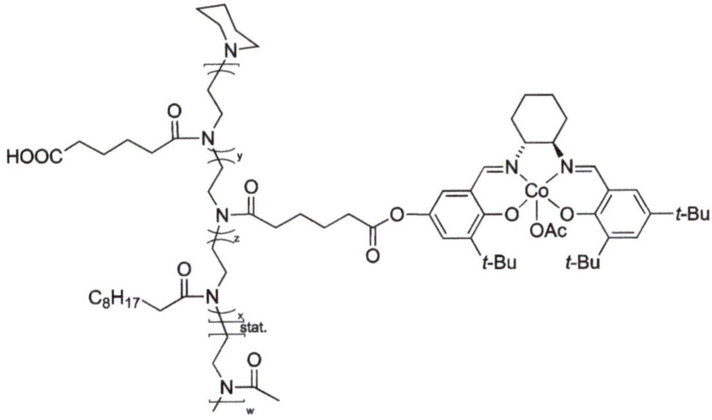

Figure 39. Amphiphilic copolymerized Co–salen catalyst **105**.

As described above, covalently linking metal–salen moieties together is one approach to obtaining multi-metallic catalysts. However, these larger motifs often require considerable synthetic efforts. In addition, many of the oligomeric complexes and polymers are prepared and used as mixtures in the catalysis, making it more complicated to elucidate the most active structure and understand how efficient catalysis is achieved. One alternative that could reduce the synthetic cost is the use of multi-metallic metal–salen assemblies based on non-covalent interactions. The most attractive case is one where the assembly can be completely controlled and only one well-defined catalyst is formed.

Following this line, Kim and coworkers [127] developed a series of dimeric catalysts where two identical Co–salen complexes were connected through the coordination of one oxygen of each salen unit to an Al(III)-containing Lewis acid (catalyst **54** in Scheme 34). The dimeric catalysts could induce high reactivities (41–46% yield for the ring-opened products and 40–46% yield for the unreacted epoxides) and enantioselectivities (up to 86% *ee* for the ring-opened products and up to 99% *ee* for the unreacted epoxides) in the HKR of epichlorohydrin, 1,2-epoxybutane, and glycidol, whereas monomeric analogues catalyzed the reactions with significantly lower reactivities and enantioselectivities. The catalysis could be performed either in THF or under solvent-free conditions. The same strategy was also extended to Lewis acids of other group 13 elements (catalyst **55** and **56** in Scheme 34) [128]. The dimeric Co–salen complex **55** (linked by $GaCl_3$) was employed in the HKR of a wide range of 3-substituted propylene oxides and glycidyl ethers, affording unreacted epoxides in 40–49% yield and 97–99.8% *ee*. Ring-opened products were obtained in 42–50% yield and >85% *ee*. The reactions were performed under solvent-free conditions in THF or in a CH_2Cl_2/THF mixture. Kinetic studies showed the participation of both intramolecular and intermolecular pathways in the HKR reactions and significantly higher reaction rates for dimeric complexes than monomeric analogues. Other Lewis acids ($ZnCl_2$, $FeCl_3$, $SnCl_4$, etc.) have also been attached to Co(III)salen complexes, and the so-formed monomeric catalysts exhibited improved reactivities in comparison with monosalen complex **64** (Figure 26) in the HKR of terminal epoxides [129–131].

Supramolecular interactions, such as aromatic donor–acceptor interactions and hydrogen bonding, have also been utilized to construct multi-metallic catalysts with controlled structures. The previously mentioned aromatic donor–acceptor complexes **42**–**44** (Figure 19) also worked very well as catalysts for the HKR of various terminal epoxides such as styrene oxide and sterically hindered *tert*-butyloxirane in $CHCl_3$ or CCl_4 [70]. The resolved epoxides were isolated in high yields (>40%) and enantioselectivities (up to 99% *ee*).

Hong and coworkers [132] designed and synthesized bis-urea-functionalized Co–salen complexes capable of forming self-assembled hydrogen-bonding structures in solution. The complex with *p*-CF_3-phenyl substituents on the urea groups gave the highest rate constant in the HKR of epichlorohydrin (catalyst **106** in Figure 40). Catalyst **106** was investigated in the HKR of epichlorohydrin, allyl glycidyl ether, 1,2-epoxybutane and 1,2-epoxyhexane under solvent-free conditions, affording unreacted epoxides in 41–43% yield and 99% *ee*. In addition, complex **106** showed significantly higher reactivity than the monosalen complex **64** (Figure 26) at the same catalyst loading. Kinetic and self-association studies supported the hypothesis that the observed rate enhancement could be attributed to the enforcement of the bimetallic mechanism by the proximal self-association of Co–salen units through urea–urea hydrogen-bonding.

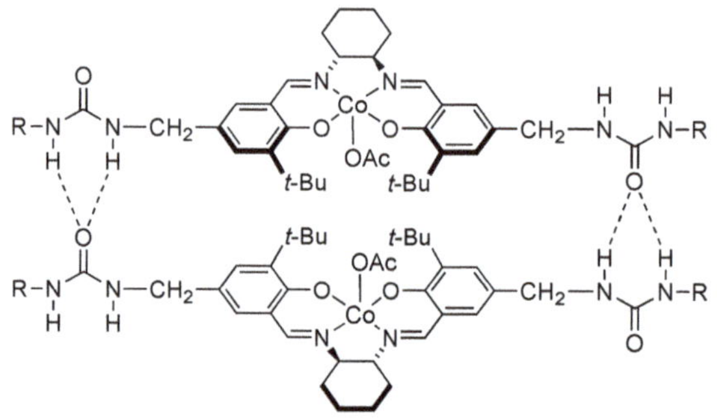

Figure 40. Bis-urea-functionalized Co–salen catalyst **106**, displayed as hydrogen-bonded dimer.

Another approach to achieving more efficient catalysis is to utilize space constraints. Li and Yang [133] placed homogenous monosalen catalyst **64** (Figure 26) in a nanocage of mesoporous silica SBA-16 (one catalyst per cage) via the *ship in a bottle* strategy. A key step was tuning the size of the pores so that the reactants used for synthesizing the catalytically active complex (salen ligand and cobalt salt) were allowed to enter the nanocage but the formed complex was hampered from escaping. The so-immobilized catalyst was easy to recycle and was effective for up to 11 successive runs, but the system gave lower enantioselectivity and reactivity than reactions performed with catalyst **64** under homogeneous conditions. To enable the bimetallic cooperative pathway, the same group described a new catalytic system where two Co–salen complexes were accommodated in one nanocage (Scheme 44) [134]. With nanoreactor **107**, the HKR of propylene oxide in CH_2Cl_2 could be completed with full conversion (50%) and excellent enantioselectivity (98% *ee* for both the unreacted epoxide and the 1,2-diol) at very low catalyst loading (1:12000 Co–salen/epoxide).

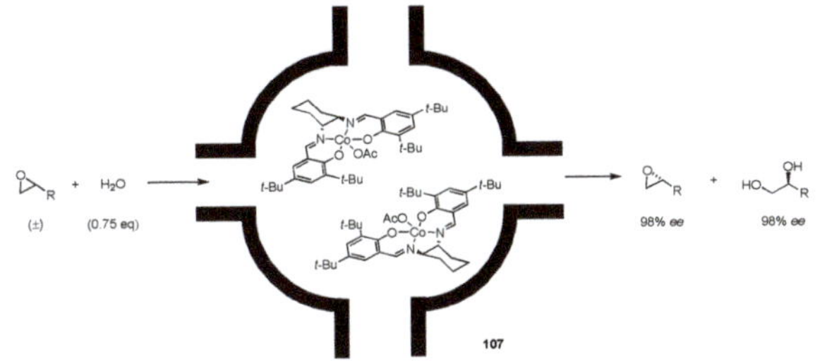

Scheme 44. The hydrolytic kinetic resolution of terminal epoxides performed in nanoreactor **107** in CH_2Cl_2.

3.3.2. Immobilized Catalysts

In the HKR of terminal epoxides catalyzed by Co(III)salen complexes, the main recycling procedure for most homogenous catalysts is to distill off all the volatile fractions and collect the solid residue. However, distillation is not an ideal separation method for industrial purposes due to the large amount of energy needed and may be limited by the stability of the catalysts at elevated temperatures. In light of this, much research has been aimed at improving the separation of the catalyst and products.

One strategy is to use a fluorous biphasic catalytic system, where employing fluorinated catalysts allows the reaction to be worked up through an organic/fluorocarbon phase separation. The products dissolve into the organic phase and the F-containing catalyst dissolves into the fluorocarbon phase. This concept was explored by Pozzi and coworkers [135,136], who synthesized a number of fluorinated Co–salen complexes for the HKR of terminal epoxides. Some of the fluorinated complexes worked well, but some complexes suffered from a compromise between the solubility in organic solvent to induce the reaction and the solubility in fluorocarbon to achieve the biphasic extraction. Other approaches to achieving improved catalyst separation include the use of ionic liquids and solvent-resistant nanofiltration, both of which have been applied successfully [137,138].

Still, the most investigated approach is the immobilization of the catalyst on polymers and inorganic materials. Following this line, Jacobsen and coworkers [139] investigated Co–salen complexes immobilized on polystyrene resin. The hydroxymethylpolystyrene-supported heterogeneous catalyst **108** (Figure 41) was applied in the HKR of epichlorohydrin and 4-hydroxy-1-butene oxide, two substrates that were not suitable for distillation. The HKR of epichlorohydrin was conducted in CH_2Cl_2, and the HKR of 4-hydroxy-1-butene oxide was run in THF. Both the ring-opened products and the unreacted epoxides were obtained in good yields (36–52% and 38–41%, respectively) and with high enantioselectivities (up to >99% *ee* for the unreacted epoxides and up to 95% *ee* for the ring-opened products). Catalyst **108** was also used for the dynamic HKR of epibromohydrin, affording the ring-opened product in 94% yield and 96% *ee*. The catalyst could be recovered by simple filtration and recycled up to five times with no apparent loss of reactivity or enantioselectivity.

Kirschning and coworkers [140] reported on a related system, where the Co–salen complex was immobilized on a chloromethylpolystyrene resin (catalyst **109** in Figure 41). Catalyst **109** was investigated in the HKR of epichlorohydrin, styrene oxide, and phenyl glycidyl ether in THF, affording ring-opened products and unreacted epoxides in similar or slightly inferior yields and enantioselectivities compared those obtained with homogeneous Co–salen catalyst **64** (Figure 26).

Catalyst **109** was also effective in catalyzing the dynamic HKR of epibromohydrin in THF, giving the corresponding diol with 94% yield and 95% *ee*. The same study also included the immobilization of Co–salen complexes on a porous glass/polymer composite material inside a "PASSflow" reactor [141] which was used to study the catalysis of the dynamic HKR of epibromohydrin in THF under continuous-flow conditions. With reactivation after each run, the modified microreactor could be used for four consecutive runs, affording the ring-opened product in 76–87% yield and 91–93% *ee* [140].

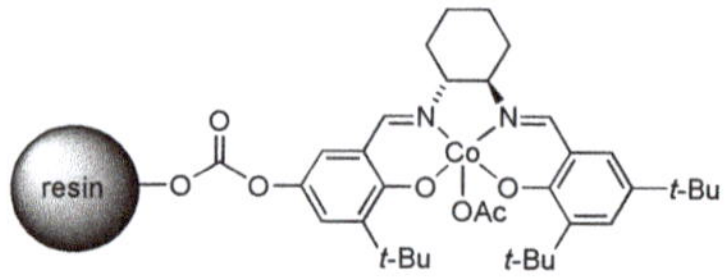

108

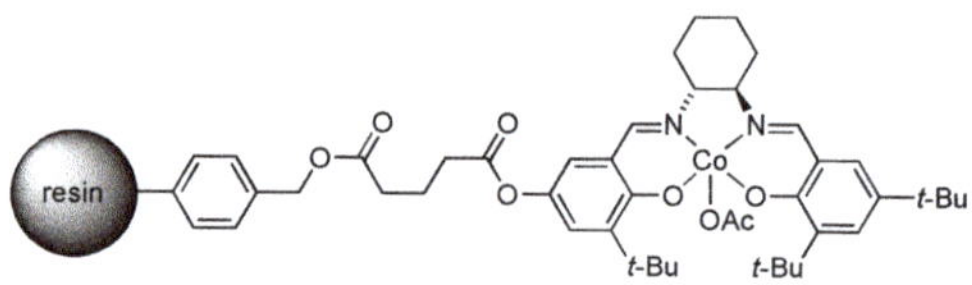

109

Figure 41. Resin-supported Co–salen catalysts **108** and **109**.

Weck and coworkers [142] synthesized polystyrene resin-supported dendronized Co–salen catalyst **110** (Figure 42) which was employed in the HKR of epichlorohydrin, 1,2-epoxyhexane, allyl glycidyl ether, and styrene oxide. Reactions performed with very low catalyst loading (0.04–0.06 mol% Co) and under solvent-free conditions afforded unreacted epoxides in high yields (40–47%) and with excellent enantioselectivities (>99% *ee*). The excellent catalytic properties were attributed to the flexible linkers and the dendronized framework which probably assisted to promote the cooperative interactions among Co–salen sites and facilitate the bimolecular pathway. Catalyst **110** could be used in five successive HKR reactions with maintained reactivity and enantioselectivity.

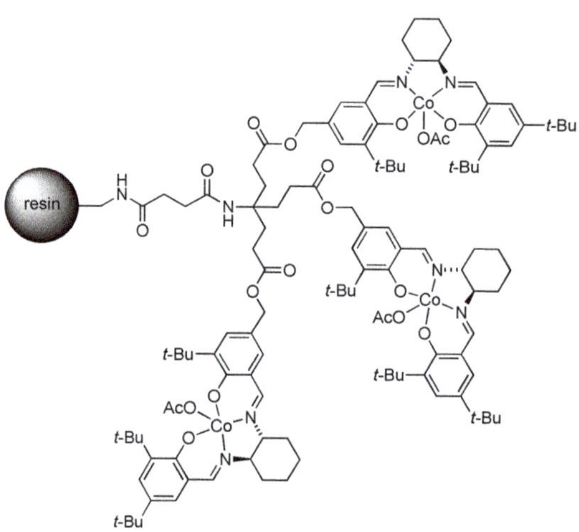

110

Figure 42. Resin-supported dendronized Co–salen catalyst **110**.

Weck and coworkers [143] grafted their macrocyclic oligomeric catalyst **76** (Figure 31) on a polystyrene resin. The resulting catalyst **111** (Figure 43) was used in the solvent-free HKR of epichlorohydrin and 1,2-epoxyhexane, achieving complete resolution (≥50% conversions and 99% *ee* of the unreacted epoxides) in 3 h or less with only 0.01 mol% Co loading. As such, catalyst **111** is the most efficient heterogeneous catalyst for the HKR of terminal epoxides to date. The catalyst could be recovered by filtration and reused up to five times with maintained enantioselectivity, although the reaction time required to reach full conversion increased after each run. A noteworthy observation was that the use of an excess of water (six equivalents instead of 0.6) led to a significant rate enhancement. The enhanced reactivity was attributed to the formation of a biphasic reaction system, in which the formed diol partitions into the aqueous phase, leaving a higher concentration of epoxide and catalyst in the organic phase.

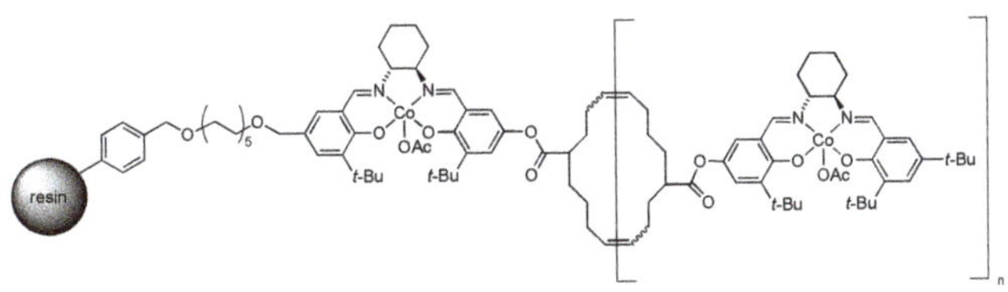

111

Figure 43. Resin-supported macrocyclic oligosalen catalyst **111** based on oligomer **76** (Figure 31).

In addition to the immobilization on polystyrene resin, Co–salen catalysts have also been immobilized on different silica-based supports. Compared to polystyrene, silica is a more rigid support and thus potentially better suited for use as a stationary phase in continuous-flow apparatus. Jacobsen and coworkers covalently immobilized a Co–salen complex on silica (Figure 44) [139]. The resulting catalyst **112** was employed as stationary phase in a continuous-flow system for the HKR of 4-hydroxy-1-butene oxide. Employing THF/H_2O as mobile phase, the reaction afforded the corresponding triol in a good yield (37%) and with high enantioselectivity (94% *ee*).

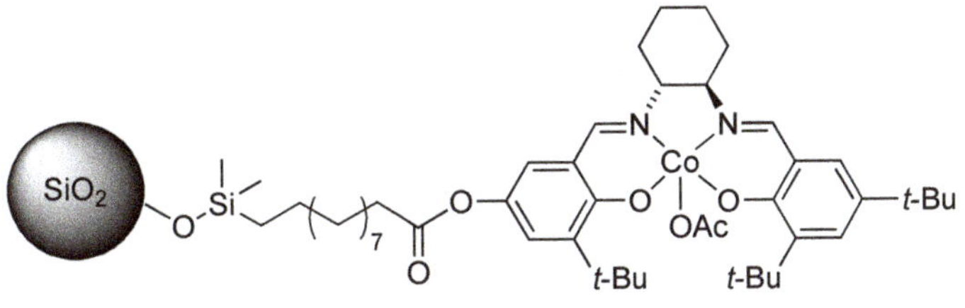

112

Figure 44. Silica-supported catalyst **112**.

Jones and coworkers [144] synthesized a series of heterogeneous polymeric catalysts by immobilizing Co–salen complexes on CAB-O-SIL silica-supported polymer brushes (Figure 45). The HKR of epichlorohydrin in CH_2Cl_2 was employed as a model reaction to compare the catalytic properties of polymeric catalyst **114** with the less flexible polymeric analogue **113**, the homogenous catalyst **64** (Figure 26), and the monosalen analogue **115**. Amongst all the catalysts, polymer **114** exhibited the highest reactivity and enantioselectivity (close to 50% conversion and > 99% *ee* for both the ring-opened product and the unreacted epoxide). The catalyst with a less flexible linker, polymer **113**, was less reactive than **114**, but still exhibited high enantioselectivity (>99% *ee* for the ring-opened product and 95% *ee* for the unreacted epoxide). Monomeric catalyst **115** performed worst, both in terms of reactivity and enantioselectivity, which was attributed to its low local concentration of Co–salen sites and thus diminished Co–salen cooperativity. Catalyst **114** was also investigated in recycling studies, where the enantioselectivity could be maintained over five catalytic runs, although the reaction rate decreased after each subsequent run. The FTIR and elemental analysis studies indicated that the loss of reactivity might be caused by cleavage of the salen ligand under the reaction condition.

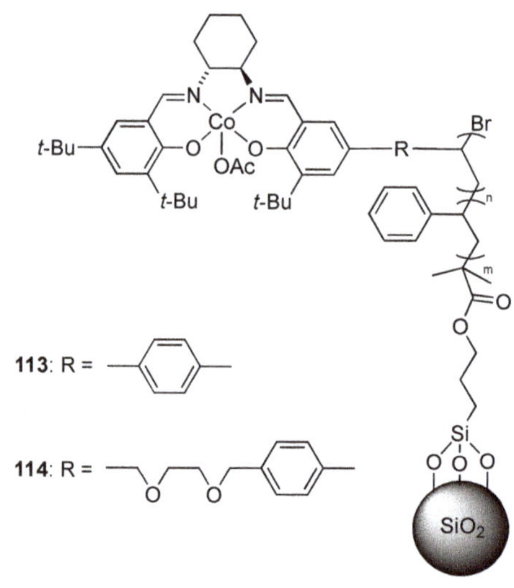

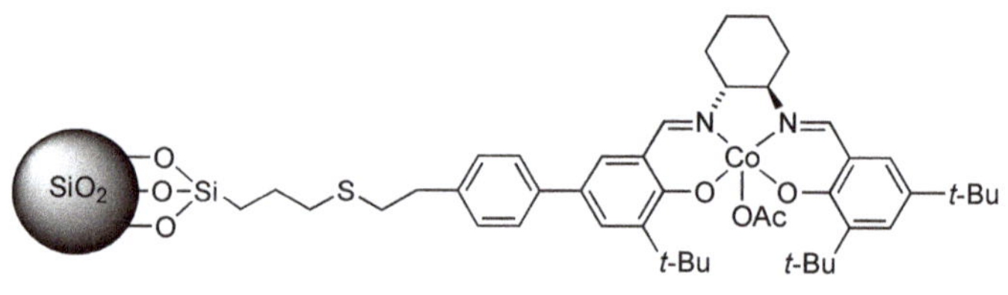

115

Figure 45. Silica-immobilized copolymeric Co–salen complexes (**113** and **114**) and silica-immobilized monomeric complex (**115**).

Jacobsen's group [145] successfully immobilized Co(III)salen complexes on gold colloids through an exchange reaction between thiol-containing Co–salen complexes and *n*-octanethiolates, the latter pre-coordinated to the gold colloid (Figure 46). With low catalyst loading (0.01 mol% Co), the so-immobilized catalyst **116** catalyzed the solvent-free HKR of 1,2-epoxyhexane ten times faster than the non-immobilized analogue **117** (Figure 46), achieving complete resolution within 5 h (≥50% conversion and >99% *ee* of the unreacted epoxide). The catalyst could be recovered by filtration through centrifugal filter units and recycled several times with maintained reactivity and enantioselectivity, only requiring re-oxidation after six consecutive reaction cycles.

Kim and coworkers [65,128] published several studies on the immobilization of salen catalysts on inorganic materials such as mesoporous silicate or alumosilicate. A series of monomeric and dimeric Co–salen catalysts were grafted onto mesoporous silica MCM-41 (Figure 47). In the solvent-free HKR of epichlorohydrin, styrene oxide, and 1,2-epoxyhexane, each of the silica-supported monosalen catalysts **118–121** exhibited decreased reactivities but comparable enantioselectivities compared to homogenous monosalen catalyst **63** (Scheme 40) [65]. In a later study, Kim and coworkers [128] immobilized dimeric Co–salen complexes **54–56** (Scheme 34) on MCM-41 (catalysts **122–124** in Figure 47). Similar to what was observed for the immobilized monosalen catalysts **118–121**, the immobilized dimeric catalysts **122–124** also exhibited decreased reactivities but comparable enantioselectivities in the HKR of epichlorohydrin compared to homogeneous analogues **54–56** (Scheme 34). In both these cases, it is clear that the immobilization hindered efficient catalysis. One explanation could be the relatively short linkers used, as previous studies clearly show that the flexibility and length of the linker is of paramount importance in enabling efficient catalysis. Another possible explanation could be the choice of mesoporous silica as solid support.

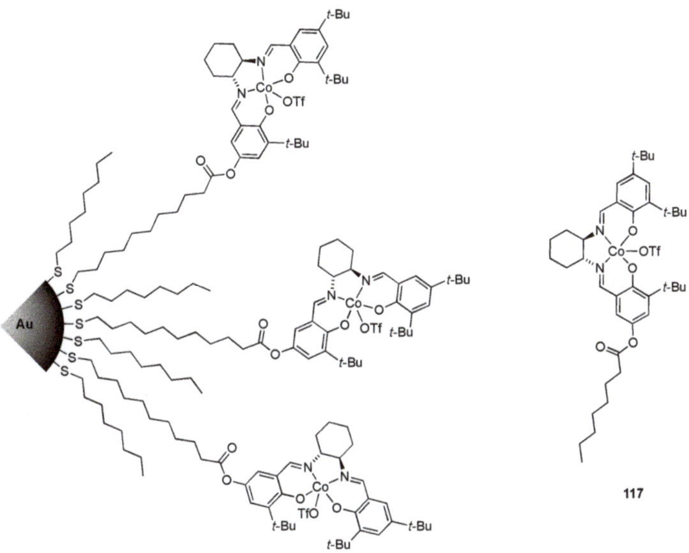

116

Figure 46. Au colloid-supported Co–salen catalyst **116** and non-immobilized analogue **117**.

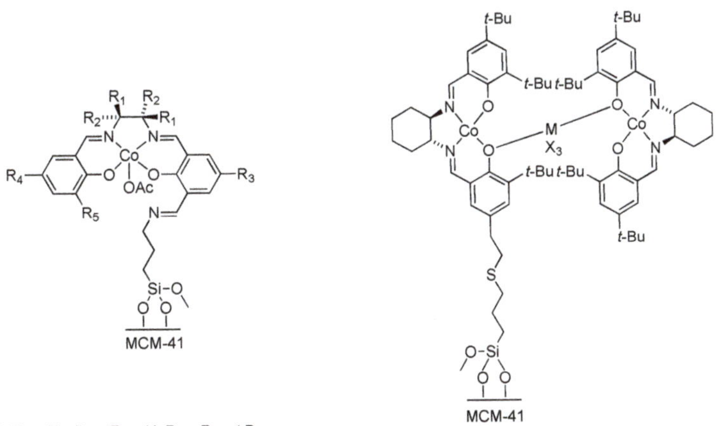

118: R_1 = Ph, R_2 = R_3 = H, R_4 = R_5 = t-Bu
119: R_1 = -$(CH_2)_4$-, R_2 = R_3 = H, R_4 = R_5 = t-Bu
120: R_1 = -$(CH_2)_4$-, R_2 = H, R_3 = t-Bu, R_4 = R_5 = H
121: R_1 = -$(CH_2)_4$-, R_2 = R_3 = H, R_4 = R_5 = t-Bu

122: M = Al, X = Cl
123: M = Ga, X = Cl
124: M = In, X = Cl

Figure 47. MCM-41-supported salen catalysts **118–124**.

To this end, Kim and coworkers [146] studied the effect of the pore size and structure of the silica support on the catalytic activity. Dimethylcarbonate (DMC) was used to partially desilylate mesoporous silica, and the resulting material was then used as a solid support for polymeric Co–salen complexes. The so-formed catalysts exhibited increased reactivities and enantioselectivities in the HKR of terminal epoxides compared to catalysts prepared without pretreatment with DMC.

Kim and coworkers [147] also investigated other mesoporous materials such as SBA-15 and SBA-16 with a wider range of pore sizes and higher stability. Co–salen complexes were attached on sulfonate functionalized SBA-16 through electrostatic interaction between the cobalt and the sulfonate (Figure 48). The so-formed catalyst **125** was employed in the HKR of epichlorohydrin in THF, affording the unreacted epoxide in 44% yield and 98% *ee*, although similar results were also obtained with homogeneous analogues.

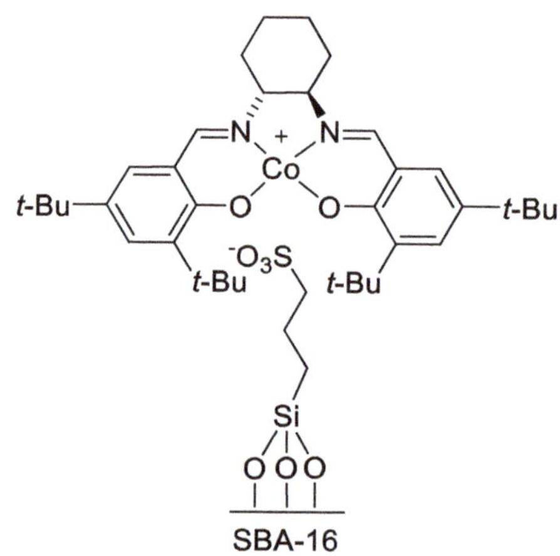

125

Figure 48. SBA-16-supported Co–salen catalyst **125**.

In a later study, the same group reported another type of non-covalent immobilization based on interactions between fluorine-bearing catalysts and different acidic sites in mesoporous Al-SBA-15 (Figure 49) [148,149]. The so-formed catalyst **126** was investigated in the HKR of terminal epoxides in THF, affording unreacted epoxides in full conversions (50%) and with good-to-excellent enantioselectivities (up to 99% *ee*). However, homogeneous analogues gave comparable results in shorter reaction times.

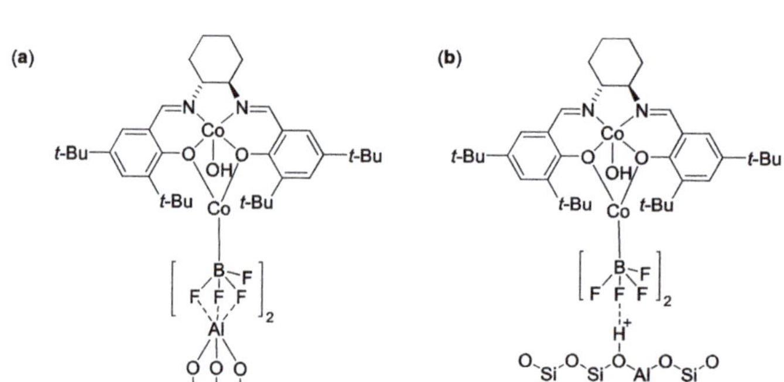

Figure 49. Al-SBA-15-supported Co–salen catalyst **126**, formed by interactions with (**a**) Lewis acidic sites and (**b**) Brønsted acidic sites.

3.3.3. Mechanistic Studies

Several practical and theoretical studies have been carried out to investigate the mechanism of the HKR of terminal epoxides by Co(III)salen complexes. Even though the cooperative mechanism has been supported by many studies [28,29], a few features about this reaction required further investigation. One of them was the influence of different counterions on the reactivity of the catalyst.

In most of the early papers regarding the HKR of epoxides using Co(III)salen catalysts, the acetate anion was employed as the counterion. However, later studies showed that the reaction could be promoted or demoted by changing the counterion [110,150]. Jacobsen and coworkers [151] performed extensive mechanistic studies in an effort to understand and explain this effect, and to identify more active catalysts for the HKR of terminal epoxides. As illustrated in Scheme 45, two possible reaction pathways that compete with each other are proposed. When the counterion (X) is completely non-nucleophilic (e.g., hexafluoroantimonate), the reaction undergoes a less selective monometallic pathway, where the epoxide is activated by the Co(III)salen(X) and attacked by water (Scheme 45, top). When the counterion is nucleophilic (e.g., chloride or acetate), a bimetallic cooperative pathway takes place (Scheme 45, bottom). In the latter pathway, the active nucleophile Co(III)salen(OH) is formed irreversibly from an initial hydrolysis of Co(III)salen(X). The bimetallic pathway being the major pathway is supported by kinetic studies by Kleij and Jacobsen [114,139] and DFT calculations by Li [152]. In the bimetallic pathway, the key to obtaining high rates in the HKR reaction is to have an equal ratio of Co(III)salen(X) (which could activate the epoxide) and Co(III)salen(OH) (which could act as nucleophile). Kinetic studies of a number Co–salen complexes with different counterions showed that Co(III)salen(OTs) achieved full conversion fastest in the HKR of 1,2-epoxyhexane. This was attributed to its slower coordination to the epoxide, thus maintaining a more favorable Co-X/Co-OH ratio throughout the reaction [151]. The experimentally observed trends of reaction rates between different counterions (OTs > OAc > Cl) were also supported by DFT calculations performed by Sherrill and coworkers [153].

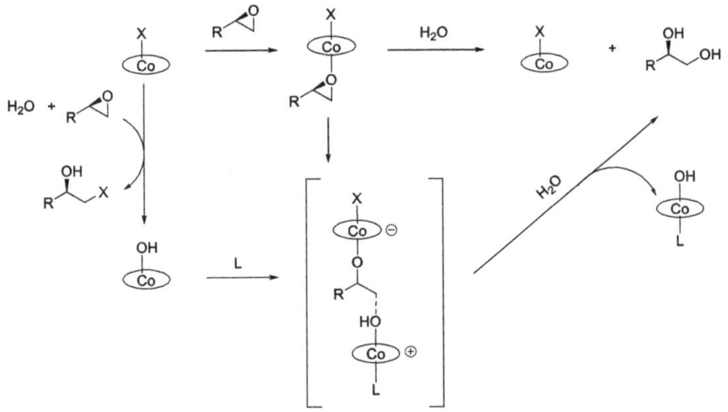

Scheme 45. Proposed reaction pathways for the HKR of terminal epoxides catalyzed by Co(III)salen complexes [151]. (top) A monometallic pathway with one Co–salen catalyst activating the epoxide and H_2O acting as a nucleophile. (bottom) A bimetallic cooperative pathway with initial formation of Co(III)salen(OH) from the reaction between Co(III)salen(X), H_2O, and epoxide, followed by a reaction between the so-formed activated nucleophile Co(III)salen(OH) and an epoxide activated by another Co(III)salen(X) catalyst.

Another mechanistic mystery concerned the deactivation of the catalyst. In the original paper where Co–salen **63** (Scheme 40) was first employed in the HKR of epoxides, the catalyst needed to be regenerated by treatment with acetic acid in air before it could be reused [109]. One of the possible explanations was that Co(III) was reduced to Co(II) through a one-electron transfer oxidation during the reaction. Although many of the later Co–salen catalysts did not require reactivation, the reason for the occasionally observed deactivation remained unclear. In a mechanistic study, Davis and coworkers ruled out the reduction hypothesis by UV-Vis and XANES (X-ray absorption near edge structure) spectroscopic studies of the catalyst before and after the HKR reaction [154]. Another suspected reason of deactivation, the formation of a catalyst dimer, was investigated by ESI-MS studies. While dimer formation was detected over time for Co–salen complex **64** (Figure 26) in dichloromethane, no difference in reactivity or enantioselectivity in the HKR reaction was observed between the dimeric material and fresh catalyst. Instead, it was proposed that the deactivation was mainly due to the formation of the less active Co(III)salen(OH) species [155]. In the proposed mechanism (Scheme 45), the Co(III)salen(OH) species is formed irreversibly and the more selective bimetallic pathway is dependent on the presence of both Co(III)salen(OH) and Co(III)salen(X), where the Co(III)salen(X) complex is good at activating the epoxide. This hypothesis was further supported by experiments where a Co(III)salen(Cl) complex was recycled in the HKR of epichlorohydrin. Without reactivation, Co(III)salen(Cl) showed a significant loss of catalytic activity after one reaction cycle, because of the rapid formation of Co(III)salen(OH). By adding a non-cooperative complex Co(III)salen(SbF_6), the catalytic activity was completely restored.

There has also been some theoretical studies focused on the mechanistic origin of the high selectivity and broad scope of the HKR of epoxides catalyzed by Co(III)salen complexes. Jacobsen and coworkers published a comprehensive study on this issue, resulting in several important findings [156]. First, effective catalysis could be induced only when the two salen moieties in the cooperative pathway had the same absolute configuration. The specific stereochemistry of the salen was not important, only the *matched* stereochemistry (i.e., the two interacting Co-complexes should either both be (*S*,*S*) or both (*R*,*R*)). Secondly, the stereochemical communication was primarily due to the stepped conformation of the entire metal–salen complex (Figure 50), and not due to the shape of the diamine backbone. The stepped conformation is the tilt of the salicylaldimine aryl rings relative to the equatorial plane of the complex. This conformation is not dependent on the presence of an enantiomerically pure diamine backbone that enforces the conformation, such as *trans*-1,2-diaminocyclohexyl. Rather, it was found that salen complexes derived from achiral diamines such as 1,2-ethylenediamine were also capable of adopting this stepped conformation. This finding was further supported by X-ray crystal structures and computational studies. Finally, the binding geometry of the terminal epoxides to the Co(III)salen complexes was independent of the substituent of the epoxides which could explain the broad scope of HKR reactions catalyzed by Co(III)salen catalysts.

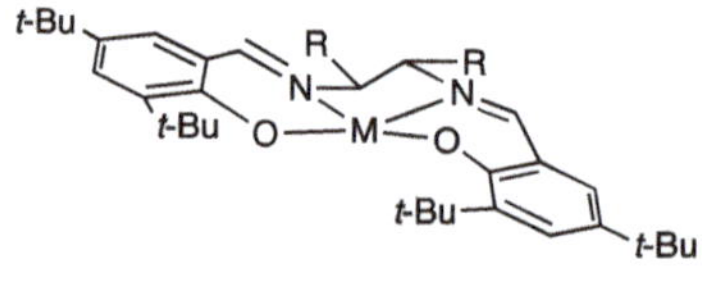

Figure 50. The stepped conformation of a metal–salen complex.

3.3.4. Other Studies

Berkessel and Ertuerk [157] reported an interesting catalyst for the HKR of terminal epoxides. They designed and synthesized DIANANE (*endo*,*endo*-2,5-diaminonorbornane)-based Cr(III) complexes **127** and **128** (Scheme 46), which exhibited high reactivity and enantioselectivity in the HKR of terminal epoxides under low catalyst loading (Scheme 46). Both catalysts were more enantioselective than the parent Cr–salen complex **1** (Scheme 4). Complex **127** with Cl^- counterion was less active than complex **128** with OTs^- counterion, which is in line with the counterion effect previously described.

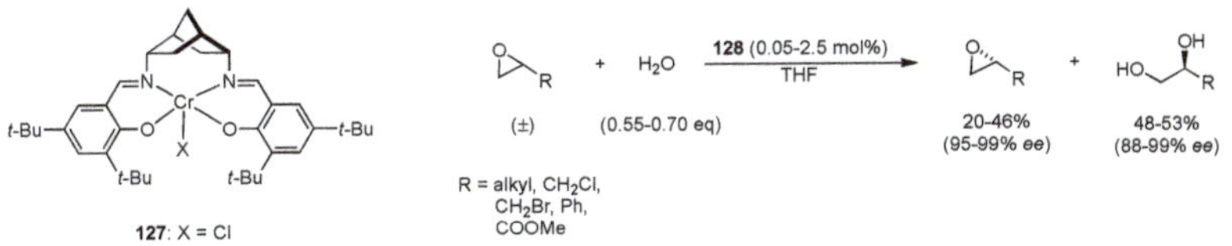

Scheme 46. The hydrolytic kinetic resolution of terminal epoxides catalyzed by Cr-DIANANE-salen complex **128**. DIANANE = *endo,endo*-2,5-diaminonorbornane.

Schulz and coworkers [67] implemented a heterobimetallic dual-catalyst system for the HKR of terminal epoxides. A 1:1 mixture of complex **129** and **130** (Figure 51) was employed in the HKR of phenyl glycidyl ether, allyl glycidyl ether, and styrene oxide in THF. The reactions with mixed catalysts afforded both products with excellent conversions (up to 57%) and enantioselectivities (up to 94% *ee* for 1,2-diols, up to 99% *ee* for the unreacted epoxides), while reactions employing only catalyst **129** were less reactive and enantioselective. Kinetic studies revealed a first order dependence on the concentration of the Co–salen complex **129**, which supported a heterobimetallic cooperative pathway similar to the cooperative mechanism in Scheme 45, where Co–salen complex **129** would be responsible for the generation of the nucleophilic Co(III)salen(OH) and Mn–salen complex **130** would be responsible for the activation of the epoxides. The study also included an investigation of different combinations of the absolute configuration of the two complexes. Higher enantioselectivity was achieved when the two complexes had the same absolute configuration which was in agreement with Jacobsen's [156] findings for a similar system.

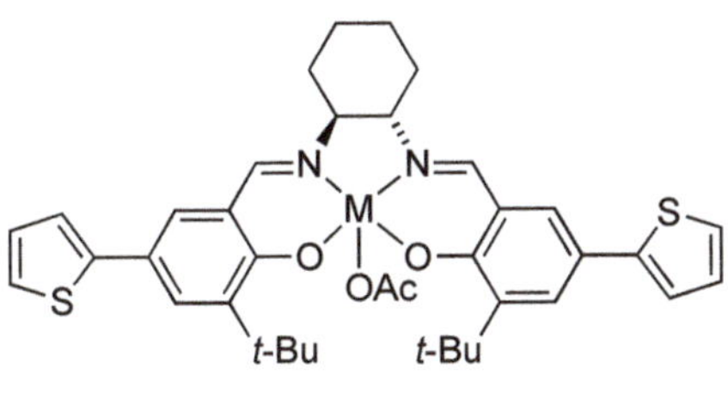

129: M = Co
130: M = Mn

Figure 51. Salen complexes **130** and **131**.

There are also examples of research that focus on more general reaction conditions. For example, Kim and coworkers [158] described the advantage of ultrasonication over mechanical stirring in the Co(III)salen-catalyzed HKR of terminal epoxides, both for homogeneous and heterogeneous catalysts. With ultrasonication, a shortening of the reaction time was observed in all cases.

3.4. With Alcohols, Phenols, and Carboxylic Acids

In addition to water (HKR), other oxygen-containing nucleophiles such as phenols, alcohols and carboxylic acids have also been employed in the KR of terminal epoxides. The ring-opened products are monoprotected enantioenriched 1,2-diols, which are versatile building blocks in the pharmaceutical industry and natural product synthesis [159]. Many of the catalysts mentioned above that are effective in the HKR of terminal epoxides are also found to work well for the KR of epoxides with phenols and alcohols as nucleophiles.

Jacobsen and coworkers [160] investigated several of their Co–salen complexes as catalysts in the KR of epoxides with phenols and alcohols as nucleophiles. In an early example, they employed catalyst **131** (Scheme 47) in the phenolic kinetic resolution of a number of different terminal epoxides.

The ring-opened products where obtained in high yields and with high enantioselectivity (Scheme 47). The protocol could also be extended to different *para-* and *meta*-substituted phenols as nucleophiles, with consistently high yields and *ee* values for the ring-opened products.

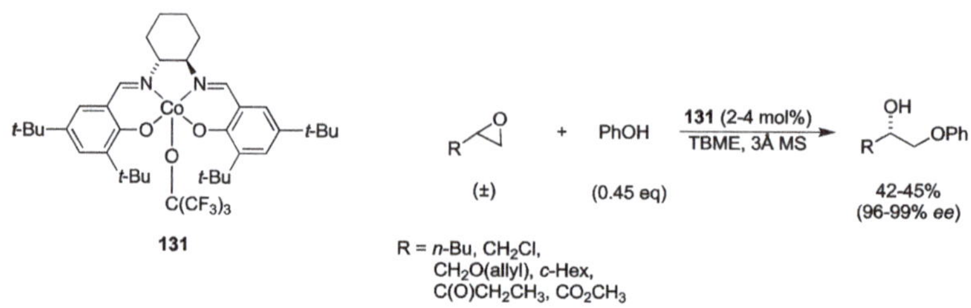

Scheme 47. The phenolic kinetic resolution catalyzed by Co–salen complex **131**.

The polystyrene resin-supported Co–salen catalyst **108** (Figure 41) was employed in the phenolic KR of terminal epoxides in the first examples of an enantioselective catalytic synthesis of parallel libraries [161]. Following this protocol, a wide range of terminal epoxides and phenol nucleophiles were employed in parallel syntheses, resulting in 110 different 1-aryloxy-2-alcohols which were obtained in high yields (80–99%) and *ee* values (up to 99% *ee*). The reactions were performed using the epoxide as a solvent. As such, an excess of epoxides was used and therefore all yields were calculated based on the nucleophile [139,161].

The oligomeric macrocyclic Co–salen catalyst **35** (Scheme 10) was proven to also be efficient in the KR of epoxides with phenols and alcohols as nucleophiles [58]. Using complex **35** as catalyst, a broad range of phenols (with electron withdrawing and -donating substituents in the *ortho-*, *meta-* and *para*-position) and primary alcohols (including benzylic and allylic) could be used as nucleophiles in the HKR of a number of different terminal epoxides. The ring-opened products were isolated in high yields (36–45%) and with excellent enantioselectivities (97–99% *ee*). All reactions were performed in acetonitrile with catalyst loadings of 0.0075–0.5 mol% Co, with more sterically hindered nucleophiles requiring the higher catalyst loadings.

Weck and coworkers applied their oligomeric Co–salen complex **77** (Figure 31) in the KR of epichlorohydrin and 1,2-epoxyhexane with a limited number of phenols and aliphatic alcohols. The reactions performed in TBME or acetonitrile and with a low catalyst loading (0.02 mol% Co) afforded the ring-opened products in 38–45% yield and 95–99% *ee* [162].

Kim and coworkers published a number of studies where different Co(III)salen catalysts were employed in the phenolic KR of terminal epoxides. Most of these studies focused on dimeric Co–salen complexes linked by group 13 elements (catalyst **54–56** in Scheme 34) [163,164] and the immobilization of these catalysts on microporous materials such as zeolite [165,166], mesoporous materials such as SBA-15 and SBA-16 [146,148], and different macroporous materials [167,168]. In all cases the catalysts exhibited good enantioselectivities, but the substrate scopes were usually limited.

Kim's group also made efforts to apply Co–salen catalysts in the KR of terminal epoxides with carboxylic acids as nucleophiles. Dimeric Co–salen complexes bridged by the Lewis acids $GaCl_3$ and $Al(NO_3)_3$ (complex **55** in Scheme 34 and complex **132** in Figure 52) were investigated in the KR of terminal epoxides with a number of different carboxylic acids in TBME [128,169]. The reactions afforded 2-hydroxy monoesters in good yields (35–43%) and with moderate to good enantioselectivities (53–86% *ee*), but most of the products needed further recrystallization or other treatment to provide practically applicable enantioenriched compounds. One of the applications of this reaction was the synthesis of highly enantioenriched (*S*)-glycidyl butyrate (Scheme 48). Hence, catalyst **55** was employed in the KR of epichlorohydrin with butyric acid. The so formed chlorohydrin (76% *ee*) was then subjected to a ring closing reaction under basic conditions, again promoted by catalyst **55**, affording (*S*)-glycidyl butyrate in 98% *ee*.

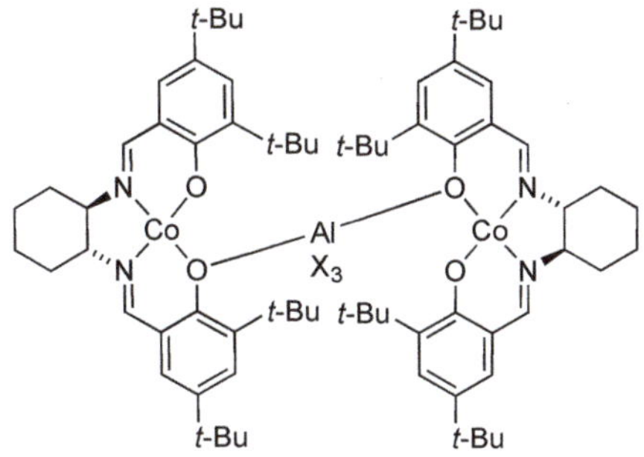

132: X = NO_3

Figure 52. Dimeric Co–salen complex **132**.

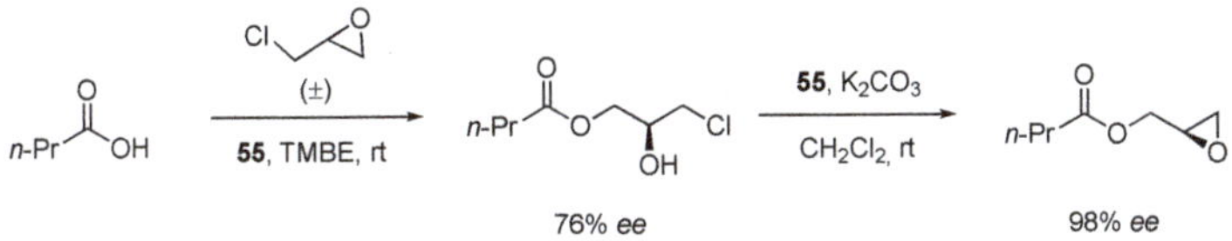

Scheme 48. The synthesis of (*S*)-glycidyl butyrate by kinetic resolution of epochlorohydrin employing catalyst **55** (Scheme 34).

The dimeric Co–salen catalyst **54** bridged by $AlCl_3$ (Scheme 34) exhibited good catalytic activity in the KR of terminal epoxides with different sulfonic acids (Scheme 49) [170]. The reactions were performed in TBME and it was discovered that the addition of tetrabutylammonium chloride (TBACl) increased the yields and *ee* values significantly (from 32% to 47% yield and from 68% to 99% *ee* for the reaction of phenyl glycidyl ether and *p*-toluenesulfonic acid). No explanation for this enhancement of the catalytic activity was given.

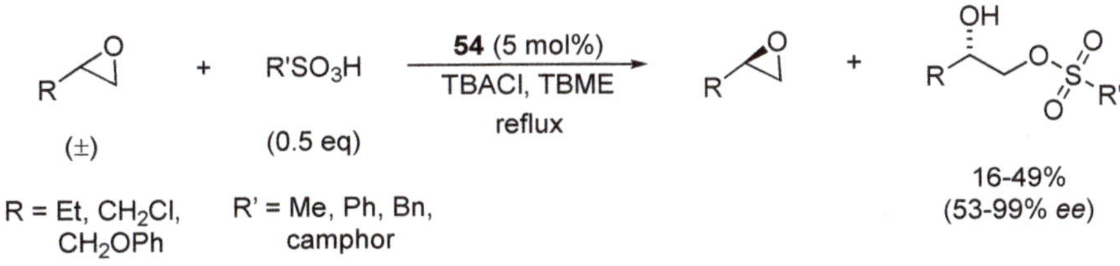

Scheme 49. The kinetic resolution of terminal epoxides with substituted sulfonic acid catalyzed by dimeric Co–salen catalyst **54** (Scheme 34). TBACl = tetrabutylammonium chloride.

3.5. With Halogens

Haufe and coworker investigated Cr–salen complex **46** (Scheme 18) as catalyst for the enantioselective fluorination of epoxides [75–77]. The studies included a limited number of examples of the KR of terminal epoxides. The reaction required a very high catalyst loading (50 mol%) and the ring-opened products were obtained in moderate yield (28–29%) and moderate to high enantioselectivity (62–90% *ee*).

A more recent example of the enantioselective fluorination of epoxide was reported by Kalow and Doyle [79]. Using dimeric Co(II)salen complex **48** (Figure 21) together with cocatalyst 1,8-diazabicyclo[5.4.0]undec-7-ene (DBU) and HF (formed in situ from benzoyl fluoride and HFIP), the ring-opened products of a number of terminal epoxides were obtained in high yields and with excellent enantioselectivities (Scheme 50). The dimeric catalyst also showed a significant rate enhancement compared to the monomeric analogue and reactions were completed in 2–4 h.

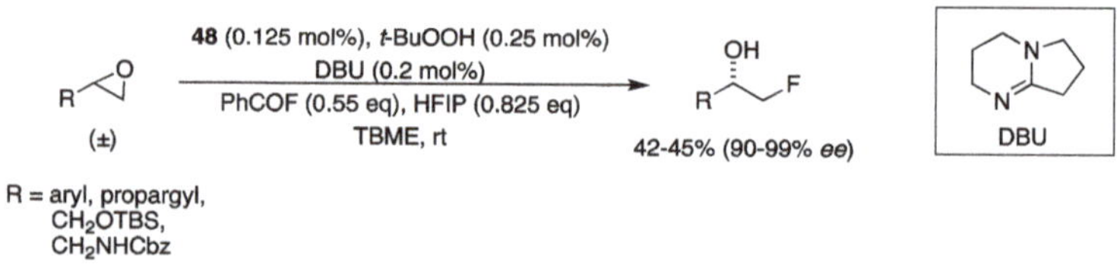

Scheme 50. The kinetic resolution of terminal epoxides with fluoride catalyzed by Co–salen complex **48** (Figure 21).

Kim and coworkers reported the KR of terminal epoxides with HCl catalyzed by dimeric Co(II)salen complexes linked by Lewis acids of group 13 metals (catalysts **54–56**, Scheme 34). Catalyst **55** afforded both ring-opened products and unreacted epoxides in high yields and enantioselectivities (up to 90% *ee* and 89% *ee*, respectively) for terminal alkyl epoxides and glycidyl ethers (Scheme 51). Aryl glycidyl ethers gave significantly lower *ee* values than alkyl glycidyl ethers and terminal alkyl epoxides. It was discovered that the *ee* value of the unreacted epoxide decreased over time, as the formed chlorohydrin was capable of performing the reverse ring-closing reaction in the presence of catalyst **55**. To prevent this, the reaction had to be terminated once a high *ee* value was reached for the unreacted epoxide. Different group 13 metals and anions were evaluated but showed similar reactivities and enantioselectivities [127,128].

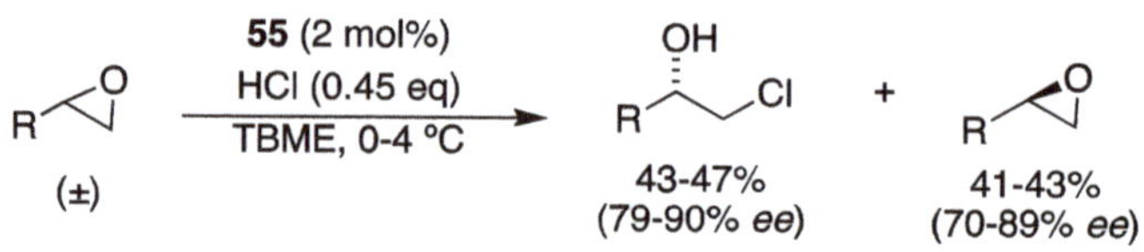

Scheme 51. The kinetic resolution of terminal epoxides with hydrochloric acid catalyzed by dimeric Co–salen complex **55** (Scheme 34).

3.6. With Carbon-Containing Nucleophiles

Cozzi and Umani-Ronchi [90] used Cr–salen-SbF_6 complex **133** (Figure 53) as a catalyst for the KR of 1,2-disubstituted aromatic epoxides with indoles. The ring-opened products were obtained in high yields and with high enantioselectivities (Scheme 52). Notably, this constitutes a rare example of a reported method for the KR of epoxides that is efficient for both *cis* and *trans* aromatic epoxides. For all substrates, the ring-opened products were obtained with complete regioselectivity (nucleophilic attack at the benzylic carbon of the epoxides). Moreover, by adjusting the amount of indole used, the reaction could be tuned to afford unreacted epoxides in satisfactory yields and with excellent enantioselectivities (91–99% *ee*).

Recently, Hajra and Roy [171] reported the first enantioselective construction of an all-carbon quaternary center from an epoxide by KR, using Co(III)salen catalysts **73** (Figure 29) and **134** (Figure 54) and spiro-epoxides with *N*-benzylindoles as nucleophiles. The desired 3,3′-bisindole methanols were obtained with complete regioselectivity, that is, with reaction of the nucleophile at the more substituted carbon of the epoxide. The excellent regioselectivity of the reaction was explained by a proposed mechanism where the Lewis acid coordinating to the epoxyindole led to the ring-opening of the epoxide to the more stabilized tertiary carbocation before the nucleophile attacks [172]. Under optimized conditions with Co(III)salen(OTf) catalyst **73**, which included the addition of 1 equivalent of water, the ring-opened products were obtained in good yields and with high enantioselectivities. The unreacted epoxides were recovered in low-to-moderate enantioselectivities. Using Co(III)salen(SbF_6) complex **134** as catalyst resulted in a DKR, giving ring-opened products in good yields and with moderate-to-good enantioselectivities, with no unreacted epoxides recovered (Scheme 53). Mechanistic studies indicated that the ring-opening reaction of spiro-epoxides was governed by an equilibration between KR and DKR processes, and that this equilibrium was further controlled by feedback inhibition caused by the formation of a catalyst–product complex. Catalyst **134** with a non-coordinating counterion was less influenced by the feedback inhibition and could therefore be operated by the dynamic kinetic process [171].

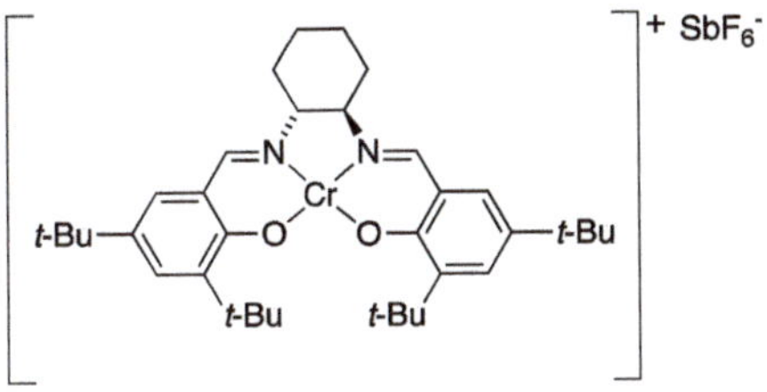

133

Figure 53. Cr–salen complex **133**.

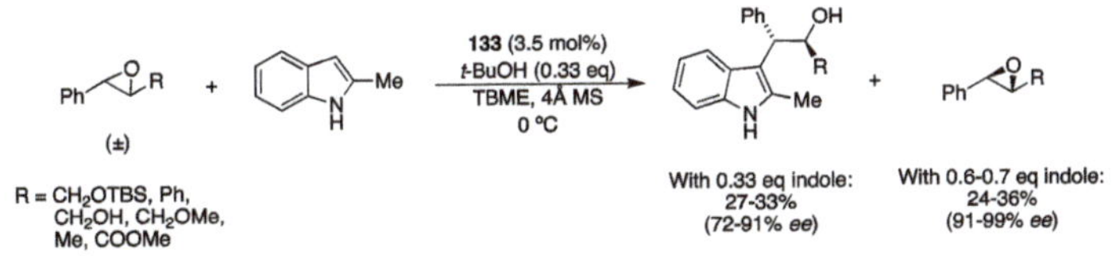

Scheme 52. The kinetic resolution of internal 1,2-disubstituted aromatic epoxides with indoles catalyzed by Cr–salen complex **133** (Figure 53).

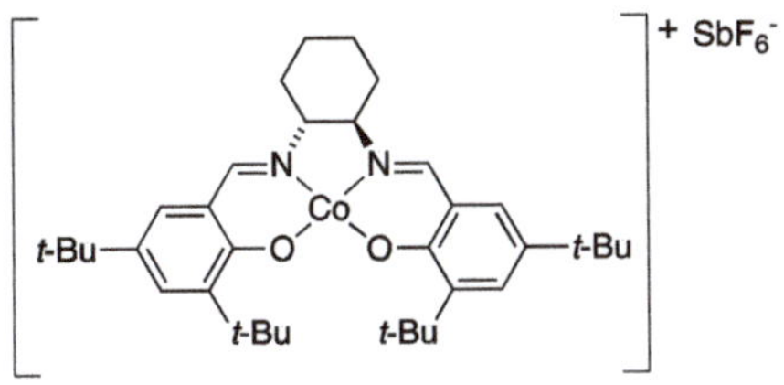

134

Figure 54. Co(II)salen complex **134**.

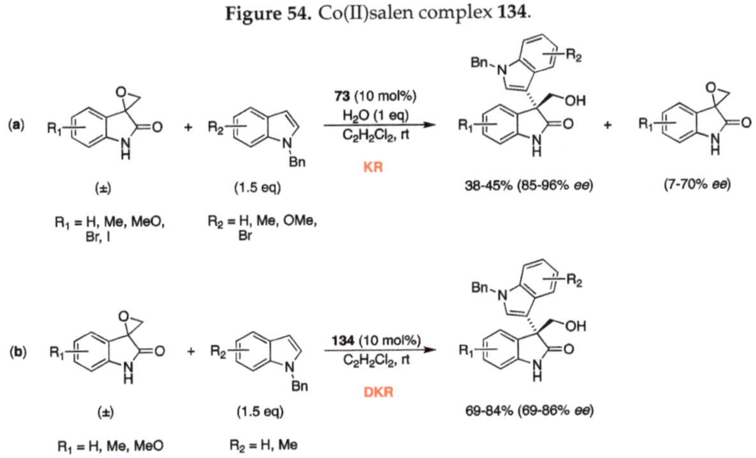

Scheme 53. (**a**) The kinetic resolution (KR) of spiro-epoxides with *N*-benzylindoles catalyzed by Co–salen complex **73** (Figure 29). (**b**) The dynamic kinetic resolution (DKR) of spiro-epoxides with *N*-benzylindoles catalyzed by Co–salen complexes **134** (Figure 54). Both processes generate quaternary all-carbon centers with high enantioselectivity.

4. Conclusions and Outlook

The catalytic asymmetric ring-opening of epoxides is a highly useful method for the preparation of synthetically important, vicinally difunctionalized organic compounds such as amino alcohols, diols, and halohydrins in high yield and with high enantiomeric purity. The use of metal–salen complexes as catalysts for ARO reactions has seen much development since the first reports some 25 years ago. From an enantioselectivity standpoint, the original monosalen catalysts reported by Jacobsen (Cr–salen for ARO with azides, Scheme 4, and Co–salen for ARO with water, Scheme 40) remain competitive to this day, but significant progress has been made in terms of catalyst loading, substrate scope, and recyclability.

One area of interest has been the development of multi-metallic catalysts capable of enforcing a bimetallic cooperative pathway. Numerous strategies have been employed, ranging from covalently linking two or more salen complexes together by installing them on dendrimers, polymers or in supramolecular assemblies. Although this approach often requires more laborious synthetic efforts, many of the resulting catalysts have shown significantly increased reaction rates for the ARO of epoxides compared to monosalen analogues, and also enabled efficient catalysis at considerably lower catalyst loadings. In addition, the development and application of multi-metallic complexes have provided further insight into the reaction mechanism and highlighted the importance of factors such as the orientation of the complexes relative to each other as well as the length and flexibility of bridging linkers.

As large-scale applications depend largely on the separation of the products and recovery and recycling of the catalyst, the development of heterogeneous catalysts has received much attention. Metal–salen complexes have been immobilized on a number of different solid surfaces and materials, providing access to catalysts with good stability that can be easily recycled without loss of catalytic properties. A further challenge in this area is to design the catalyst so that the cooperative interactions are still possible in order to achieve high reactivity and enantioselectivity. In terms of recyclability, the use of ionic liquids has also received some attention and shows much promise. Looking forward, one interesting area that has so far only been sparingly explored is the development of catalysts bridging homogeneous and heterogeneous catalysis, for example, by using soluble supports or encapsulating metal–salen complexes inside porous materials. The facile synthesis and easy modification of metal–salen complexes has also made them attractive catalysts to use for demonstrating the design and efficacy of new catalytic concepts and systems in general.

Different catalytic systems based on increasing the local concentration of catalyst or improving the cooperative interactions have allowed for significantly decreased catalyst loadings. This includes the development of polymeric salen complexes and the incorporation of metal–salen complexes in nanoreactors. Very low catalyst loadings (< 0.01 mol%) have mainly been realized for the hydrolytic kinetic resolution (HKR) of epoxides catalyzed by Co–salen complexes, with a few examples for the ARO of *meso*-epoxides with azides catalyzed by Cr–salen. However, for other nucleophiles there is still a lot of room for improvement.

Regarding the substrate scope, many of the reported catalysts have only been investigated for a limited number of substrates, both in terms of nucleophiles and epoxides. With regard to the nucleophile, one desirable development would be a wider application and investigation of different carbon-based nucleophiles in the metal–salen catalyzed ARO of epoxides, in order to produce asymmetric C–C bonds. Another class of nucleophiles that merits further investigation is oxygen-containing nucleophiles other than water, as this could result in the regioselective preparation of monoprotected diols, something which can be difficult to achieve by other methods.

Another aspect of the substrate scope is the epoxide. For the KR of epoxides, the vast majority of the research has been focused on terminal epoxides. This can be explained by the limited availability of enantiopure terminal epoxides by other methods, making the KR of terminal epoxides highly attractive as the unreacted epoxide can be recovered in high *ee*. As such, the metal–salen-catalyzed HKR have been successfully applied to a broad range of terminal epoxides with a wide variety of substituents, although no examples of the HKR of internal epoxides have been published to date. There are a few examples of the KR of *trans*-epoxides with anilines and indoles, and one example of the KR of *cis*-epoxides with indoles, but other than that there is a distinct lack of protocols for the KR of internal epoxides.

Extending the metal–salen catalyzed ARO of epoxides to include internal epoxides would also make it a good complement to other synthetic protocols. For example, the successful HKR of *trans*-epoxides would yield *anti*-diols, making it complementary to Sharpless dihydroxylation of olefins which works well for *trans*-olefins and gives the corresponding *syn*-diols with high regio-, diastereo- and enantioselectivity [64]. In addition, since asymmetric epoxidation methods such as Jacobsen's Mn–salen epoxidation usually work poorly for *trans*-alkenes [24,173], achieving an efficient KR of *trans*-epoxides would give access to otherwise difficult to obtain enantiopure *trans*-epoxides.

There is also a lack of examples of the ARO of more substituted epoxides, with only a handful reported examples of the ARO of trisubstituted and 2,2-disubstituted epoxides. Extending the substrate scope to more substituted epoxides would also enable further studies of the regioselectivity of this type of catalysis. So far, most studies report complete regioselectivity with the investigated substrate scope, but further studies are needed to better understand, and by extension tune, this selectivity.

There are also several examples of innovative ways of utilizing metal–salen-catalyzed ARO reactions to solve general challenging synthetic problems. This includes the enantioselective construction of all-carbon quaternary centers from epoxides and the development of sequential asymmetric alkene epoxidation/ring-opening reactions. Both of these reactions are potentially of high importance and we hope to see much further research in this direction. Another aspect that merits further study is the dynamic kinetic resolution, which allows the preparation of enantiopure ring-opened products from racemic starting materials in 100% theoretical yield. There are a few examples of the successful application of this highly attractive strategy, mainly for the HKR of terminal epoxides. However, it is an area where much progress remains to be made.

To conclude, catalytic systems for the asymmetric ring opening of epoxides based on metal–salen complexes have been extensively studied and developed. With excellent catalytic activity and selectivity, this class of catalysts is a valuable tool in the area of asymmetric catalysis. As such, it remains an active field of research and we expect further advances are still forthcoming.

Author Contributions: Conceptualization, A.L., Y.L. and K.W.; writing—original draft preparation, A.L. and Y.L.; writing—review and editing, A.L., Y.L. and K.W. All authors have read and agreed to the published version of the manuscript.

Funding: K.W. thanks the Swedish Research Council, the LMK foundation, the Swedish Foundation for Strategic Research, and the Knut and Alice Wallenbergs Stiftelse for generous grants. Y.L. thanks the Chinese Scholarship Council for the scholarship.

Conflicts of Interest: The authors declare no conflict of interest.

References

1. Walsh, P.J.; Kozlowski, M.C. *Fundamentals of Asymmetric Catalysis*; University Science Books: Sausalito, CA, USA, 2009.
2. Halpern, J.; Trost, B.M. Asymmetric Catalysis. *Proc. Natl. Acad. Sci. USA* **2004**, *101*, 5347. [CrossRef]
3. Yoon, T.P.; Jacobsen, E.N. Privileged Chiral Catalysts. *Science* **2003**, *299*, 1691. [CrossRef]
4. Shaw, S.; White, J.D. Asymmetric Catalysis Using Chiral Salen–Metal Complexes: Recent Advances. *Chem. Rev.* **2019**, *119*, 9381–9426. [CrossRef]
5. Cozzi, P.G. Metal–Salen Schiff base complexes in catalysis: Practical aspects. *Chem. Soc. Rev.* **2004**, *33*, 410–421. [CrossRef]
6. Zhang, W.-Z.; Lu, X.-B. Chiral Salen Complexes. In *Privileged Chiral Ligands and Catalysts*; Zhou, Q.L., Ed.; Wiley: Weinheim, Germany, 2011; pp. 257–293.
7. Trost, B.; Fleming, I. *Comprehensive Organic Synthesis*; Pergamon: Oxford, UK, 1991.

8. Xia, Q.H.; Ge, H.Q.; Ye, C.P.; Liu, Z.M.; Su, K.X. Advances in Homogeneous and Heterogeneous Catalytic Asymmetric Epoxidation. *Chem. Rev.* **2005**, *105*, 1603–1662. [CrossRef]
9. Pastor, I.M.; Yus, M. Asymmetric Ring Opening of Epoxides. *Curr. Org. Chem.* **2005**, *9*, 1–29. [CrossRef]
10. Schneider, C. Synthesis of 1, 2-difunctionalized fine chemicals through catalytic, enantioselective ring-opening reactions of epoxides. *Synthesis* **2006**, *2006*, 3919–3944. [CrossRef]
11. Clarke, R.M.; Storr, T. The chemistry and applications of multimetallic salen complexes. *Dalton Trans.* **2014**, *43*, 9380–9391. [CrossRef]
12. Matsunaga, S.; Shibasaki, M. Multimetallic Schiff Base Complexes as Cooperative Asymmetric Catalysts. *Synthesis* **2013**, *45*, 421–437. [CrossRef]
13. Haak, R.M.; Wezenberg, S.J.; Kleij, A.W. Cooperative multimetallic catalysis using metallosalens. *Chem. Commun.* **2010**, *46*, 2713–2723. [CrossRef]
14. Abd El Sater, M.; Jaber, N.; Schulz, E. Chiral Salen Complexes for Asymmetric Heterogeneous Catalysis: Recent Examples for Recycling and Cooperativity. *ChemCatChem* **2019**, *11*, 3662–3687. [CrossRef]
15. Baleizão, C.; Garcia, H. Chiral Salen Complexes: An Overview to Recoverable and Reusable Homogeneous and Heterogeneous Catalysts. *Chem. Rev.* **2006**, *106*, 3987–4043. [CrossRef]
16. Shioiri, T.; Hamada, Y. Natural product syntheses utilizing 4-alkoxycarbonyloxazoles as β-hydroxy-α-amino acid synthons. *Heterocycles* **1988**, *27*, 1035–1050. [CrossRef]
17. Kikelj, D.; Kidrič, J.; Pristovšek, P.; Pečar, S.; Urleb, U.; Krbavčič, A.; Hönig, H. Preparation of diastereomerically pure immunologically active carbocyclic nor-muramyldipeptide analogues. *Tetrahedron* **1992**, *48*, 5915–5932. [CrossRef]
18. Vacca, J.P.; Dorsey, B.; Schleif, W.; Levin, R.; McDaniel, S.; Darke, P.; Zugay, J.; Quintero, J.; Blahy, O.; Roth, E. L-735,524: An orally bioavailable human immunodeficiency virus type 1 protease inhibitor. *Proc. Natl. Acad. Sci. USA* **1994**, *91*, 4096–4100. [CrossRef]
19. Wang, R.; Wong, C.-H. Synthesis of sialyl Lewis X mimetics: Use of O-α-fucosyl-(1R, 2R)-2-aminocyclohexanol as core structure. *Tetrahedron Lett.* **1996**, *37*, 5427–5430. [CrossRef]
20. Martinez, L.E.; Leighton, J.L.; Carsten, D.H.; Jacobsen, E.N. Highly enantioselective ring opening of epoxides catalyzed by (salen) Cr (III) complexes. *J. Am. Chem. Soc.* **1995**, *117*, 5897–5898. [CrossRef]
21. Jacobsen, E.N. Asymmetric catalysis of epoxide ring-opening reactions. *Acc. Chem. Res.* **2000**, *33*, 421–431. [CrossRef]
22. Pakulski, Z.; Pietrusiewicz, K.M. Enantioselective desymmetrization of phospholene meso-epoxide by nucleophilic opening of the epoxide. *Tetrahedron Asymmetry* **2004**, *15*, 41–45. [CrossRef]
23. Saha, B.; Lin, M.-H.; RajanBabu, T. Exceptionally Active Yttrium−Salen Complexes for the Catalyzed Ring Opening of Epoxides by TMSCN and TMSN3. *J. Org. Chem.* **2007**, *72*, 8648–8655. [CrossRef]
24. Zhang, W.; Loebach, J.L.; Wilson, S.R.; Jacobsen, E.N. Enantioselective epoxidation of unfunctionalized olefins catalyzed by salen manganese complexes. *J. Am. Chem. Soc.* **1990**, *112*, 2801–2803. [CrossRef]
25. Jacobsen, E.N.; Zhang, W.; Muci, A.R.; Ecker, J.R.; Deng, L. Highly enantioselective epoxidation catalysts derived from 1, 2-diaminocyclohexane. *J. Am. Chem. Soc.* **1991**, *113*, 7063–7064. [CrossRef]
26. Brandes, B.D.; Jacobsen, E.N. Highly enantioselective, catalytic epoxidation of trisubstituted olefins. *J. Org. Chem.* **1994**, *59*, 4378–4380. [CrossRef]
27. Schaus, S.E.; Larrow, J.F.; Jacobsen, E.N. Practical synthesis of enantiopure cyclic 1, 2-amino alcohols via catalytic asymmetric ring opening of meso epoxides. *J. Org. Chem.* **1997**, *62*, 4197–4199. [CrossRef]
28. Hansen, K.B.; Leighton, J.L.; Jacobsen, E.N. On the mechanism of asymmetric nucleophilic ring-opening of epoxides catalyzed by (salen) CrIII complexes. *J. Am. Chem. Soc.* **1996**, *118*, 10924–10925. [CrossRef]
29. Guillaneux, D.; Zhao, S.-H.; Samuel, O.; Rainford, D.; Kagan, H.B. Nonlinear effects in asymmetric catalysis. *J. Am. Chem. Soc.* **1994**, *116*, 9430–9439. [CrossRef]
30. Konsler, R.G.; Karl, J.; Jacobsen, E.N. Cooperative asymmetric catalysis with dimeric salen complexes. *J. Am. Chem. Soc.* **1998**, *120*, 10780–10781. [CrossRef]
31. Ma, D.Y.; Xiao, Z.Y.; Etxabe, J.; Wärnmark, K. Pseudo-C2-Symmetric Bimetallic Bissalen Catalysts for Efficient and Enantioselective Ring-Opening of meso-Epoxides. *ChemCatChem* **2012**, *4*, 1321–1329. [CrossRef]
32. Li, Y.; Lidskog, A.; Armengol-Relats, H.; Pham, T.H.; Favraud, A.; Nicolas, M.; Dawaigher, S.; Xiao, Z.; Ma, D.; Lindbäck, E.; et al. Enantiotopic Discrimination by Coordination-Desymmetrized meso-Ligands. *ChemCatChem* **2020**, *12*, 1575–1579. [CrossRef]

33. Gianneschi, N.C.; Bertin, P.A.; Nguyen, S.T.; Mirkin, C.A.; Zakharov, L.N.; Rheingold, A.L. A supramolecular approach to an allosteric catalyst. *J. Am. Chem. Soc.* **2003**, *125*, 10508–10509. [CrossRef]
34. Gianneschi, N.C.; Cho, S.H.; Nguyen, S.T.; Mirkin, C.A. Reversibly addressing an allosteric catalyst in situ: Catalytic molecular tweezers. *Angew. Chem. Int. Ed.* **2004**, *43*, 5503–5507. [CrossRef] [PubMed]
35. Ma, D.Y.; Norouzi-Arasi, H.; Sheibani, E.; Wärnmark, K. Dynamic Supramolecular [(Salen) CrCl] Complexes as Efficient Catalysts for Ring Opening of Epoxides. *ChemCatChem* **2010**, *2*, 629–632. [CrossRef]
36. Lindbäck, E.; Norouzi-Arasi, H.; Sheibani, E.; Ma, D.; Dawaigher, S.; Wärnmark, K. Synthesis of Cr (III) Salen Complexes as Supramolecular Catalytic Systems for Ring-Opening Reactions of Epoxides. *ChemistrySelect* **2016**, *1*, 1789–1794. [CrossRef]
37. Odille, F.G.J.; Jónsson, S.; Stjernqvist, S.; Rydén, T.; Wärnmark, K. On the Characterization of Dynamic Supramolecular Systems: A General Mathematical Association Model for Linear Supramolecular Copolymers and Application on a Complex Two-Component Hydrogen-Bonding System. *Chem. Eur. J.* **2007**, *13*, 9617–9636. [CrossRef] [PubMed]
38. Zulauf, A.; Mellah, M.; Schulz, E. New Chiral Calixsalen Chromium Complexes: Recyclable Asymmetric Catalysts. *Chem. Eur. J.* **2010**, *16*, 11108–11114. [CrossRef]
39. Canali, L.; Sherrington, D.C. Utilisation of homogeneous and supported chiral metal (salen) complexes in asymmetric catalysis. *Chem. Soc. Rev.* **1999**, *28*, 85–93. [CrossRef]
40. Angelino, M.D.; Laibinis, P.E. Polymer-supported salen complexes for heterogeneous asymmetric synthesis: Stability and selectivity. *J. Polym. Sci. Part A Polym. Chem.* **1999**, *37*, 3888–3898. [CrossRef]
41. Gigante, B.; Corma, A.; García, H.; Sabater, M.J. Assessment of the negative factors responsible for the decrease in the enantioselectivity for the ring opening of epoxides catalyzed by chiral supported Cr (III)-salen complexes. *Catal. Lett.* **2000**, *68*, 113–119. [CrossRef]
42. Baleizão, C.; Gigante, B.; Sabater, M.J.; Garcia, H.; Corma, A. On the activity of chiral chromium salen complexes covalently bound to solid silicates for the enantioselective epoxide ring opening. *Appl. Catal. A* **2002**, *228*, 279–288. [CrossRef]
43. Dioos, B.M.; Geurts, W.A.; Jacobs, P.A. Coordination of Cr III (salen) on functionalised silica for asymmetric ring opening reactions of epoxides. *Catal. Lett.* **2004**, *97*, 125–129. [CrossRef]
44. Dioos, B.M.L.; Jacobs, P.A. Impregnation of dimeric CrIII(salen) on silica and its application in epoxide asymmetric ring opening reactions. *Appl. Catal. A* **2005**, *282*, 181–188. [CrossRef]
45. Dioos, B.M.L.; Jacobs, P.A. CrIII(salen) impregnated on silica for asymmetric ring opening reactions and its recovery via desorption/re-impregnation. *Tetrahedron Lett.* **2003**, *44*, 8815–8817. [CrossRef]
46. Dioos, B.M.L.; Jacobs, P.A. Heterogenisation of dimeric Cr(salen) with supported ionic liquids. *J. Catal.* **2006**, *243*, 217–219. [CrossRef]
47. Keilitz, J.; Haag, R. Intramolecular Acceleration of Asymmetric Epoxide Ring-Opening by Dendritic Polyglycerol Salen–CrIII Complexes. *Eur. J. Org. Chem.* **2009**, *2009*, 3272–3278. [CrossRef]
48. Zheng, X.; Jones, C.W.; Weck, M. Ring-Expanding Olefin Metathesis: A Route to Highly Active Unsymmetrical Macrocyclic Oligomeric Co-Salen Catalysts for the Hydrolytic Kinetic Resolution of Epoxides. *J. Am. Chem. Soc.* **2007**, *129*, 1105–1112. [CrossRef]
49. Kinslow, K.; Sevde, A.M.; Liang, J.; Liu, Y. Enantioselective ring opening of epoxides with TMSN3 by macrocyclic oligomer-supported Cr (III)–salen catalysts. *Tetrahedron Asymmetry* **2015**, *26*, 385–392. [CrossRef]
50. Song, C.E.; Oh, C.R.; Roh, E.J.; Choo, D.J. Cr (salen) catalysed asymmetric ring opening reactions of epoxides in room temperature ionic liquids. *Chem. Commun.* **2000**, 1743–1744. [CrossRef]
51. Jiao, J.; Tan, C.; Li, Z.; Liu, Y.; Han, X.; Cui, Y. Design and assembly of chiral coordination cages for asymmetric sequential reactions. *J. Am. Chem. Soc.* **2018**, *140*, 2251–2259. [CrossRef]
52. Tan, C.; Han, X.; Li, Z.; Liu, Y.; Cui, Y. Controlled exchange of achiral linkers with chiral linkers in Zr-based UiO-68 metal–organic framework. *J. Am. Chem. Soc.* **2018**, *140*, 16229–16236. [CrossRef]
53. Zhang, X.; Hou, Y.; Ettelaie, R.; Guan, R.; Zhang, M.; Zhang, Y.; Yang, H. Pickering Emulsion-Derived Liquid–Solid Hybrid Catalyst for Bridging Homogeneous and Heterogeneous Catalysis. *J. Am. Chem. Soc.* **2019**, *141*, 5220–5230. [CrossRef]
54. Bartoli, G.; Bosco, M.; Carlone, A.; Locatelli, M.; Massaccesi, M.; Melchiorre, P.; Sambri, L. Asymmetric Aminolysis of Aromatic Epoxides: A Facile Catalytic Enantioselective Synthesis of anti-β-Amino Alcohols. *Org. Lett.* **2004**, *6*, 2173–2176. [CrossRef] [PubMed]

55. Kureshy, R.I.; Kumar, M.; Agrawal, S.; Khan, N.U.H.; Dangi, B.; Abdi, S.H.; Bajaj, H.C. Enantioselective desymmetrization of meso-epoxides with anilines catalyzed by polymeric and monomeric Ti (IV) salen complexes. *Chirality* **2011**, *23*, 76–83. [CrossRef]
56. Kureshy, R.I.; Prathap, K.J.; Kumar, M.; Bera, P.K.; Khan, N.-u.H.; Abdi, S.H.R.; Bajaj, H.C. Synthesis of enantiopure β-amino alcohols via AKR/ARO of epoxides using recyclable macrocyclic Cr (III) salen complexes. *Tetrahedron* **2011**, *67*, 8300–8307. [CrossRef]
57. Birrell, J.A.; Jacobsen, E.N. A practical method for the synthesis of highly enantioenriched trans-1, 2-amino alcohols. *Org. Lett.* **2013**, *15*, 2895–2897. [CrossRef] [PubMed]
58. White, D.E.; Tadross, P.M.; Lu, Z.; Jacobsen, E.N. A broadly applicable and practical oligomeric (salen)Co catalyst for enantioselective epoxide ring-opening reactions. *Tetrahedron* **2014**, *70*, 4165–4180. [CrossRef] [PubMed]
59. Sharma, A.; Agarwal, J.; Peddinti, R.K. Direct access to the optically active VAChT inhibitor vesamicol and its analogues via the asymmetric aminolysis of meso-epoxides with secondary aliphatic amines. *Org. Biomol. Chem.* **2017**, *15*, 1913–1920. [CrossRef]
60. Roy, S.; Bhanja, P.; Islam, S.S.; Bhaumik, A.; Islam, S.M. A new chiral Fe (iii)–salen grafted mesoporous catalyst for enantioselective asymmetric ring opening of racemic epoxides at room temperature under solvent-free conditions. *Chem. Commun.* **2016**, *52*, 1871–1874. [CrossRef]
61. Islam, M.M.; Bhanja, P.; Halder, M.; Kundu, S.K.; Bhaumik, A.; Islam, S.M. Chiral Co(iii)–salen complex supported over highly ordered functionalized mesoporous silica for enantioselective aminolysis of racemic epoxides. *RSC Adv.* **2016**, *6*, 109315–109321. [CrossRef]
62. Sun, Z.; Chen, J.; Liu, Y.; Tu, T. Chiral Titanium Coordination Assemblies: Robust Cooperative Self-Supported Catalysts for Asymmetric Ring Opening of meso-Epoxides with Aliphatic Amines. *Adv. Synth. Catal.* **2017**, *359*, 494–505. [CrossRef]
63. Xi, W.; Liu, Y.; Xia, Q.; Li, Z.; Cui, Y. Direct and Post-Synthesis Incorporation of Chiral Metallosalen Catalysts into Metal–Organic Frameworks for Asymmetric Organic Transformations. *Chem. Eur. J.* **2015**, *21*, 12581–12585. [CrossRef]
64. Kolb, H.C.; VanNieuwenhze, M.S.; Sharpless, K.B. Catalytic Asymmetric Dihydroxylation. *Chem. Rev.* **1994**, *94*, 2483–2547. [CrossRef]
65. Kim, G.-J.; Park, D.-W. The catalytic activity of new chiral salen complexes immobilized on MCM-41 in the asymmetric hydrolysis of epoxides to diols. *Catal. Today* **2000**, *63*, 537–547. [CrossRef]
66. Haak, R.M.; Martinez Belmonte, M.; Escudero-Adan, E.C.; Benet-Buchholz, J.; Kleij, A.W. Olefin metathesis as a tool for multinuclear Co(III)salen catalyst construction: Access to cooperative catalysts. *Dalton Trans.* **2010**, *39*, 593–602. [CrossRef] [PubMed]
67. Hong, X.; Mellah, M.; Schulz, E. Heterobimetallic dual-catalyst systems for the hydrolytic kinetic resolution of terminal epoxides. *Catal. Sci. Technol.* **2014**, *4*, 2608–2617. [CrossRef]
68. Dandachi, H.; Zaborova, E.; Kolodziej, E.; David, O.R.P.; Hannedouche, J.; Mellah, M.; Jaber, N.; Schulz, E. Mixing and matching chiral cobalt- and manganese-based calix-salen catalysts for the asymmetric hydrolytic ring opening of epoxides. *Tetrahedron Asymmetry* **2016**, *27*, 246–253. [CrossRef]
69. Ready, J.M.; Jacobsen, E.N. Highly active oligomeric (salen)Co catalysts for asymmetric epoxide ring-opening reactions. *J. Am. Chem. Soc.* **2001**, *123*, 2687–2688. [CrossRef]
70. Liang, J.; Soucie, L.N.; Blechschmidt, D.R.; Yoder, A.; Gustafson, A.; Liu, Y. Aromatic Donor-Acceptor Interaction-Based Co(III)-salen Self-Assemblies and Their Applications in Asymmetric Ring Opening of Epoxides. *Org. Lett.* **2019**, *21*, 513–518. [CrossRef]
71. Jacobsen, E.N.; Kakiuchi, F.; Konsler, R.G.; Larrow, J.F.; Tokunaga, M. Enantioselective catalytic ring opening of epoxides with carboxylic acids. *Tetrahedron Lett.* **1997**, *38*, 773–776. [CrossRef]
72. Müller, K.; Faeh, C.; Diederich, F. Fluorine in pharmaceuticals: Looking beyond intuition. *Science* **2007**, *317*, 1881–1886. [CrossRef]
73. Purser, S.; Moore, P.R.; Swallow, S.; Gouverneur, V. Fluorine in medicinal chemistry. *Chem. Soc. Rev.* **2008**, *37*, 320–330. [CrossRef]
74. Banks, R.E.; Smart, B.E.; Tatlow, J. *Organofluorine Chemistry: Principles and Commercial Applications*; Springer Science & Business Media: New York, NY, USA, 2013.
75. Bruns, S.; Haufe, G. Enantioselective introduction of fluoride into organic compounds: First asymmetric ring opening of epoxides by hydrofluorinating reagents. *J. Fluor. Chem.* **2000**, *104*, 247–254. [CrossRef]

76. Haufe, G.; Bruns, S.; Runge, M. Enantioselective ring-opening of epoxides by HF-reagents: Asymmetric synthesis of fluoro lactones. *J. Fluor. Chem.* **2001**, *112*, 55–61. [CrossRef]
77. Haufe, G.; Bruns, S. (Salen)chromium Complex Mediated Asymmetric Ring Opening of meso- and Racemic Epoxides with Different Fluoride Sources. *Adv. Synth. Catal.* **2002**, *344*, 165–171. [CrossRef]
78. Kalow, J.A.; Doyle, A.G. Enantioselective ring opening of epoxides by fluoride anion promoted by a cooperative dual-catalyst system. *J. Am. Chem. Soc.* **2010**, *132*, 3268–3269. [CrossRef]
79. Kalow, J.A.; Doyle, A.G. Mechanistic Investigations of Cooperative Catalysis in the Enantioselective Fluorination of Epoxides. *J. Am. Chem. Soc.* **2011**, *133*, 16001–16012. [CrossRef]
80. Wu, M.H.; Jacobsen, E.N. Asymmetric ring opening of meso epoxides with thiols: Enantiomeric enrichment using a bifunctional nucleophile. *J. Org. Chem.* **1998**, *63*, 5252–5254. [CrossRef]
81. Wu, J.; Hou, X.-L.; Dai, L.-X.; Xia, L.-J.; Tang, M.-H. Enantioselective ring opening of meso-epoxides with thiols catalyzed by a chiral (salen) Ti (IV) complex. *Tetrahedron Asymmetry* **1998**, *9*, 3431–3436. [CrossRef]
82. Li, Z.; Zhou, Z.; Li, K.; Wang, L.; Zhou, Q.; Tang, C. Regio-and stereoselective ring-opening of epoxides using organic dithiophosphorus acids as nucleophiles. *Tetrahedron Lett.* **2002**, *43*, 7609–7611. [CrossRef]
83. Zhou, Z.; Li, Z.; Quanyong, W.; Liu, B.; Li, K.; Zhao, G.; Zhou, Q.; Tang, C. (Salen) Ti (IV) complex catalyzed asymmetric ring-opening of epoxides using dithiophosphorus acid as the nucleophile. *J. Organomet. Chem.* **2006**, *691*, 5790–5797. [CrossRef]
84. Yang, M.; Zhu, C.; Yuan, F.; Huang, Y.; Pan, Y. Enantioselective Ring-Opening Reaction of m eso-Epoxides with ArSeH Catalyzed by Heterometallic Ti–Ga–Salen System. *Org. Lett.* **2005**, *7*, 1927–1930. [CrossRef]
85. Sun, J.; Yang, M.; Yuan, F.; Jia, X.; Yang, X.; Pan, Y.; Zhu, C. Catalytic Asymmetric Ring-Opening Reaction of meso-Epoxides with Aryl Selenols and Thiols Catalyzed by a Heterobimetallic Gallium-Titanium-Salen Complex. *Adv. Synth. Catal.* **2009**, *351*, 920–930. [CrossRef]
86. Sun, J.; Yuan, F.; Yang, M.; Pan, Y.; Zhu, C. Enantioselective ring-opening reaction of meso-epoxides with ArSH catalyzed by heterobimetallic Ti–Ga–Salen system. *Tetrahedron Lett.* **2009**, *50*, 548–551. [CrossRef]
87. Tiecco, M.; Testaferri, L.; Marini, F.; Sternativo, S.; Del Verme, F.; Santi, C.; Bagnoli, L.; Temperini, A. Synthesis of enantiomerically enriched β-hydroxy selenides by catalytic asymmetric ring opening of meso-epoxides with (phenylseleno) silanes. *Tetrahedron* **2008**, *64*, 3337–3342. [CrossRef]
88. Scheffler, U.; Mahrwald, R. Recent Advances in Organocatalytic Methods for Asymmetric C–C Bond Formation. *Chem. Eur. J.* **2013**, *19*, 14346–14396. [CrossRef] [PubMed]
89. Brahmachari, G. Design for carbon–carbon bond forming reactions under ambient conditions. *RSC Adv.* **2016**, *6*, 64676–64725. [CrossRef]
90. Bandini, M.; Cozzi, P.G.; Melchiorre, P.; Umani-Ronchi, A. Kinetic Resolution of Epoxides by a C-C Bond-Forming Reaction: Highly Enantioselective Addition of Indoles to cis, trans, and meso Aromatic Epoxides Catalyzed by [Cr (salen)] Complexes. *Angew. Chem. Int. Ed.* **2004**, *43*, 84–87. [CrossRef] [PubMed]
91. Keith, J.M.; Larrow, J.F.; Jacobsen, E.N. Practical considerations in kinetic resolution reactions. *Adv. Synth. Catal.* **2001**, *343*, 5–26. [CrossRef]
92. Larrow, J.F.; Schaus, S.E.; Jacobsen, E.N. Kinetic resolution of terminal epoxides via highly regioselective and enantioselective ring opening with TMSN3. An efficient, catalytic route to 1, 2-amino alcohols. *J. Am. Chem. Soc.* **1996**, *118*, 7420–7421. [CrossRef]
93. Schaus, S.E.; Jacobsen, E.N. Dynamic kinetic resolution of epichlorohydrin via enantioselective catalytic ring opening with TMSN3. Practical synthesis of aryl oxazolidinone antibacterial agents. *Tetrahedron Lett.* **1996**, *37*, 7937–7940. [CrossRef]
94. Lebel, H.; Jacobsen, E.N. Chromium catalyzed kinetic resolution of 2,2-disubstituted epoxides. *Tetrahedron Lett.* **1999**, *40*, 7303–7306. [CrossRef]
95. Lebel, H.; Jacobsen, E.N. Enantioselective Total Synthesis of Taurospongin A. *J. Org. Chem.* **1998**, *63*, 9624–9625. [CrossRef]
96. Dioos, B.M.L.; Jacobs, P.A. CrIII(salen) catalysed asymmetric ring opening of monocyclic terpene-epoxides. *Tetrahedron Lett.* **2003**, *44*, 4715–4717. [CrossRef]
97. Dioos, B.M.L.; Jacobs, P.A. Microwave-assisted Cr(salen)-catalysed asymmetric ring opening of epoxides. *J. Catal.* **2005**, *235*, 428–430. [CrossRef]
98. Chen, S.-W.; Thakur, S.S.; Li, W.; Shin, C.-K.; Kawthekar, R.B.; Kim, G.-J. Efficient catalytic synthesis of optically pure 1,2-azido alcohols through enantioselective epoxide ring opening with HN3. *J. Mol. Catal. A Chem.* **2006**, *259*, 116–120. [CrossRef]

99. Bai, S.; Li, B.; Peng, J.; Zhang, X.; Yang, Q.; Li, C. Promoted activity of Cr(Salen) in a nanoreactor for kinetic resolution of terminal epoxides. *Chem. Sci.* **2012**, *3*, 2864–2867. [CrossRef]
100. Bartoli, G.; Bosco, M.; Carlone, A.; Locatelli, M.; Melchiorre, P.; Sambri, L. Asymmetric Catalytic Synthesis of Enantiopure N-Protected 1,2-Amino Alcohols. *Org. Lett.* **2004**, *6*, 3973–3975. [CrossRef]
101. Baudequin, C.; Baudoux, J.; Levillain, J.; Cahard, D.; Gaumont, A.-C.; Plaquevent, J.-C. Ionic liquids and chirality: Opportunities and challenges. *Tetrahedron Asymmetry* **2003**, *14*, 3081–3093. [CrossRef]
102. Kureshy, R.I.; Prathap, K.J.; Agrawal, S.; Kumar, M.; Khan, N.-u.H.; Abdi, S.H.R.; Bajaj, H.C. Highly Efficient Recyclable CoIII–salen Complexes in the Catalyzed Asymmetric Aminolytic Kinetic Resolution of Aryloxy/Terminal Epoxides for the Simultaneous Production of N-Protected 1,2-Amino Alcohols and the Corresponding Epoxides in High Optical Purity. *Eur. J. Org. Chem.* **2009**, *2009*, 2863–2871. [CrossRef]
103. Kureshy, R.I.; Kumar, M.; Agrawal, S.; Khan, N.-u.H.; Abdi, S.H.R.; Bajaj, H.C. Aminolytic kinetic resolution of trans epoxides for the simultaneous production of chiral trans β-amino alcohols in the presence of chiral Cr(III) salen complex using an ionic liquid as a green reaction media. *Tetrahedron Asymmetry* **2010**, *21*, 451–456. [CrossRef]
104. Kureshy, R.I.; Singh, S.; Khan, N.-u.H.; Abdi, S.H.R.; Agrawal, S.; Jasra, R.V. Enantioselective aminolytic kinetic resolution (AKR) of epoxides catalyzed by recyclable polymeric Cr(III) salen complexes. *Tetrahedron Asymmetry* **2006**, *17*, 1638–1643. [CrossRef]
105. Kureshy, R.I.; Prathap, K.J.; Singh, S.; Agrawal, S.; Khan, N.-U.H.; Abdi, S.H.R.; Jasra, R.V. Chiral recyclable dimeric and polymeric Cr(III) salen complexes catalyzed aminolytic kinetic resolution of trans-aromatic epoxides under microwave irradiation. *Chirality* **2007**, *19*, 809–815. [CrossRef] [PubMed]
106. Kureshy, R.I.; Prathap, K.J.; Roy, T.; Maity, N.C.; Khan, N.-u.H.; Abdi, S.H.R.; Bajaj, H.C. Reusable Chiral Dicationic Chromium(III) Salen Catalysts for Aminolytic Kinetic Resolution of trans-Epoxides. *Adv. Synth. Catal.* **2010**, *352*, 3053–3060. [CrossRef]
107. Tak, R.; Kumar, M.; Menapara, T.; Choudhary, M.K.; Kureshy, R.I.; Khan, N.-u.H. Asymmetric Catalytic Syntheses of Pharmaceutically Important β-Amino-α-Hydroxyl Esters by Enantioselective Aminolysis of Methyl Phenylglycidate. *ChemCatChem* **2017**, *9*, 322–328. [CrossRef]
108. Kumar, M.; Kureshy, R.I.; Shah, A.K.; Das, A.; Khan, N.-u.H.; Abdi, S.H.R.; Bajaj, H.C. Asymmetric Aminolytic Kinetic Resolution of Racemic Epoxides Using Recyclable Chiral Polymeric Co(III)-Salen Complexes: A Protocol for Total Utilization of Racemic Epoxide in the Synthesis of (R)-Naftopidil and (S)-Propranolol. *J. Org. Chem.* **2013**, *78*, 9076–9084. [CrossRef]
109. Tokunaga, M.; Larrow, J.F.; Kakiuchi, F.; Jacobsen, E.N. Asymmetric Catalysis with Water: Efficient Kinetic Resolution of Terminal Epoxides by Means of Catalytic Hydrolysis. *Science* **1997**, *277*, 936–938. [CrossRef]
110. Schaus, S.E.; Brandes, B.D.; Larrow, J.F.; Tokunaga, M.; Hansen, K.B.; Gould, A.E.; Furrow, M.E.; Jacobsen, E.N. Highly Selective Hydrolytic Kinetic Resolution of Terminal Epoxides Catalyzed by Chiral (salen)CoIII Complexes. Practical Synthesis of Enantioenriched Terminal Epoxides and 1,2-Diols. *J. Am. Chem. Soc.* **2002**, *124*, 1307–1315. [CrossRef]
111. Kureshy, R.I.; Khan, N.H.; Abdi, S.H.R.; Patel, S.T.; Jasra, R.V. Simultaneous production of chirally enriched epoxides and 1,2-diols from racemic epoxides via hydrolytic kinetic resolution (HKR). *J. Mol. Catal. A Chem.* **2002**, *179*, 73–77. [CrossRef]
112. Kureshy, R.I.; Singh, S.; Khan, N.U.H.; Abdi, S.H.R.; Ahmad, I.; Bhatt, A.; Jasra, R.V. Improved catalytic activity of homochiral dimeric cobalt-salen complex in hydrolytic kinetic resolution of terminal racemic epoxides. *Chirality* **2005**, *17*, 590–594. [CrossRef]
113. Venkatasubbaiah, K.; Gill, C.S.; Takatani, T.; Sherrill, C.D.; Jones, C.W. A Versatile Co(bisalen) Unit for Homogeneous and Heterogeneous Cooperative Catalysis in the Hydrolytic Kinetic Resolution of Epoxides. *Chem. Eur. J.* **2009**, *15*, 3951–3955. [CrossRef]
114. Wezenberg, S.J.; Kleij, A.W. Cooperative Activation in the Hydrolytic Kinetic Resolution of Epoxides by a Bis-Cobalt(III)salen-Calix 4 arene Hybrid. *Adv. Synth. Catal.* **2010**, *352*, 85–91. [CrossRef]
115. Ready, J.M.; Jacobsen, E.N. A practical oligomeric (salen)Co catalyst for asymmetric epoxide ring-opening reactions. *Angew. Chem. Int. Ed.* **2002**, *41*, 1374–1377. [CrossRef]
116. Kahn, M.G.C.; Weck, M. Highly crosslinked polycyclooctyl-salen cobalt (III) for the hydrolytic kinetic resolution of terminal epoxides. *Catal. Sci. Technol.* **2012**, *2*, 386–389. [CrossRef]

117. Sadhukhan, A.; Khan, N.-u.H.; Roy, T.; Kureshy, R.I.; Abdi, S.H.R.; Bajaj, H.C. Asymmetric Hydrolytic Kinetic Resolution with Recyclable Macrocyclic CoIII-Salen Complexes: A Practical Strategy in the Preparation of (R)-Mexiletine and (S)-Propranolol. *Chem. Eur. J.* **2012**, *18*, 5256–5260. [CrossRef]
118. Roy, T.; Barik, S.; Kumar, M.; Kureshy, R.I.; Ganguly, B.; Khan, N.-u.H.; Abdi, S.H.R.; Bajaj, H.C. Asymmetric hydrolytic kinetic resolution with recyclable polymeric Co(III)-salen complexes: A practical strategy in the preparation of (S)-metoprolol, (S)-toliprolol and (S)-alprenolol: Computational rationale for enantioselectivity. *Catal. Sci. Technol.* **2014**, *4*, 3899–3908. [CrossRef]
119. Hong, X.; Mellah, M.; Bordier, F.; Guillot, R.; Schulz, E. Electrogenerated Polymers as Efficient and Robust Heterogeneous Catalysts for the Hydrolytic Kinetic Resolution of Terminal Epoxides. *ChemCatChem* **2012**, *4*, 1115–1121. [CrossRef]
120. Dandachi, H.; Nasrallah, H.; Ibrahim, F.; Hong, X.; Mellah, M.; Jaber, N.; Schulz, E. Chiral calix-salen cobalt complexes, catalysts for the enantioselective dynamic hydrolytic kinetic resolution of epibromohydrin. *J. Mol. Catal. A Chem.* **2014**, *395*, 457–462. [CrossRef]
121. Breinbauer, R.; Jacobsen, E.N. Cooperative Asymmetric Catalysis with Dendrimeric [Co(salen)] Complexes. *Angew. Chem. Int. Ed.* **2000**, *39*, 3604–3607. [CrossRef]
122. Song, Y.; Yao, X.; Chen, H.; Bai, C.; Hu, X.; Zheng, Z. Highly enantioselective resolution of terminal epoxides using polymeric catalysts. *Tetrahedron Lett.* **2002**, *43*, 6625–6627. [CrossRef]
123. Song, Y.M.; Chen, H.L.; Hu, X.Q.; Bai, C.M.; Zheng, Z. Highly enantioselective resolution of terminal epoxides with cross-linked polymeric salen-Co(III) complexes. *Tetrahedron Lett.* **2003**, *44*, 7081–7085. [CrossRef]
124. Zheng, X.; Jones, C.W.; Weck, M. Poly(styrene)-Supported Co–Salen Complexes as Efficient Recyclable Catalysts for the Hydrolytic Kinetic Resolution of Epichlorohydrin. *Chem. Eur. J.* **2006**, *12*, 576–583. [CrossRef]
125. Zheng, X.; Jones, C.W.; Weck, M. Engineering Polymer-Enhanced Bimetallic Cooperative Interactions in the Hydrolytic Kinetic Resolution of Epoxides. *Adv. Synth. Catal.* **2008**, *350*, 255–261. [CrossRef]
126. Rossbach, B.M.; Leopold, K.; Weberskirch, R. Self-Assembled Nanoreactors as Highly Active Catalysts in the Hydrolytic Kinetic Resolution (HKR) of Epoxides in Water. *Angew. Chem. Int. Ed.* **2006**, *45*, 1309–1312. [CrossRef] [PubMed]
127. Thakur, S.S.; Li, W.J.; Kim, S.J.; Kim, G.J. Highly reactive and enantioselective kinetic resolution of terminal epoxides with H_2O and HCl catalyzed by new chiral (salen)Co complex linked with Al. *Tetrahedron Lett.* **2005**, *46*, 2263–2266. [CrossRef]
128. Thakur, S.S.; Chen, S.W.; Li, W.J.; Shin, C.K.; Kim, S.J.; Koo, Y.M.; Kim, G.J. A new dinuclear chiral salen complexes for asymmetric ring opening and closing reactions: Synthesis of valuable chiral intermediates. *J. Organomet. Chem.* **2006**, *691*, 1862–1872. [CrossRef]
129. Jiang, C.J.; Chen, Z.R. Chiral cobalt salen complexes containing Lewis acid: A highly reactive and enantioselective catalyst for the hydrolytic kinetic resolution of terminal epoxides. *Kinet. Catal.* **2008**, *49*, 474–478. [CrossRef]
130. Kawthekar, R.B.; Bi, W.-t.; Kim, G.-J. Asymmetric ring opening of epoxides catalyzed by novel heterobimetallic Schiff-bases containing transition metal salts. *Bull. Korean Chem. Soc.* **2008**, *29*, 313–318. [CrossRef]
131. Jiang, C. Asymmetric ring opening of terminal epoxides via kinetic resolution catalyzed by chiral (salen)Co mixture. *Kinet. Catal.* **2011**, *52*, 691–696. [CrossRef]
132. Park, J.; Lang, K.; Abboud, K.A.; Hong, S. Self-Assembly Approach toward Chiral Bimetallic Catalysts: Bis-Urea-Functionalized (Salen)Cobalt Complexes for the Hydrolytic Kinetic Resolution of Epoxides. *Chem. Eur. J.* **2011**, *17*, 2236–2245. [CrossRef]
133. Yang, H.; Zhang, L.; Su, W.; Yang, Q.; Li, C. Asymmetric ring-opening of epoxides on chiral Co(Salen) catalyst synthesized in SBA-16 through the "ship in a bottle" strategy. *J. Catal.* **2007**, *248*, 204–212. [CrossRef]
134. Yang, H.; Zhang, L.; Zhong, L.; Yang, Q.; Li, C. Enhanced Cooperative Activation Effect in the Hydrolytic Kinetic Resolution of Epoxides on [Co(salen)] Catalysts Confined in Nanocages. *Angew. Chem. Int. Ed.* **2007**, *46*, 6861–6865. [CrossRef]
135. Cavazzini, M.; Quici, S.; Pozzi, G. Hydrolytic kinetic resolution of terminal epoxides catalyzed by fluorous chiral Co(salen) complexes. *Tetrahedron* **2002**, *58*, 3943–3949. [CrossRef]
136. Shepperson, I.; Cavazzini, M.; Pozzi, G.; Quici, S. Fluorous biphasic hydrolytic kinetic resolution of terminal epoxides. *J. Fluorine Chem.* **2004**, *125*, 175–180. [CrossRef]

137. Rim Oh, C.; Joon Choo, D.; Ho Shim, W.; Hoon Lee, D.; Joo Roh, E.; Lee, S.-g.; Eui Song, C. Chiral Co(iii)(salen)-catalysed hydrolytic kinetic resolution of racemic epoxides in ionic liquids. *Chem. Commun.* **2003**, 1100–1101. [CrossRef] [PubMed]
138. Aerts, S.; Weyten, H.; Buekenhoudt, A.; Gevers, L.E.M.; Vankelecom, I.F.J.; Jacobs, P.A. Recycling of the homogeneous Co-Jacobsen catalyst through solvent-resistent nanofiltration (SRNF). *Chem. Commun.* **2004**, 710–711. [CrossRef]
139. Annis, D.A.; Jacobsen, E.N. Polymer-supported chiral Co(salen) complexes: Synthetic applications and mechanistic investigations in the hydrolytic kinetic resolution of terminal epoxides. *J. Am. Chem. Soc.* **1999**, *121*, 4147–4154. [CrossRef]
140. Solodenko, W.; Jas, G.; Kunz, U.; Kirschning, A. Continuous enantioselective kinetic resolution of terminal epoxides using immobilized chiral cobalt-salen complexes. *Synthesis* **2007**, 583–589. [CrossRef]
141. Kunz, U.; Schönfeld, H.; Solodenko, W.; Jas, G.; Kirschning, A. Manufacturing and Construction of PASSflow Flow Reactors and Their Utilization in Suzuki−Miyaura Cross-Coupling Reactions. *Ind. Eng. Chem. Res.* **2005**, *44*, 8458–8467. [CrossRef]
142. Goyal, P.; Zheng, X.; Weck, M. Enhanced Cooperativity in Hydrolytic Kinetic Resolution of Epoxides using Poly(styrene) Resin-Supported Dendronized Co-(Salen) Catalysts. *Adv. Synth. Catal.* **2008**, *350*, 1816–1822. [CrossRef]
143. Kahn, M.G.C.; Stenlid, J.H.; Weck, M. Poly(styrene) Resin-Supported Cobalt(III) Salen Cyclic Oligomers as Active Heterogeneous HKR Catalysts. *Adv. Synth. Catal.* **2012**, *354*, 3016–3024. [CrossRef]
144. Gill, C.S.; Venkatasubbaiah, K.; Phan, N.T.S.; Weck, M.; Jones, C.W. Enhanced Cooperativity through Design: Pendant CoIII-Salen Polymer Brush Catalysts for the Hydrolytic Kinetic Resolution of Epichlorohydrin (Salen=N,N′-Bis(salicylidene)ethylenediamine Dianion). *Chem. Eur. J.* **2008**, *14*, 7306–7313. [CrossRef]
145. Belser, T.; Jacobsen, E.N. Cooperative catalysis in the hydrolytic kinetic resolution of epoxides by chiral (salen)Co(III) complexes immobilized on gold colloids. *Adv. Synth. Catal.* **2008**, *350*, 967–971. [CrossRef]
146. Kawthekar, R.B.; Lee, Y.-H.; Kim, G.-J. Asymmetric catalysis by chiral salen complexes immobilized on mesoporous materials having modified pore channel system by dimethylcarbonate. *J. Porous Mater.* **2009**, *16*, 367–378. [CrossRef]
147. Kim, Y.-S.; Guo, X.-F.; Kim, G.-J. Highly active new chiral Co(III) salen catalysts immobilized by electrostatic interaction with sulfonic acid linkages on ordered mesoporous SBA-16 silica. *Chem. Commun.* **2009**, 4296–4298. [CrossRef] [PubMed]
148. Kim, Y.-S.; Lee, C.-Y.; Kim, G.-J. Asymmetric Ring Opening of Terminal Epoxides Catalyzed by Chiral Co(III)-BF3 Salen Complex Immobilized on SBA-16. *Bull. Korean Chem. Soc.* **2009**, *30*, 1771–1777. [CrossRef]
149. Kim, Y.-S.; Lee, C.-Y.; Kim, G.-J. Synthesis of New Bimetallic Chiral Salen Catalyst Bearing Co(BF4)(2) Salt and Its Application in Asymmetric Ring Opening of Epoxide. *Bull. Korean Chem. Soc.* **2010**, *31*, 2973–2979. [CrossRef]
150. Kim, G.-J.; Lee, H.; Kim, S.-J. Catalytic activity and recyclability of new enantioselective chiral Co–salen complexes in the hydrolytic kinetic resolution of epichlorohydrine. *Tetrahedron Lett.* **2003**, *44*, 5005–5008. [CrossRef]
151. Nielsen, L.P.C.; Stevenson, C.P.; Blackmond, D.G.; Jacobsen, E.N. Mechanistic investigation leads to a synthetic improvement in the hydrolytic kinetic resolution of terminal epoxides. *J. Am. Chem. Soc.* **2004**, *126*, 1360–1362. [CrossRef] [PubMed]
152. Sun, K.; Li, W.-X.; Feng, Z.; Li, C. Cooperative activation in ring-opening hydrolysis of epoxides by Co-salen complexes: A first principle study. *Chem. Phys. Lett.* **2009**, *470*, 259–263. [CrossRef]
153. Kennedy, M.R.; Burns, L.A.; Sherrill, C.D. Counterion and Substrate Effects on Barrier Heights of the Hydrolytic Kinetic Resolution of Terminal Epoxides Catalyzed by Co(III)-salen. *J. Phys. Chem. A* **2015**, *119*, 403–409. [CrossRef]
154. Jain, S.; Zheng, X.; Jones, C.W.; Weck, M.; Davis, R.J. Importance of counterion reactivity on the deactivation of co-salen catalysts in the hydrolytic kinetic resolution of epichlorohydrin. *Inorg. Chem.* **2007**, *46*, 8887–8896. [CrossRef]
155. Jain, S.; Venkatasubbaiah, K.; Jones, C.W.; Davis, R.J. Factors influencing recyclability of Co(III)-salen catalysts in the hydrolytic kinetic resolution of epichlorohydrin. *J. Mol. Catal. A Chem.* **2010**, *316*, 8–15. [CrossRef]

156. Ford, D.D.; Nielsen, L.P.C.; Zuend, S.J.; Musgrave, C.B.; Jacobsen, E.N. Mechanistic Basis for High Stereoselectivity and Broad Substrate Scope in the (salen)Co(III)-Catalyzed Hydrolytic Kinetic Resolution. *J. Am. Chem. Soc.* **2013**, *135*, 15595–15608. [CrossRef] [PubMed]
157. Berkessel, A.; Ertuerk, E. Hydrolytic kinetic resolution of epoxides catalyzed by chromium(III)-endo,endo-2,5-diaminonorbornane-salen Cr(III)-DIANANE-salen complexes. Improved activity, low catalyst loading. *Adv. Synth. Catal.* **2006**, *348*, 2619–2625. [CrossRef]
158. Lee, Y.; Yu, J.; Park, K.; Kim, G.J. Superior Effect of Ultrasonic Homogenization to Mechanical Agitation on Accelerating Reaction Rates in Asymmetric Ring Opening of Epoxides. *Bull. Korean Chem. Soc.* **2017**, *38*, 795–803. [CrossRef]
159. Wright, J.L.; Gregory, T.F.; Heffner, T.G.; MacKenzie, R.G.; Pugsley, T.A.; Meulen, S.V.; Wise, L.D. Discovery of selective dopamine D4 receptor antagonists: 1-Aryloxy-3-(4-aryloxypiperidinyl)-2-propanols. *Bioorgan. Med. Chem. Lett.* **1997**, *7*, 1377–1380. [CrossRef]
160. Ready, J.M.; Jacobsen, E.N. Asymmetric Catalytic Synthesis of α-Aryloxy Alcohols: Kinetic Resolution of Terminal Epoxides via Highly Enantioselective Ring-Opening with Phenols. *J. Am. Chem. Soc.* **1999**, *121*, 6086–6087. [CrossRef]
161. Peukert, S.; Jacobsen, E.N. Enantioselective Parallel Synthesis Using Polymer-Supported Chiral Co(salen) Complexes. *Org. Lett.* **1999**, *1*, 1245–1248. [CrossRef]
162. Zhu, X.; Venkatasubbaiah, K.; Weck, M.; Jones, C.W. Highly active oligomeric Co(salen) catalysts for the asymmetric synthesis of alpha-aryloxy or alpha-alkoxy alcohols via kinetic resolution of terminal epoxides. *J. Mol. Catal. A Chem.* **2010**, *329*, 1–6. [CrossRef]
163. Kawthekar, R.B.; Ahn, C.-H.; Kim, G.-J. Kinetic resolution of terminal epoxides with phenols promoted by heterometallic Co-Al and Co-Ga salen complexes. *Catal. Lett.* **2007**, *115*, 62–69. [CrossRef]
164. Lee, K.-Y.; Lee, C.-Y.; Kim, G.-J. Asymmetric synthesis of alpha-aryloxy alcohols via kinetic resolution of terminal epoxides catalyzed by (R,R)-N,N ′-bis(2-hydroxy-5-tert-butylsalisilidine)-1,2-cyclohexenediamino cobalt. *React. Kinet. Catal. Lett.* **2008**, *93*, 75–83. [CrossRef]
165. Guo, X.-F.; Kim, Y.-S.; Kim, G.-J. Chiral (Salen) Cobalt Complexes Encapsulated in Mesoporous MFI as an Enantioselective Catalyst for Asymmetric Ring Opening of Terminal Epoxides. *Top. Catal.* **2009**, *52*, 153–160. [CrossRef]
166. Lee, K.-Y.; Lee, C.-Y.; Kim, G.-J. Non-Covalent Immobilization of Chiral (Salen) Complexes on HF-treated Mesoporous MFI-type Zeolite for Asymmetric Catalysis. *Bull. Korean Chem. Soc.* **2009**, *30*, 389–396. [CrossRef]
167. Kim, Y.-S.; Guo, X.-f.; Kim, G.-J. Asymmetric Ring Opening Reaction of Catalyst Immobilized on Silica Monolith with Bimodal Meso/Macroscopic Pore Structure. *Top. Catal.* **2009**, *52*, 197–204. [CrossRef]
168. Jeon, H.-G.; Choi, S.D.; Jeon, S.K.; Park, G.W.; Yang, J.Y.; Kim, G.-J. Application of Mesoporous Silica Foam for Immobilization of Salen Complexes in Chiral Intermediates Synthesis. *Bull. Korean Chem. Soc.* **2015**, *36*, 1396–1404. [CrossRef]
169. Li, W.J.; Thakur, S.S.; Chen, S.W.; Shin, C.K.; Kawthekar, R.B.; Kim, G.J. Synthesis of optically active 2-hydroxy monoesters via-kinetic resolution and asymmetric cyclization catalyzed by heterometallic chiral (salen) Co complex. *Tetrahedron Lett.* **2006**, *47*, 3453–3457. [CrossRef]
170. Lee, Y.W.; Yang, H.C.; Kim, G.-J. Synthesis of Highly Enantiomerically Enriched Arenesulfonic Acid 2-Hydroxy Esters via Kinetic Resolution of Terminal Epoxides. *Appl. Chem. Eng.* **2016**, *27*, 490–494. [CrossRef]
171. Hajra, S.; Roy, S. Feedback Inhibition in Chemical Catalysis Leads the Dynamic Kinetic to Kinetic Resolution in C3-Indolylation of Spiro-epoxyoxindoles. *Org. Lett.* **2020**, *22*, 1458–1463. [CrossRef]
172. Hajra, S.; Maity, S.; Maity, R. Efficient Synthesis of 3,3′-Mixed Bisindoles via Lewis Acid Catalyzed Reaction of Spiro-epoxyoxindoles and Indoles. *Org. Lett.* **2015**, *17*, 3430–3433. [CrossRef]
173. Linker, T. The Jacobsen–Katsuki Epoxidation and Its Controversial Mechanism. *Angew. Chem. Int. Ed.* **1997**, *36*, 2060–2062. [CrossRef]

catalysts

MDPI

Review

α-Functionalization of Imines via Visible Light Photoredox Catalysis

Alberto F. Garrido-Castro [1], M. Carmen Maestro [1,*] and José Alemán [1,2,*]

1 Department of Organic Chemistry, Universidad Autónoma de Madrid, 28049 Madrid, Spain; albertof.garrido@uam.es
2 Institute for Advanced Research in Chemical Sciences (IAdChem), Universidad Autónoma de Madrid, 28049 Madrid, Spain
* Correspondence: carmen.maestro@uam.es (M.C.M.); jose.aleman@uam.es (J.A.); Tel.: +34914973875 (J.A.)

Received: 6 May 2020; Accepted: 16 May 2020; Published: 19 May 2020

Abstract: The innate electrophilicity of imine building blocks has been exploited in organic synthetic chemistry for decades. Inspired by the resurgence in photocatalysis, imine reactivity has now been redesigned through the generation of unconventional and versatile radical intermediates under mild reaction conditions. While novel photocatalytic approaches have broadened the range and applicability of conventional radical additions to imine acceptors, the possibility to use these imines as latent nucleophiles via single-electron reduction has also been uncovered. Thus, multiple research programs have converged on this issue, delivering creative and practical strategies to achieve racemic and asymmetric α-functionalizations of imines under visible light photoredox catalysis.

Keywords: amines; imines; photoredox catalysis; radical additions; radical–radical couplings; stereoselectivity; umpolung chemistry; visible light

1. Introduction

Visible light photoredox catalysis has been at the forefront of organic chemistry research for over a decade, establishing itself as a sustainable and multifaceted synthetic tool [1]. Irradiation of catalytic amounts of polypyridyl complexes and organic sensitizers under mild conditions has proven to be an excellent activation pathway to access a wide variety of radical intermediates (Figure 1). Spurred by this resurgence, long-standing challenges in the field have been resolved, while a plethora of transformations continue to be developed in an effort to revamp organic synthesis.

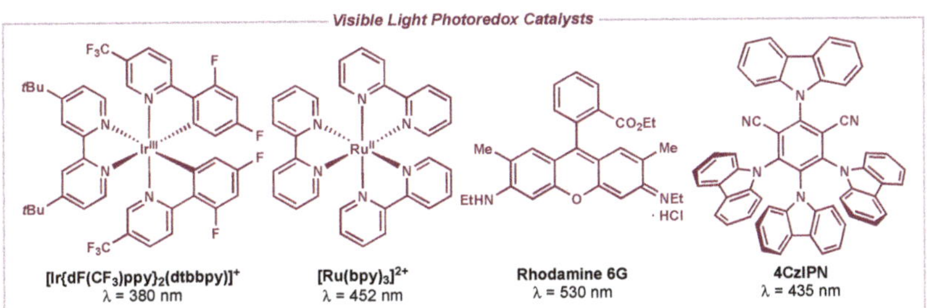

Figure 1. Visible light photoredox catalysts (λ = local absorbance maximum for lowest energy absorption).

The reactivity of imines has certainly undergone a complete makeover, as different strategies involving these key building blocks have been developed (Scheme 1). The classical approach still relies on the innate electrophilic nature of imines to undergo standard alkyl radical addition (pathway A in

Scheme 1, left). Thanks to photoredox catalysis, the generation of nucleophilic radicals starting from mild alkylating reagents [2] has provided a broader range to a severely limited transformation in the past due to hazardous reagents and impractical conditions.

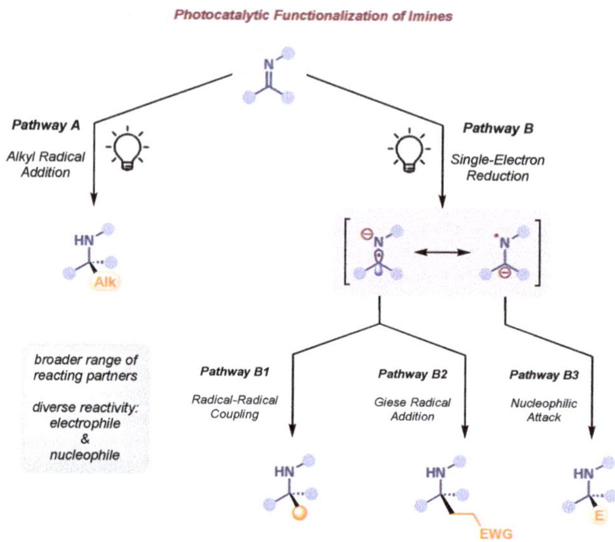

Scheme 1. Photocatalytic functionalization of imines: pathway A, alkyl radical addition (left); pathway B, single-electron reduction (right).

Alternatively, the photocatalytic single-electron reduction of imines has emerged as a powerful technology to generate radical anion intermediates which exist as two different resonant forms (pathway B in Scheme 1, right) [3,4]. The α-amino radical species can engage in radical–radical couplings with a large pool of reacting partners (pathway B1 in Scheme 1), while also displaying a complementary nucleophilic behavior to their corresponding electrophilic imine precursors. Indeed, they can be trapped by electron-deficient π-systems, a combination which would not be feasible using polar chemistry (pathway B2 in Scheme 1). Interestingly, the *N*-centered radical species can be quickly quenched by an H atom donor to yield a stable carbanion capable of reacting with a traditional electrophile in polar fashion (pathway B3 in Scheme 1).

The single-electron reduction event (pathway B in Scheme 1, right) can be a challenging redox process which often requires assistance [4]. While some electron-poor imines can undergo a straightforward photocatalytic reduction (such as *N*-sulfonyl- or α-keto-imines), other neutral imines feature a reduction potential which falls out of range of most photocatalysts. The addition of an external Lewis acid can increase the reduction potential of the imine (less negative) through coordination. Moreover, hydrogen-bonding via Brønsted acid can make this reduction a thermodynamically favorable process thanks to proton-coupled electron transfer (PCET), wherein an electron transfer from the photocatalyst to the imine takes place in concert with a proton transfer from the Brønsted acid to the imine.

2. Photocatalytic Radical Additions to Imines—*Pathway A*

2.1. Racemic Photocatalytic Radical Additions to Imines

Racemic radical additions to imines under visible light photocatalysis began to appear in 2016, when Bode reported the cyclization of silicon amine protocol (SLAP) reagents with an imine moiety

(Scheme 2, left) [5]. These α-silyl amine precursors could undergo mild single-electron oxidation to render α-amino radicals, which could then engage with the imine to yield a wide variety of piperazine derivatives. The protocol was expanded further with α-silyl ether and thioether precursors to access morpholines, oxazepanes, thiomorpholines and thiazepanes (Scheme 2, right) [6,7]. It should be noted that, in this case, a Lewis acid was required to activate the imine, and the photocatalytic cycle could start with an initial Lewis acid-assisted single-electron reduction of the imine.

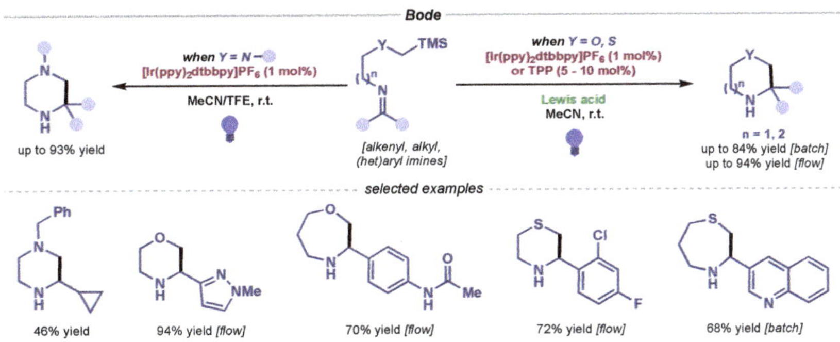

Scheme 2. Cyclization of SLAP reagents developed by Bode for the synthesis of piperazines (left) and morpholines, oxazepanes, thiomorpholines and thiazepanes (right).

The first intermolecular photocatalytic radical addition was published in 2017 by Molander's group [8]. The design of a general and modular approach based on the swift single-electron oxidation of ammonium alkyl bis(catecholato)silicates enabled the alkylation of different *N*-sulfonyl- and *N*-aryl-imines (Scheme 3, top left) [9]. In addition, Friestad employed these silicon reagents to perform the alkylation of *N*-acyl hydrazones in the presence of a Lewis acid (Scheme 3, top right) [10]. Alkyl silicates have also been utilized by Kelly and Molander to achieve the synthesis of various saturated *N*-heterocycles via radical alkylation and subsequent cyclization in a radical polar crossover (RPC) process (Scheme 3, bottom) [11].

More radical precursors have also been deployed in an attempt to expand the synthetic prowess of this transformation. For instance, Hanna, Jr. and Molander disclosed the photocatalytic activation of alkyl trifluoroborates, enabling the radical alkylation of non-activated imines (Scheme 4) [12,13].

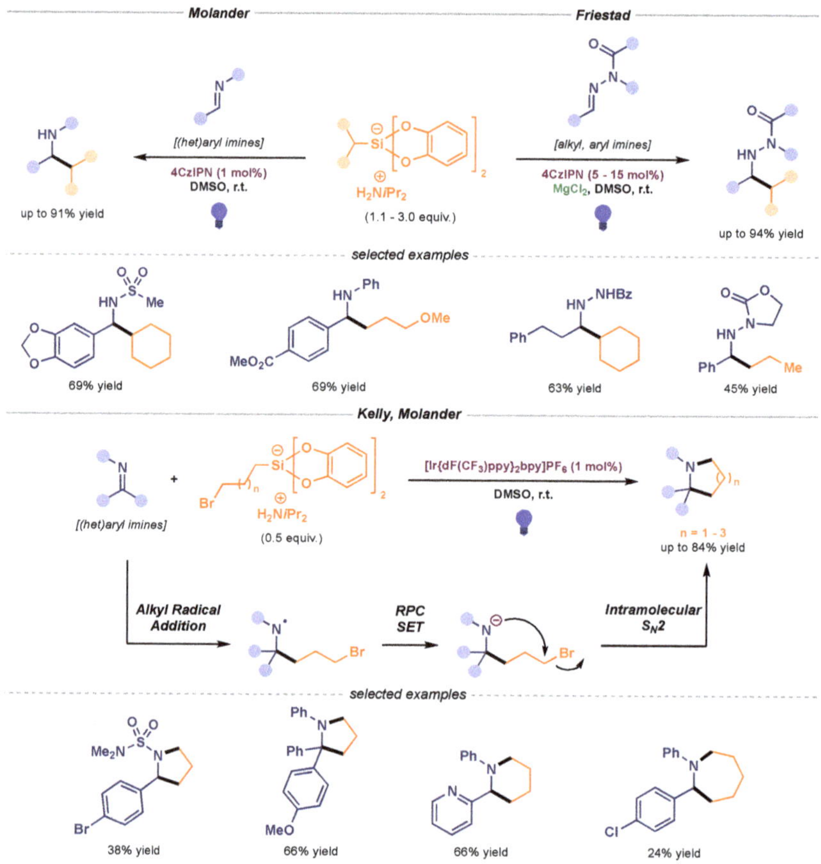

Scheme 3. Intermolecular photocatalytic radical additions to imines using alkyl silicates developed by Molander (top left), Friestad (top right) and Kelly and Molander (bottom).

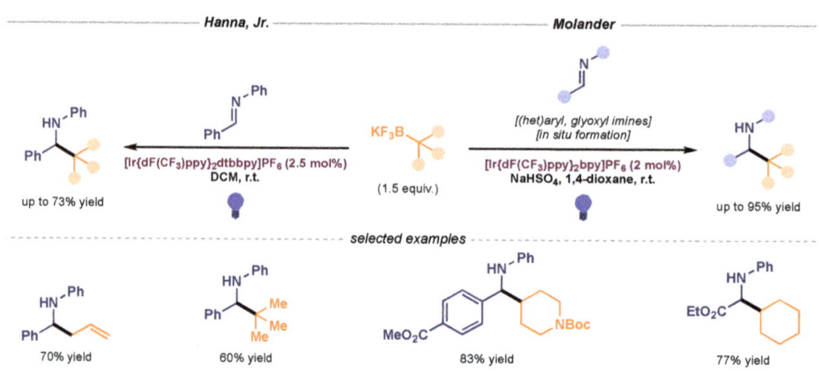

Scheme 4. Photocatalytic radical additions to imines using alkyl trifluoroborates developed by Hanna, Jr (left) and Molander (right).

Alkyl carboxylic acids hold a preferred position among radical precursors due to their versatility and ubiquity. Indeed, these alkylating agents can be implemented into mechanistically distinct photoredox pathways. Deprotonation of the acid can render a carboxylate species which can then undergo single-electron oxidation and subsequent decarboxylation to afford the alkyl radical intermediate. Alternatively, these acids can be activated with *N*-hydroxyphthalimide (NHPI) or its tetrachlorinated derivative (TCNHPI) through a simple esterification process to provide redox-active esters (RAEs). In this case, single-electron reduction can deliver the alkyl radical intermediate. This flexible behavior has been exploited by several research groups attempting to perform the alkyl radical addition to imines (Scheme 5). Weng and Lu reported the decarboxylative benzylation process following the oxidative pathway (Scheme 5, left) [14,15], while Mariano and Wang published a reductive version. In this later case, the decarboxylative glycosylation of imines was featured, although a Hantzsch ester (HEH) derivative was needed as a stoichiometric photosensitizer (Scheme 5, right) [16,17].

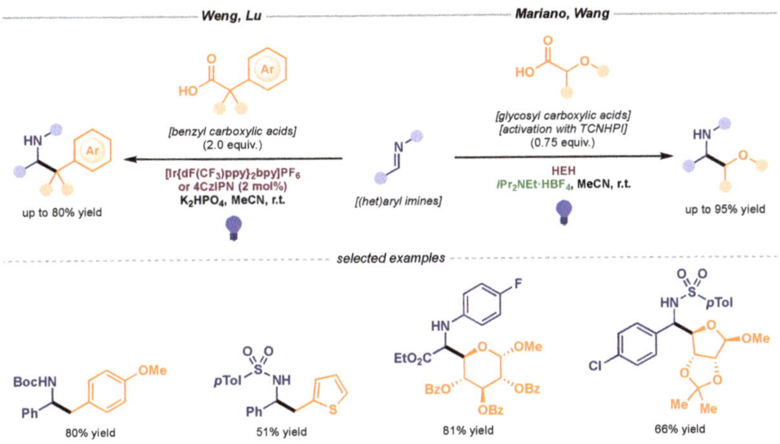

Scheme 5. Photocatalytic radical additions to imines using alkyl carboxylic acids developed by Weng and Lu (left) and Mariano and Wang (right).

Notably, the radical fluoroalkylation of imines had remained inaccessible in the field until Maestro and Alemán recently reported the direct difluoromethylation of imine moieties (Scheme 6) [18]. This general procedure was predicated on the single-electron oxidation of readily available zinc difluoromethane sulfinate (DFMS) in the presence of an organophotoredox catalyst (Rhodamine 6G).

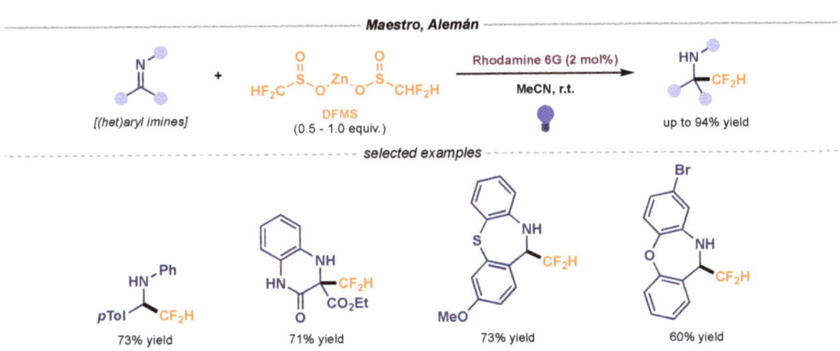

Scheme 6. Photocatalytic difluoromethyl radical addition to imines using DFMS developed by Maestro and Alemán.

Lastly, hydrogen atom transfer (HAT) has also been used in order to perform the C-H activation of different alkyl radical precursors and perform the desired alkylation reaction with activated imines (Scheme 7). Lu and Gong reported the α-oxyalkyl radical addition of 1,3-dioxolane to fluoroalkyl imines (Scheme 7, top) [19,20], while Dilman managed to install different alkyl and acyl radicals into *N*-sulfonyl imines (Scheme 7, bottom) [21–23].

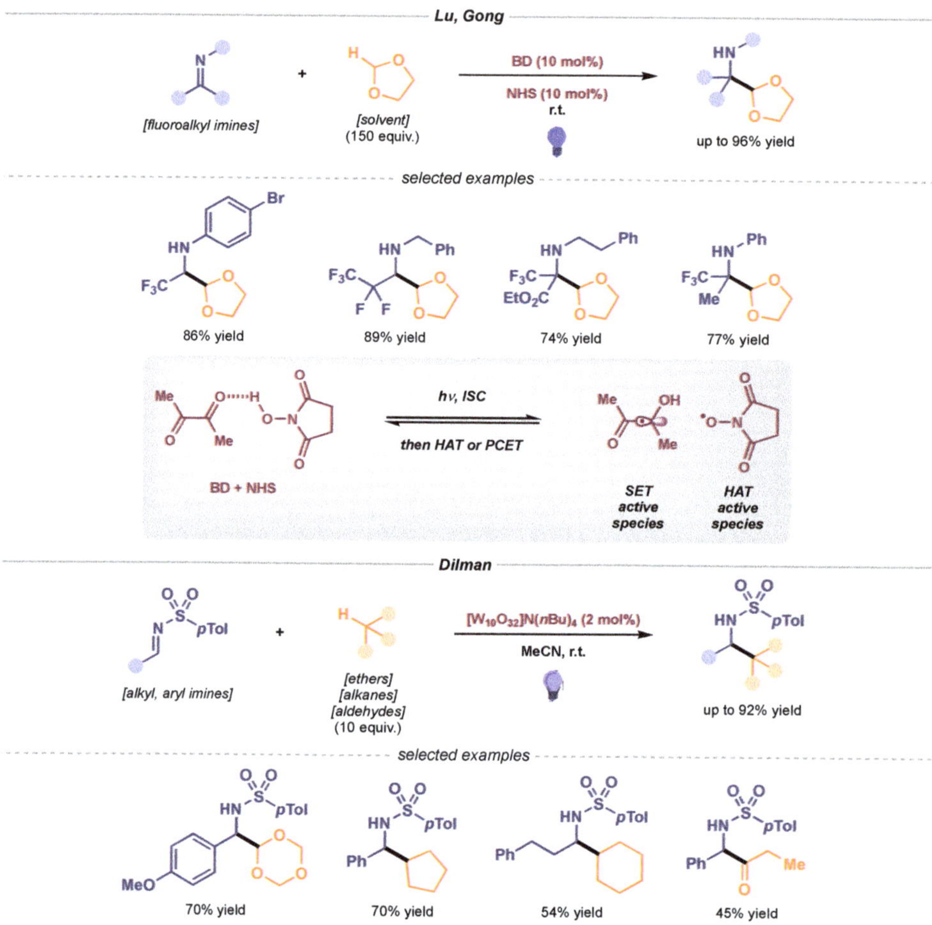

Scheme 7. Photocatalytic radical additions to imines using HAT developed by Lu and Gong (top) and Dilman (bottom).

2.2. *Stereoselective Photocatalytic Radical Additions to Imines*

In the field of asymmetric photocatalytic additions to imines, Knowles first described in 2013 an elegant intramolecular example in which a hydrazone trapped a ketyl radical intermediate—generated by PCET—in enantioselective fashion thanks to the chiral induction exerted by a chiral phosphoric acid (Scheme 8) [24].

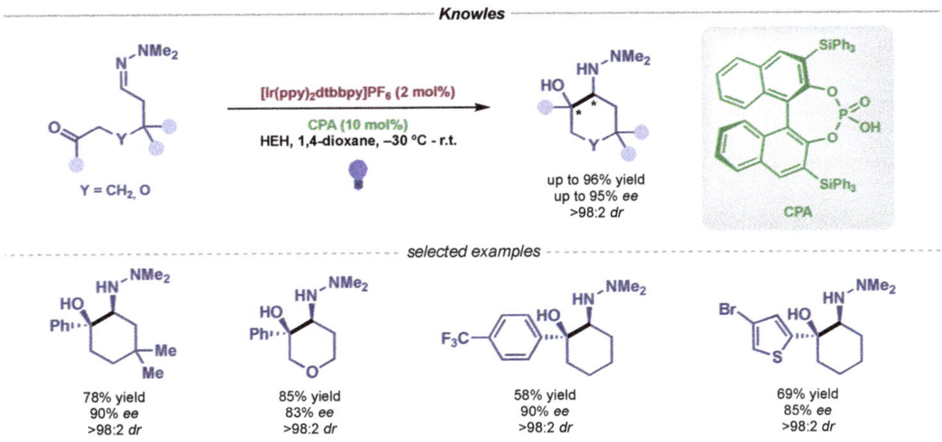

Scheme 8. Asymmetric intramolecular photocatalytic radical addition to hydrazones developed by Knowles.

No further reports were published on this topic until Maestro and Alemán disclosed in 2017 an asymmetric intermolecular radical alkylation of imines based on the use of chiral sulfoxides (Scheme 9) [25]. The photocatalytic reduction of NHPI-derived RAEs delivered the alkyl radical, which then engaged with the enantiopure *N*-sulfinimine in diastereoselective fashion to afford α-branched benzyl amine derivatives.

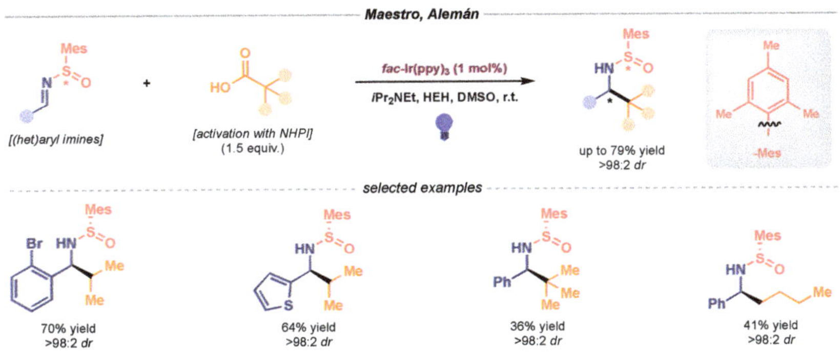

Scheme 9. Asymmetric intermolecular photocatalytic radical addition to *N*-sulfinimines developed by Maestro and Alemán.

Moreover, Gong's research group developed a series of transformations in 2018 and 2019 based on chiral Lewis acid-catalyzed radical alkylations of different imine scaffolds (Scheme 10, top) [26,27]. When using redox-active alkyl trifluoroborates and silanes, the Cu-BOX complexes acted as bifunctional chiral photocatalysts, performing both the asymmetric induction and the single-electron oxidation of the radical precursors, while suppressing the need for an external photocatalyst. In the latest report published by Gong, an HAT-photocatalyst (5,7,12,14-pentacenetetrone, PT) was required to perform the C-H activation of benzyl and allyl positions, as well as non-activated alkanes (Scheme 10, bottom) [28].

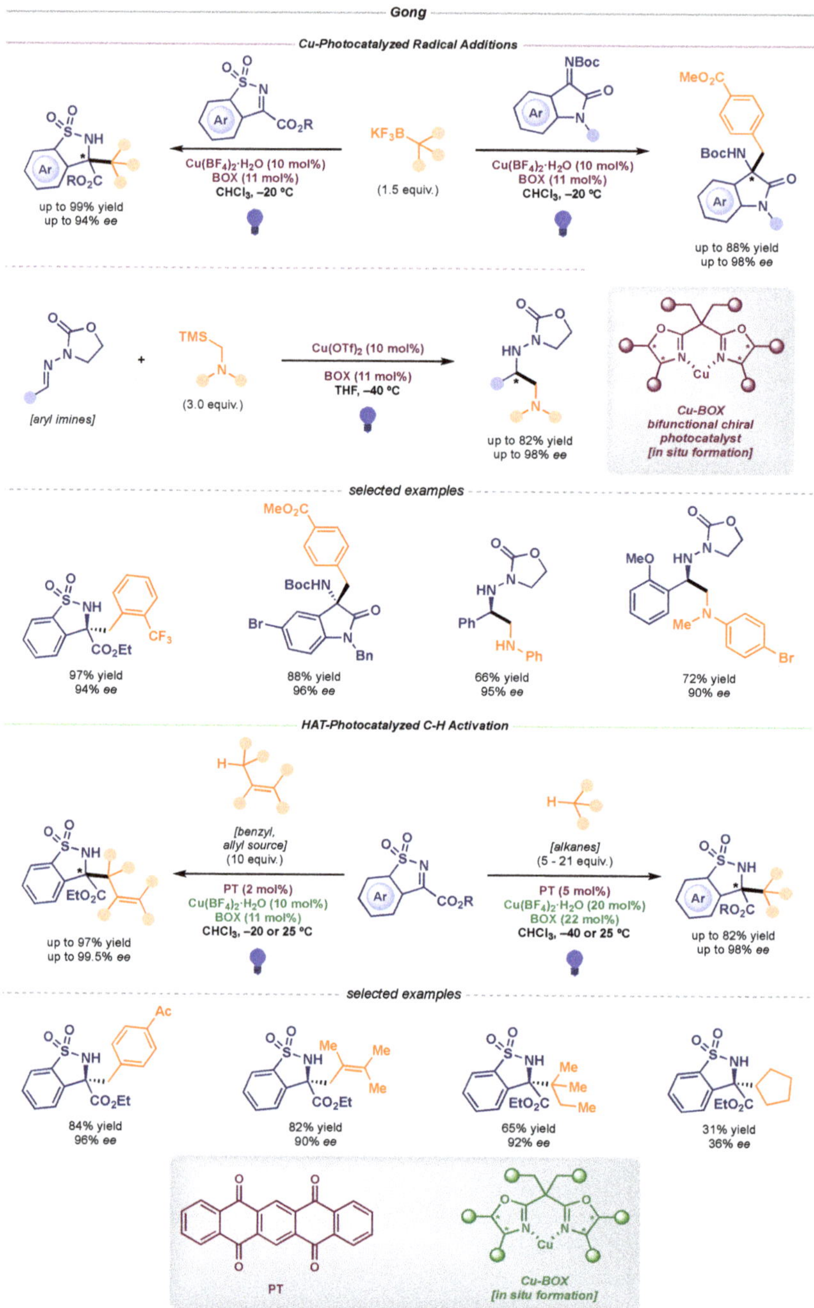

Scheme 10. Asymmetric intermolecular photocatalytic reactions developed by Gong: Cu-photocatalyzed radical additions (top) and HAT-photocatalyzed C-H activation (bottom).

The generality observed throughout this section noticeably stands out, wherein an assortment of radical precursors has been inserted into mechanistically similar protocols based on their photocatalytic activation to render the nucleophilic radical intermediate. Most notably, the range of imine building blocks employed in these reactions is quite impressive, as both activated and non-activated substrates have proven to be suitable acceptors to the different radical additions.

3. Photocatalytic α-Amino Radical Reactivity via Single-Electron Reduction of Imine Derivatives—*Pathway B*

3.1. Racemic α-Amino Radical–Radical Couplings—Pathway B1

The generation of α-amino radical intermediates derived from imine building blocks was first reported by Kisch (Scheme 11, top) [29,30]. By means of a family of heterogeneous photocatalyst semiconductors, their group developed a series of transformations involving the coupling of α-amino radicals and allyl radicals (pathway B1 in Scheme 1) [31]. CdS powder as well as supported versions on SiO_2, ZnS, and Al_2O_3 can behave as the photocatalyst semiconductor which features a surface with the ability to engage in interfacial electron transfer (IFET). Upon visible light absorption, the semiconductor can generate an electron-hole pair—essentially reducing and oxidizing surface centers. These sites can then perform IFET with the adsorbed substrates, delivering the two radical intermediates that eventually afford the recombination product (known as semiconductor photocatalysis B). Later on, Pu and Shen used CdSe/CdS core/shell quantum dots (QDs) as photocatalysts for the transfer hydrogenation of diaryl imines with a thiophenol as H atom donor (Scheme 11, bottom) [32,33].

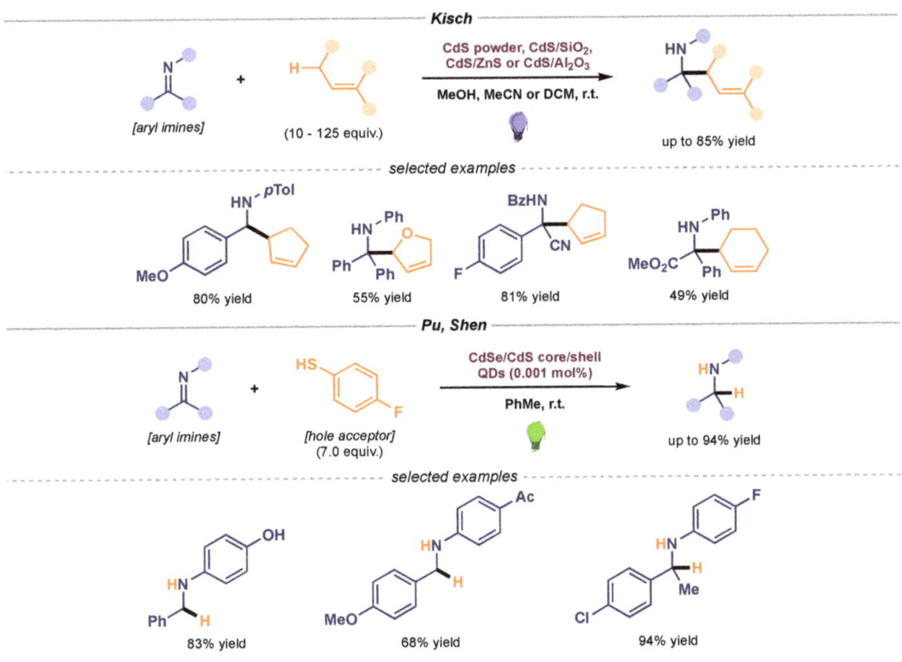

Scheme 11. Heterogeneous photocatalytic α-amino radical–radical couplings developed by Kisch (top) and Pu and Shen (bottom).

In the field of homogenous photocatalysis, MacMillan first reported the coupling of α-amino radicals with α-oxybenzyl and β-enaminyl radicals, giving access to pinacol-type products and the formal β-Mannich reaction, respectively (Scheme 12, top) [34,35]. Notably, the generation of the radical reacting partners required elegant multicatalytic approaches. The pinacol-type coupling reaction relied on the initial photocatalytic oxidation and deprotonation of methyl thioglycolate to produce a thiyl radical. Then, this *S*-centered radical could perform the H atom abstraction from the benzyl ether to afford the α-oxybenzyl radical and regenerate the thiol catalyst. On the other hand, the β-enaminyl radical could be accessed following: i) initial condensation of a cyclic ketone with a simple aminocatalyst (azepane), ii) subsequent oxidation of the catalytic enamine, and iii) final allylic deprotonation (Scheme 12, bottom).

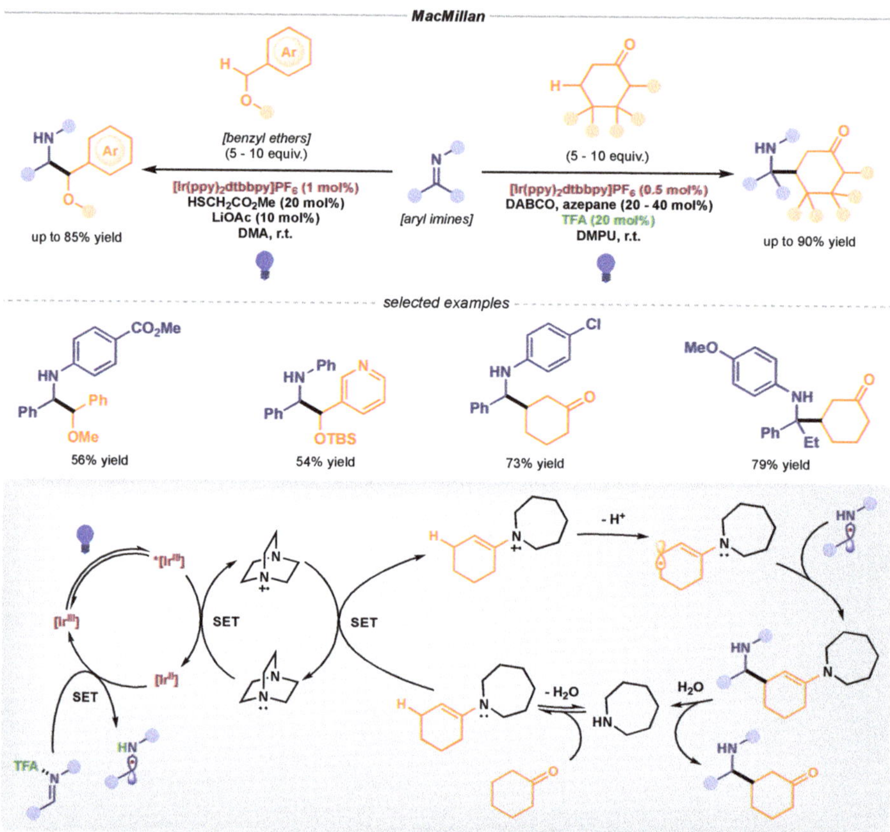

Scheme 12. Photocatalytic α-amino radical–radical couplings (top) and mechanistic proposal for the formal β-Mannich reaction (bottom) developed by MacMillan.

Concurrently, Rueping displayed the ability of these α-amino radicals to react with each other to render symmetrical and unsymmetrically substituted 1,2-diamines (Scheme 13) [36,37].

Following the development of these transformations, different variants of the radical–radical coupling of α-amino radicals began to appear. Sudo successfully employed an organophotoredox catalyst in a similar symmetrical coupling [38], whereas Gilmore obtained vicinal primary diamine products through *in situ* formation of the imines with aldehydes and ammonia [39]. Regarding unsymmetrical adducts, Wang reported the coupling of imine-derived α-amino radicals with tetrahydroisoquinoline-derived α-amino radicals [40]. In most cases, the imine reduction is believed to be assisted by coordination to an external acidic species.

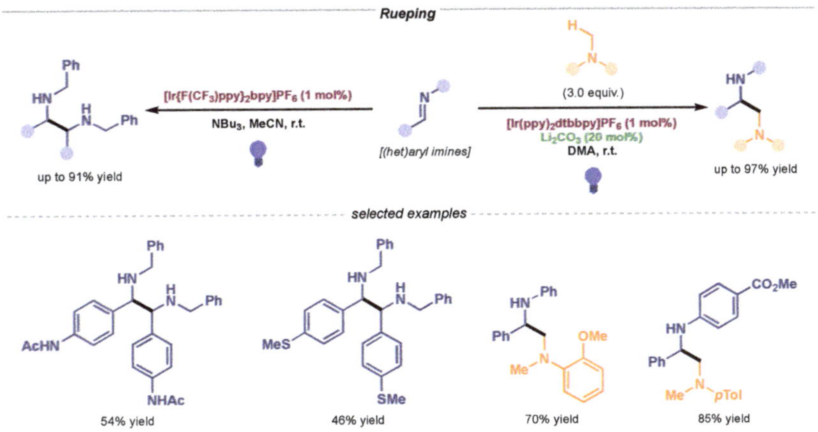

Scheme 13. Photocatalytic α-amino radical–radical couplings developed by Rueping.

A unique example based on the reactivity of these α-amino radicals was reported by Opatz featuring a four-component reaction which gave access to structurally diverse products (Scheme 14) [41]. The protocol involved the simultaneous construction of three new bonds: C-N (via *in situ* formation of the imine), C-S (via sulfonyl radical addition to styrene) and C-C (via α-amino radical-benzyl radical coupling).

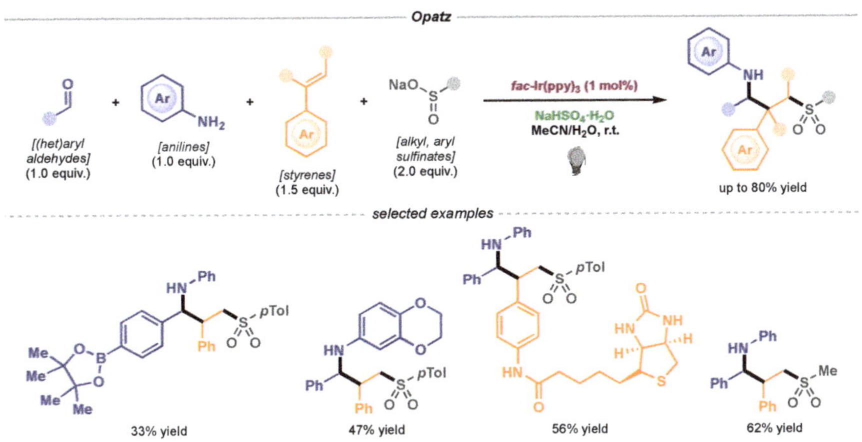

Scheme 14. Photocatalytic α-amino radical-benzyl radical coupling developed by Opatz.

Arylation reactions, which had remained elusive in the context of radical chemistry with imines, were achieved by Xia and Lehnherr and Rovis (Scheme 15) [42,43]. Through generation of stabilized aryl radical intermediates, the radical–radical coupling became feasible with 1,4-dicyanobenzene and 4-cyanopyridine precursors, respectively.

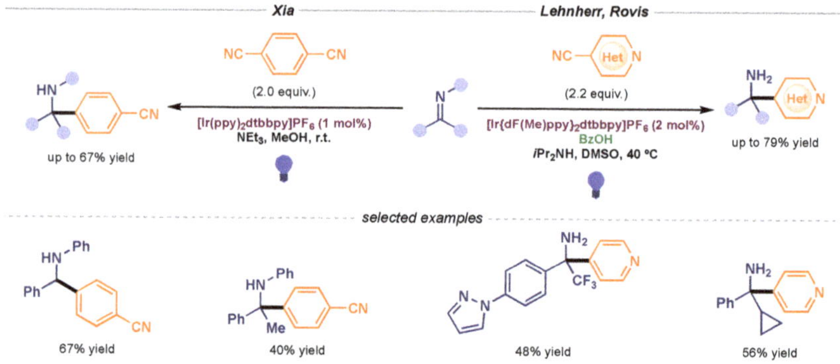

Scheme 15. Photocatalytic α-amino radical-aryl radical couplings developed by Xia (left) and Lehnherr and Rovis (right).

3.2. Racemic α-Amino Radical Additions to Activated Olefins—Pathway B2

The development of the single-electron reduction of imines to produce new radical–radical couplings has undoubtedly revamped this area. Furthermore, the polarity reversal displayed in this redox process served as a platform upon which an even greater challenge could be tackled. The new nucleophilic character of the α-amino radical was exploited in a series of Giese radical additions (pathway B2 in Scheme 1) reported by several groups. Indeed, Chen [44], Dixon [45,46], Ngai [47], and Rueping [48] independently worked on the addition of these intermediates to different activated olefins (Scheme 16).

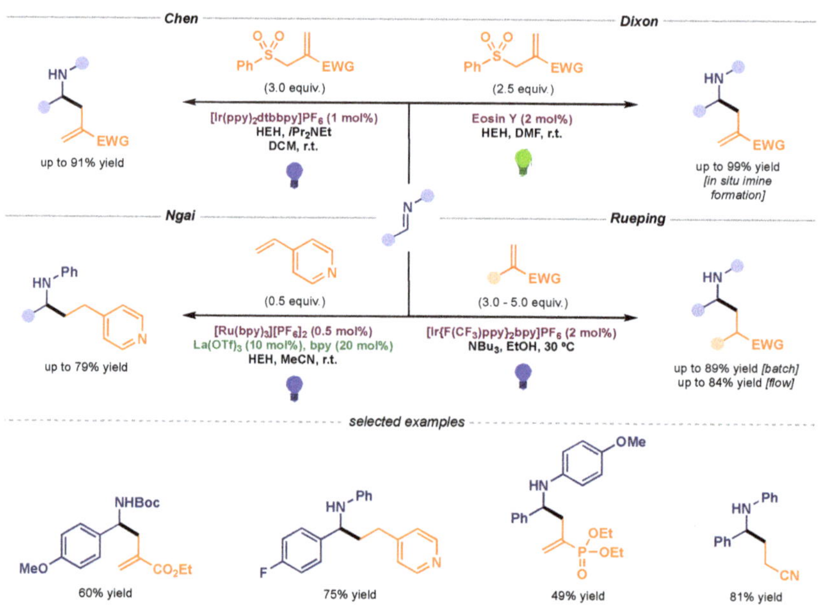

Scheme 16. Photocatalytic α-amino radical additions to activated olefins developed by Chen (top left), Dixon (top right), Ngai (bottom left), and Rueping (bottom right).

In addition, Dixon's lab described two procedures built on these radical additions to electrophilic partners followed by cyclization events, thus granting access to molecules of higher structural complexity (Scheme 17) [49,50]. Notably, the addition of the α-amino radical to vinyl sulfones led to a reverse polarity Povarov reaction (Scheme 17, top), while bridged 1,3-diazepanes could be prepared via α-amino radical addition to the 4-position of a quinoline core and subsequent ring closure at the 2-position (Scheme 17, bottom).

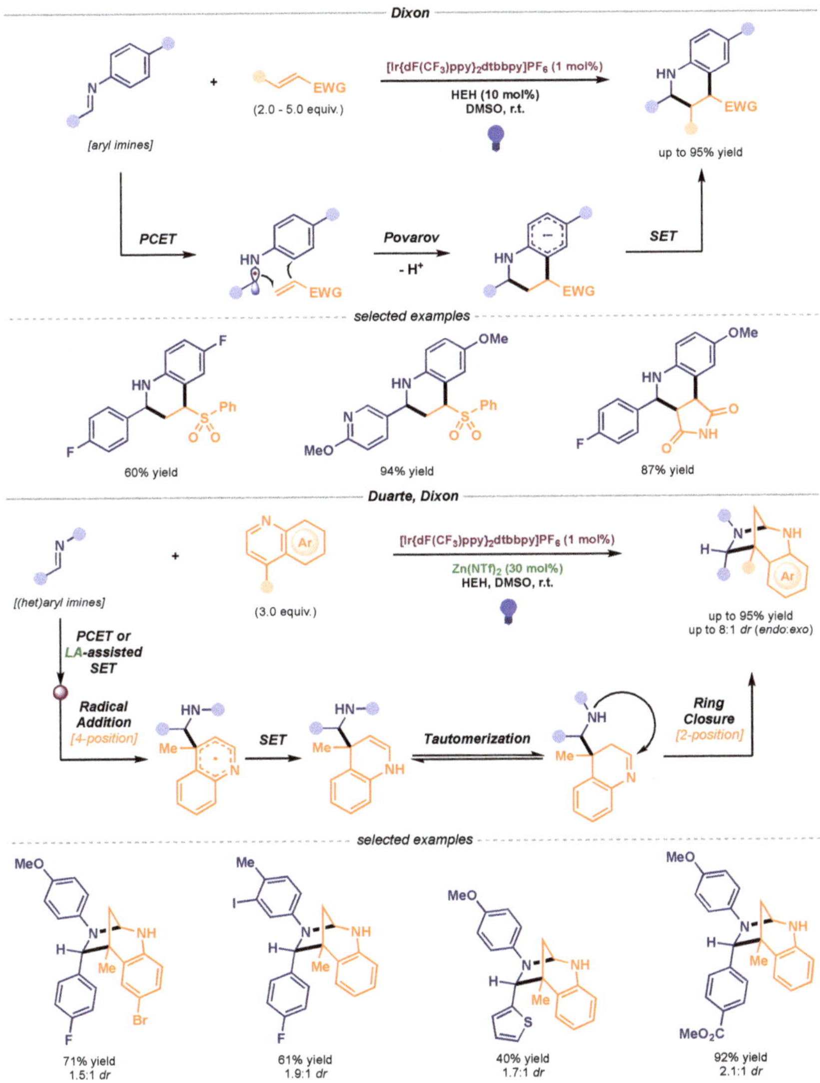

Scheme 17. Photocatalytic reactions of α-amino radicals for the synthesis of polycyclic structures developed by Dixon: reverse polarity Povarov reaction (top) and bridged 1,3-diazepane construction (bottom).

3.3. Racemic α-Amino Carbanion Nucleophilic Attacks—Pathway B3

After thorough evaluation of the results outlined above, it could be assumed that the single-electron reduction of imine scaffolds usually delivers a radical anion that can exist as two different resonant forms. As mentioned previously, the N-radical resonant form can engage in an HAT to yield a stable carbanion (pathway B3 in Scheme 1). This behavior has been exploited by Yu and Fan and Walsh to achieve polar nucleophilic attacks on numerous electrophiles, such as CO_2 and aldehydes, as well as the direct hydrolysis of these anionic intermediates to afford the formal reduction to benzyl amines (Scheme 18) [51–54].

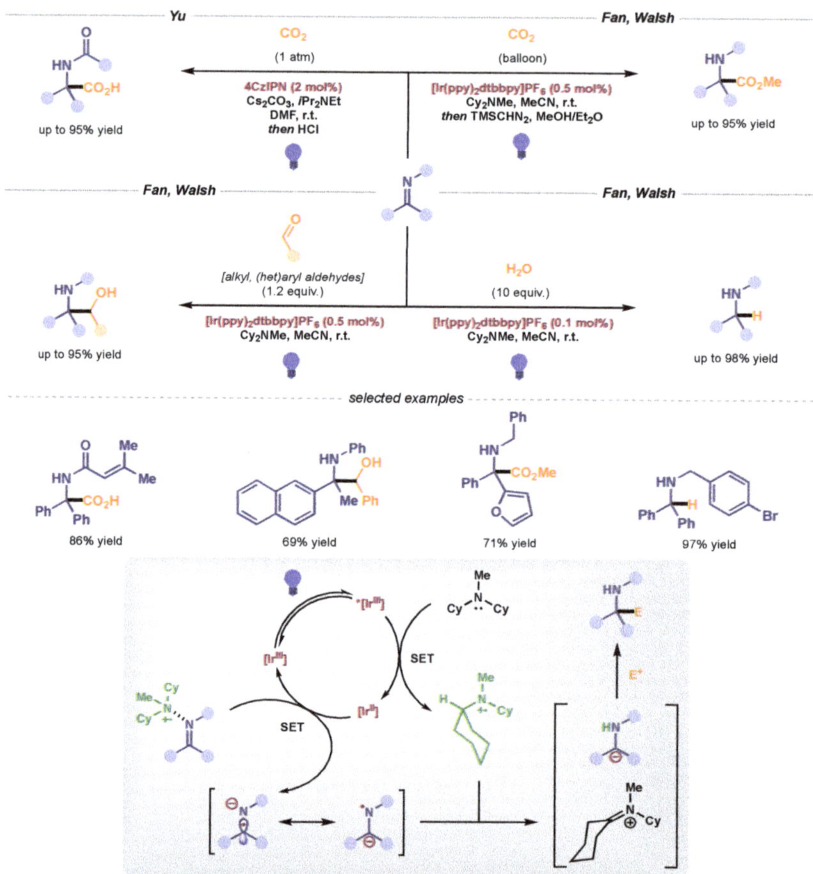

Scheme 18. Photocatalytic carbanion additions to electrophiles developed by Yu and Fan and Walsh: carboxylation (top), reaction with aldehydes (middle left) and hydrolysis (middle right).

3.4. Stereoselective Photocatalytic α-Amino Radical Reactivity via Single-Electron Reduction of Imine Derivatives—Pathway B

Controlling the stereochemical outcome of transformations involving the single-electron reduction of imines has proven to be an outstanding challenge. An exceptional solution to this problem was developed by Ooi's lab in 2015 and 2016 (Scheme 19) [55,56], wherein the radical anion derived from this redox process was ion-paired with a chiral phosphonium salt, rendering an asymmetric radical–radical coupling with α-amino radicals (pathway B1 in Scheme 1).

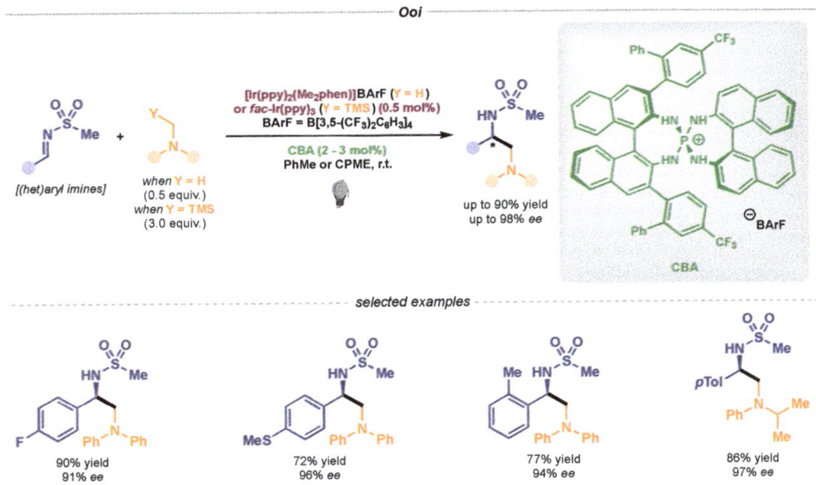

Scheme 19. Asymmetric photocatalytic α-amino radical–radical couplings developed by Ooi.

Moreover, Jiang's lab reported two transformations involving the use of traditional organocatalysts (Scheme 20) [57,58]. In the first one, the formal reduction of activated ketimines was achieved via coordination of Takemoto's urea catalyst to the radical anion and ensuing asymmetric HAT (pathway B1 in Scheme 1). In the second one, a chiral Brønsted acid managed to activate vinyl pyridines—acting as acceptors in a Giese radical addition—while exerting stereocontrol through H-bond interactions (pathway B2 in Scheme 1).

Finally, Ward and Wenger published an interesting asymmetric reduction of cyclic imines in which enzymatic catalysis played a fundamental role (Scheme 21) [59]. Initial photocatalytic reduction and HAT would deliver a racemic mixture of both amines (pathway B1 in Scheme 1), yet, in the presence of monoamine oxidase MAO-N-9, only one enantiomer could undergo subsequent re-oxidation (enantiomer recycling). Therefore, the combination of photocatalysis and biocatalysis yielded an elegant dynamic kinetic resolution (DKR) en route to chiral amines.

In this section, the pool of radical precursors employed in reactions following pathway B1 may not be as diverse when compared to conventional radical additions (pathway A). However, thanks to the nucleophilic character of the imine-derived radical anion, pathways B2 and B3 have widened the scope of reacting partners since they no longer require photocatalytic activation. Remarkably, common electrophiles used in polar methodologies find a smooth transition into these new radical processes thanks to the inspired polarity reversal enforced upon the imine substrates.

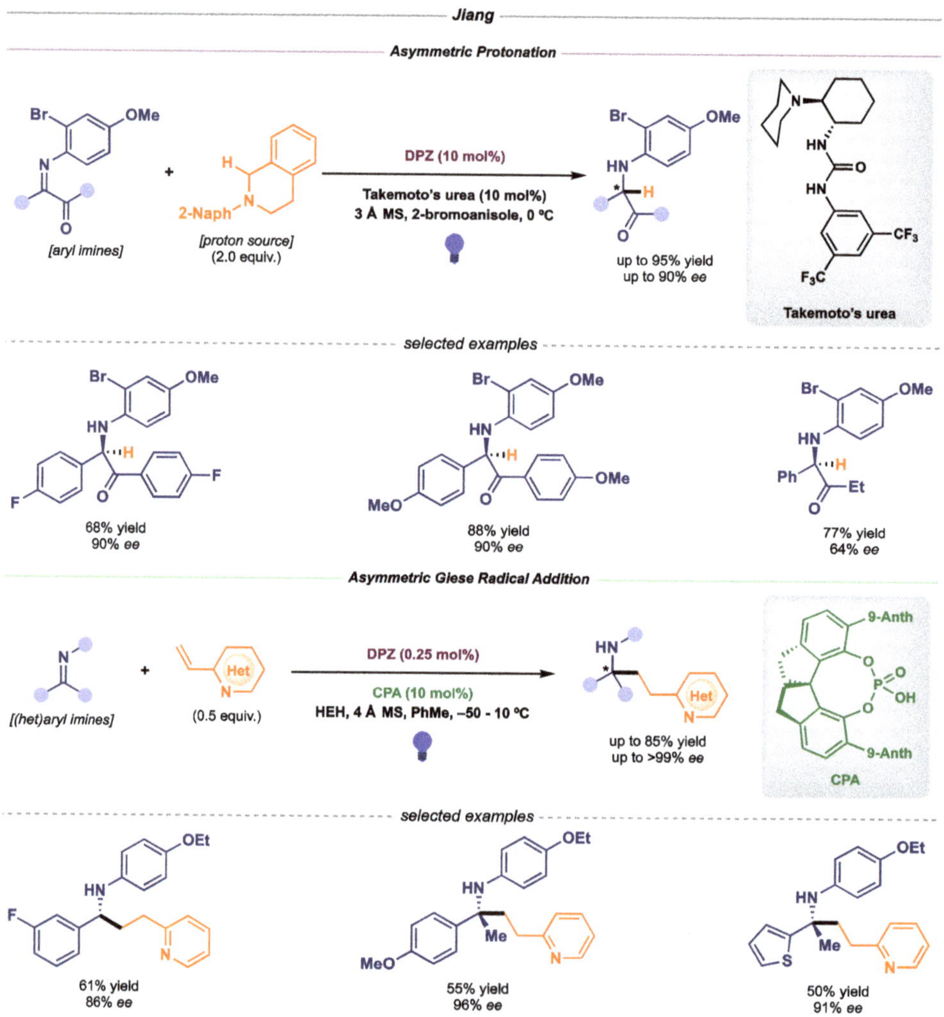

Scheme 20. Asymmetric photocatalytic α-amino radical-mediated reactions developed by Jiang: asymmetric protonation (top) and asymmetric Giese radical addition (bottom).

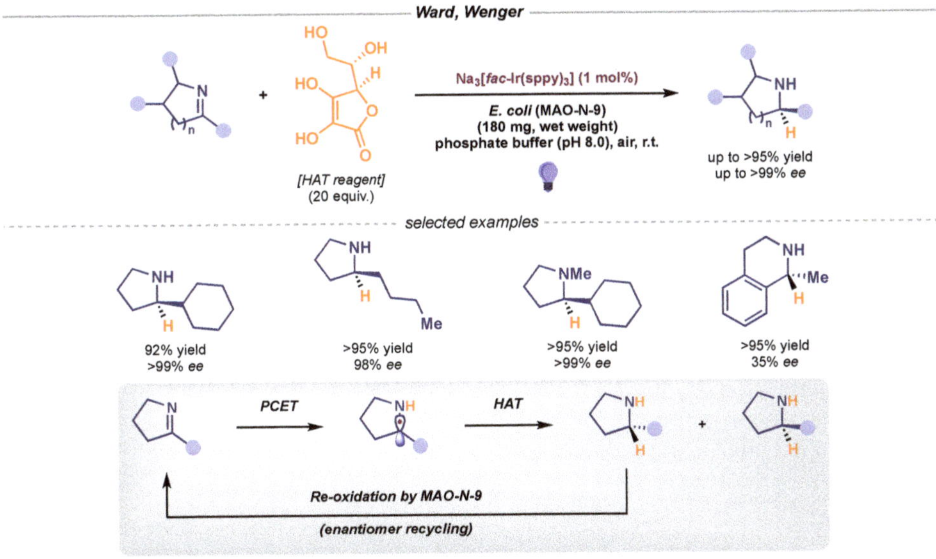

Scheme 21. Asymmetric bio- and photo-catalytic reduction of cyclic imines developed by Ward and Wenger.

4. Conclusions

The development of new methodologies to perform the α-functionalization of imine building blocks has proven to be a subject of intense research during the past decade, as numerous approaches featuring outstanding versatility have surfaced in the context of visible light photoredox catalysis. Most importantly, the complimentary nature to the different photocatalytic strategies employed in imine chemistry has provided immense flexibility, since the imine reagent can now be used as both an electrophile and, strikingly, a nucleophile. This multifaceted behavior has delivered a wide array of racemic transformations in this area. However, asymmetric functionalization of the C=N moiety has remained a great challenge in photoredox catalysis. Nevertheless, brilliant activation strategies have been deployed to achieve stereoselectivity, although an increase in generality and modularity can be expected as the field continues to grow.

Author Contributions: Conceptualization, A.F.G.-C. and M.C.M.; literature search, A.F.G.-C. and M.C.M.; writing—original draft preparation, A.F.G.-C.; writing—review and editing, A.F.G.-C., M.C.M., and J.A.; supervision, M.C.M. and J.A.; project administration, M.C.M. and J.A.; funding acquisition, M.C.M. and J.A. All authors have read and agreed to the published version of the manuscript.

Funding: This research was funded by the European Research Council, grant number 647550; the Spanish Government, grant number RTI2018-095038-B-I00; the "Comunidad de Madrid" and European Structural Funds, grant number S2018/NMT-4367.

Conflicts of Interest: The authors declare no conflict of interest.

Abbreviations

For uncommon abbreviations outlined throughout the manuscript, see the following list:

abbreviation	full description
4CzIPN	2,4,5,6-tetra(9H-carbazol-9-yl)isophthalonitrile
Ac	acyl
Alk	alkyl
Anth	anthracenyl
Ar	aryl
BArF	$B[3,5-(CF_3)_2C_6H_3]_4$
BD	2,3-butanedione
Boc	*tert*-butyloxycarbonyl
BOX	bis(oxazoline)
bpy	2,2′-bipyridine
Bu	butyl
Bz	benzoyl
CBA	chiral Brønsted acid
CPA	chiral phosphoric acid
CPME	cyclopentyl methyl ether
Cy	cyclohexyl
DABCO	1,4-diazabicyclo[2.2.2]octane
DCM	dichloromethane
DFMS	zinc difluoromethanesulfinate
DKR	dynamic kinetic resolution
DMA	*N,N*-dimethylacetamide
DMF	*N,N*-dimethylformamide
DMPU	*N,N′*-dimethylpropyleneurea
DMSO	dimethylsulfoxide
DPZ	5,6-bis(5-methoxythiophen-2-yl)pyrazine-2,3-dicarbonitrile
dr	diastereomeric ratio
dtbbpy	4,4′-di-*tert*-butyl-2,2′-dipyridyl
E	electrophile
ee	enantiomeric excess
Et	ethyl
EWG	electron withdrawing group
HAT	hydrogen atom transfer
HEH	Hantzsch ester
Het	heteroaryl
IFET	interfacial electron transfer
*i*Pr	*iso*-propyl
MAO	monoamine oxidase
Me	methyl
Mes	mesityl
MS	molecular sieves
Naph	naphthyl
NHPI	*N*-hydroxyphthalimide
NHS	*N*-hydroxysuccinimide
PCET	proton-coupled electron transfer
Ph	phenyl
phen	1,10-phenanthroline
ppy	2-phenylpyridine
PT	5,7,12,14-pentacenetetrone
*p*Tol	*para*-tolyl
QD	quantum dot
RAE	redox-active ester
RPC	radical polar crossover

SET	single-electron transfer
SLAP	silicon amine protocol
S_N2	bimolecular nucleophilic substitution
sppy	3-(pyridin-2-yl)benzenesulfonate
TBS	*tert*-butyldimethylsilyl
TCNHPI	*N*-hydroxytetrachlorophthalimide
Tf	trifluoromethanesulfonyl
TFA	trifluoroacetic acid
TFE	2,2,2-trifluoroethanol
THF	tetrahydrofuran
TMS	trimethylsilyl
TPP	2,4,6-triphenylpyrylium tetrafluoroborate

References

1. *Visible Light Photocatalysis in Organic Chemistry*, 1st ed; Stephenson, C.R.J.; Yoon, T.P.; MacMillan, D.W.C. (Eds.) Wiley-VCH: Weinheim, Germany, 2018.
2. Matsui, J.K.; Lang, S.B.; Heitz, D.R.; Molander, G.A. Photoredox-Mediated Routes to Radicals: The Value of Catalytic Radical Generation in Synthetic Methods Development. *ACS Catal.* **2017**, *7*, 2563–2575. [CrossRef]
3. Lee, K.N.; Ngai, M.-Y. Recent developments in transition-metal photoredox-catalysed reactions of carbonyl derivatives. *Chem. Commun.* **2017**, *53*, 13093–13112. [CrossRef] [PubMed]
4. Leitch, J.A.; Rossolini, T.; Rogova, T.; Maitland, J.A.P.; Dixon, D.J. α-Amino Radicals via Photocatalytic Single-Electron Reduction of Imine Derivatives. *ACS Catal.* **2020**, *10*, 2009–2025. [CrossRef]
5. Hsieh, S.-Y.; Bode, J.W. Silicon Amine Reagents for the Photocatalytic Synthesis of Piperazines from Aldehydes and Ketones. *Org. Lett.* **2016**, *18*, 2098–2101. [CrossRef] [PubMed]
6. Jackl, M.K.; Legnani, L.; Morandi, B.; Bode, J.W. Continuous Flow Synthesis of Morpholines and Oxazepanes with Silicon Amine Protocol (SLAP) Reagents and Lewis Acid Facilitated Photoredox Catalysis. *Org. Lett.* **2017**, *19*, 4696–4699. [CrossRef] [PubMed]
7. Hsieh, S.-Y.; Bode, J.W. Lewis Acid Induced Toggle from Ir(II) to Ir(IV) Pathways in Photocatalytic Reactions: Synthesis of Thiomorpholines and Thiazepanes from Aldehydes and SLAP Reagents. *ACS Cent. Sci.* **2017**, *3*, 66–72. [CrossRef]
8. Patel, N.R.; Kelly, C.B.; Siegenfeld, A.P.; Molander, G.A. Mild, Redox-Neutral Alkylation of Imines Enabled by an Organic Photocatalyst. *ACS Catal.* **2017**, *7*, 1766–1770. [CrossRef]
9. Prieto, A.; Bouyssi, D.; Monteiro, N. Radical-Mediated Formal C(sp^2)–H Functionalization of Aldehyde-Derived *N,N*-Dialkylhydrazones. *Eur. J. Org. Chem.* **2018**, 2378–2393. [CrossRef]
10. Cullen, S.T.J.; Friestad, G.K. Alkyl Radical Addition to Aliphatic and Aromatic *N*-Acylhydrazones Using an Organic Photoredox Catalyst. *Org. Lett.* **2019**, *21*, 8290–8294. [CrossRef]
11. Pantaine, L.R.E.; Milligan, J.A.; Matsui, J.K.; Kelly, C.B.; Molander, G.A. Photoredox Radical/Polar Crossover Enables Construction of Saturated Nitrogen Heterocycles. *Org. Lett.* **2019**, *21*, 2317–2321. [CrossRef]
12. Plasko, D.P.; Jordan, C.J.; Ciesa, B.E.; Merrill, M.A.; Hanna, J.M., Jr. Visible light-promoted alkylation of imines using potassium organotrifluoroborates. *Photochem. Photobiol. Sci.* **2018**, *17*, 534–538. [CrossRef] [PubMed]
13. Yi, J.; Badir, S.O.; Alam, R.; Molander, G.A. Photoredox-Catalyzed Multicomponent Petasis Reaction with Alkyltrifluoroborates. *Org. Lett.* **2019**, *21*, 4853–4858. [CrossRef] [PubMed]
14. Guo, J.; Wu, Q.-L.; Xie, Y.; Weng, J.; Lu, G. Visible-Light-Mediated Decarboxylative Benzylation of Imines with Arylacetic Acids. *J. Org. Chem.* **2018**, *83*, 12559–12567, The reaction pathway could not be confirmed; both the radical addition and the radical–radical coupling are most likely taking place. [CrossRef] [PubMed]
15. Zhang, H.-H.; Yu, S. Radical Alkylation of Imines with 4-Alkyl-1,4-dihydropyridines Enabled by Photoredox/Brønsted Acid Cocatalysis. *J. Org. Chem.* **2017**, *82*, 9995–10006, The reaction pathway could not be confirmed; both the radical addition and the radical–radical coupling are most likely taking place. [CrossRef]
16. Ji, P.; Zhang, Y.; Wei, Y.; Huang, H.; Hu, W.; Mariano, P.A.; Wang, W. Visible-Light-Mediated, Chemo- and Stereoselective Radical Process for the Synthesis of C-Glycoamino Acids. *Org. Lett.* **2019**, *21*, 3086–3092. [CrossRef]

17. Jia, J.; Lefebvre, Q.; Rueping, M. Reductive coupling of imines with redox-active esters by visible light photoredox organocatalysis. *Org. Chem. Front.* **2020**, *7*, 602–608. [CrossRef]
18. Garrido-Castro, A.F.; Gini, A.; Maestro, M.C.; Alemán, J. Unlocking the direct photocatalytic difluoromethylation of C=N bonds. *Chem. Commun.* **2020**, *56*, 3769–3772. [CrossRef]
19. Yang, S.; Zhu, S.; Lu, D.; Gong, Y. Formylation of Fluoroalkyl Imines through Visible-Light-Enabled H-Atom Transfer Catalysis: Access to Fluorinated α-Amino Aldehydes. *Org. Lett.* **2019**, *21*, 2019–2024. [CrossRef]
20. For a photoinduced α-oxyalkylation of imines generated in situ, see: Zhang, L.; Deng, Y.; Shi, F. Light promoted aqueous phase amine synthesis via three-component coupling reactions. *Tetrahedron Lett.* **2013**, *54*, 5217–5219.
21. Supranovich, V.I.; Levin, V.V.; Dilman, A.D. Radical Addition to *N*-Tosylimines via C−H Activation Induced by Decatungstate Photocatalyst. *Org. Lett.* **2019**, *21*, 4271–4274. [CrossRef]
22. For an inverse hydroboration of imines wherein the boryl radical is generated from *N*-Heterocyclic Carbene (NHC) boranes via HAT, see: Zhou, N.; Yuan, X.-A.; Zhao, Y.; Xie, J.; Zhu, C. Synergistic Photoredox Catalysis and Organocatalysis for Inverse Hydroboration of Imines. *Angew. Chem., Int. Ed.* **2018**, *57*, 3990–3994, The reaction pathway could not be confirmed; both the radical addition and the radical–radical coupling are most likely taking place.
23. Zhang, H.-H.; Yu, S. Visible-Light-Induced Radical Acylation of Imines with α-Ketoacids Enabled by Electron-Donor−Acceptor Complexes. *Org. Lett.* **2019**, *21*, 3711–3715, A radical–radical coupling is most likely taking place. [CrossRef] [PubMed]
24. Rono, L.J.; Yayla, H.G.; Wang, D.Y.; Armstrong, M.F.; Knowles, R.R. Enantioselective Photoredox Catalysis Enabled by Proton-Coupled Electron Transfer: Development of an Asymmetric Aza-Pinacol Cyclization. *J. Am. Chem. Soc.* **2013**, *135*, 17735–17738. [CrossRef] [PubMed]
25. Garrido-Castro, A.F.; Choubane, H.; Daaou, M.; Maestro, M.C.; Alemán, J. Asymmetric radical alkylation of *N*-sulfinimines under visible light photocatalytic conditions. *Chem. Commun.* **2017**, *53*, 7764–7767. [CrossRef]
26. Li, Y.; Zhou, K.; Wen, Z.; Cao, S.; Shen, X.; Lei, M.; Gong, L. Copper(II)-Catalyzed Asymmetric Photoredox Reactions: Enantioselective Alkylation of Imines Driven by Visible Light. *J. Am. Chem. Soc.* **2018**, *140*, 15850–15858. [CrossRef]
27. Han, B.; Li, Y.; Yu, Y.; Gong, L. Photocatalytic enantioselective α-aminoalkylation of acyclic imine derivatives by a chiral copper catalyst. *Nat. Commun.* **2019**, *10*, 3804–3812. [CrossRef]
28. Li, Y.; Lei, M.; Gong, L. Photocatalytic regio- and stereoselective C(sp^3)–H functionalization of benzylic and allylic hydrocarbons as well as unactivated alkanes. *Nat. Catal.* **2019**, *2*, 1016–1026, The reaction pathway could not be confirmed; both the radical addition and the radical–radical coupling are most likely taking place. [CrossRef]
29. Schindler, W.; Knoch, F.; Kisch, H. Semiconductor-Catalysed Photoaddition: γ,δ-Unsaturated Amines from Cyclopentene and Schiff Bases. *Chem. Ber.* **1996**, *129*, 925–932. [CrossRef]
30. Keck, H.; Schindler, W.; Knoch, F.; Kisch, H. Type B Semiconductor Photocatalysis: The Synthesis of Homoallyl Amines by Cadmium Sulfide-Catalyzed Linear Photoaddition of Olefins and Enol/Allyl Ethers to *N*-Phenylbenzophenone Imine. *Chem. Eur. J.* **1997**, *3*, 1638–1645. [CrossRef]
31. Kisch, H. Semiconductor Photocatalysis for Chemoselective Radical Coupling Reactions. *Acc. Chem. Res.* **2017**, *50*, 1002–1010. [CrossRef]
32. Xi, Z.-W.; Yang, L.; Wang, D.-Y.; Pu, C.-D.; Shen, Y.-M.; Wu, C.-D.; Peng, X.-G. Visible-Light Photocatalytic Synthesis of Amines from Imines via Transfer Hydrogenation Using Quantum Dots as Catalysts. *J. Org. Chem.* **2018**, *83*, 11886–11895. [CrossRef] [PubMed]
33. Van As, D.J.; Connell, T.U.; Brzozowski, M.; Scully, A.D.; Polyzos, A. Photocatalytic and Chemoselective Transfer Hydrogenation of Diarylimines in Batch and Continuous Flow. *Org. Lett.* **2018**, *20*, 905–908. [CrossRef] [PubMed]
34. Hager, D.; MacMillan, D.W.C. Activation of C−H Bonds via the Merger of Photoredox and Organocatalysis: A Coupling of Benzylic Ethers with Schiff Bases. *J. Am. Chem. Soc.* **2014**, *136*, 16986–16989. [CrossRef] [PubMed]
35. Jeffrey, J.L.; Petronijević, F.R.; MacMillan, D.W.C. Selective Radical−Radical Cross-Couplings: Design of a Formal β-Mannich Reaction. *J. Am. Chem. Soc.* **2015**, *137*, 8404–8407. [CrossRef]
36. Nakajima, M.; Fava, E.; Loescher, S.; Jiang, Z.; Rueping, M. Photoredox-Catalyzed Reductive Coupling of Aldehydes, Ketones, and Imines with Visible Light. *Angew. Chem., Int. Ed.* **2015**, *54*, 8828–8832. [CrossRef]

37. Fava, E.; Millet, A.; Nakajima, M.; Loescher, S.; Rueping, M. Reductive Umpolung of Carbonyl Derivatives with Visible-Light Photoredox Catalysis: Direct Access to Vicinal Diamines and Amino Alcohols via α-Amino Radicals and Ketyl Radicals. *Angew. Chem., Int. Ed.* **2016**, *55*, 6776–6779. [CrossRef]
38. Okamoto, S.; Kojiyama, K.; Tsujioka, H.; Sudo, A. Metal-free reductive coupling of C=O and C=N bonds driven by visible light: Use of perylene as a simple photoredox catalyst. *Chem. Commun.* **2016**, *52*, 11339–11342. [CrossRef]
39. Rong, J.; Seeberger, P.H.; Gilmore, K. Chemoselective Photoredox Synthesis of Unprotected Primary Amines Using Ammonia. *Org. Lett.* **2018**, *20*, 4081–4085. [CrossRef]
40. Xia, Q.; Tian, H.; Dong, J.; Qu, Y.; Li, L.; Song, H.; Liu, Y.; Wang, Q. *N*-Arylamines Coupled with Aldehydes, Ketones, and Imines by Means of Photocatalytic Proton-Coupled Electron Transfer. *Chem. Eur. J.* **2018**, *24*, 9269–9273. [CrossRef]
41. Kammer, L.M.; Krumb, M.; Spitzbarth, B.; Lipp, B.; Kühlborn, J.; Busold, J.; Mulina, O.M.; Terentev, A.O.; Opatz, T. Photoredox-Catalyzed Four-Component Reaction for the Synthesis of Complex Secondary Amines. *Org. Lett.* **2020**, *22*, 3318–3322. [CrossRef]
42. Chen, M.; Zhao, X.; Yang, C.; Xia, W. Visible-Light-Triggered Directly Reductive Arylation of Carbonyl/Iminyl Derivatives through Photocatalytic PCET. *Org. Lett.* **2017**, *19*, 3807–3810. [CrossRef] [PubMed]
43. Nicastri, M.C.; Lehnherr, D.; Lam, Y.; DiRocco, D.A.; Rovis, T. Synthesis of Sterically Hindered Primary Amines by Concurrent Tandem Photoredox Catalysis. *J. Am. Chem. Soc.* **2020**, *142*, 987–998. [CrossRef] [PubMed]
44. Qi, L.; Chen, Y. Polarity-Reversed Allylations of Aldehydes, Ketones, and Imines Enabled by Hantzsch Ester in Photoredox Catalysis. *Angew. Chem., Int. Ed.* **2016**, *55*, 13312–13315. [CrossRef] [PubMed]
45. Fuentes de Arriba, A.L.; Urbitsch, F.; Dixon, D.J. Umpolung synthesis of branched α-functionalized amines from imines via photocatalytic three-component reductive coupling reactions. *Chem. Commun.* **2016**, *52*, 14434–14437. [CrossRef]
46. Rossolini, T.; Leitch, J.A.; Grainger, R.; Dixon, D.J. Photocatalytic Three-Component Umpolung Synthesis of 1,3-Diamines. *Org. Lett.* **2018**, *20*, 6794–6798. [CrossRef]
47. Lee, K.N.; Lei, Z.; Ngai, M.-Y. β-Selective Reductive Coupling of Alkenylpyridines with Aldehydes and Imines via Synergistic Lewis Acid/Photoredox Catalysis. *J. Am. Chem. Soc.* **2017**, *139*, 5003–5006. [CrossRef]
48. Lefebvre, Q.; Porta, R.; Millet, A.; Jia, J.; Rueping, M. One Amine–3 Tasks: Reductive Coupling of Imines with Olefins in Batch and Flow. *Chem. Eur. J.* **2020**, *26*, 1363–1367. [CrossRef]
49. Leitch, J.A.; Fuentes de Arriba, A.L.; Tan, J.; Hoff, O.; Martínez, C.M.; Dixon, D.J. Photocatalytic reverse polarity Povarov reaction. *Chem. Sci.* **2018**, *9*, 6653–6658. [CrossRef]
50. Leitch, J.A.; Rogova, T.; Duarte, F.; Dixon, D.J. Dearomative Photocatalytic Construction of Bridged 1,3-Diazepanes. *Angew. Chem., Int. Ed.* **2020**, *59*, 4121–4130. [CrossRef]
51. Ju, T.; Fu, Q.; Ye, J.-H.; Zhang, Z.; Liao, L.-L.; Yan, S.-S.; Tian, X.-Y.; Luo, S.-P.; Li, J.; Yu, D.-G. Selective and Catalytic Hydrocarboxylation of Enamides and Imines with CO_2 to Generate α,α-Disubstituted α-Amino Acids. *Angew. Chem., Int. Ed.* **2018**, *57*, 13897–13901. [CrossRef]
52. Fan, X.; Gong, X.; Ma, M.; Wang, R.; Walsh, P.J. Visible light-promoted CO_2 fixation with imines to synthesize diaryl α-amino acids. *Nat. Commun.* **2018**, *9*, 4936–4943. [CrossRef] [PubMed]
53. Wang, R.; Ma, M.; Gong, X.; Fan, X.; Walsh, P.J. Reductive Cross-Coupling of Aldehydes and Imines Mediated by Visible Light Photoredox Catalysis. *Org. Lett.* **2019**, *21*, 27–31. [CrossRef] [PubMed]
54. Wang, R.; Ma, M.; Gong, X.; Panetti, G.B.; Fan, X.; Walsh, P.J. Visible-Light-Mediated Umpolung Reactivity of Imines: Ketimine Reductions with Cy_2NMe and Water. *Org. Lett.* **2018**, *20*, 2433–2436. [CrossRef] [PubMed]
55. Uraguchi, D.; Kinoshita, N.; Kizu, T.; Ooi, T. Synergistic Catalysis of Ionic Brønsted Acid and Photosensitizer for a Redox Neutral Asymmetric α-Coupling of *N*-Arylaminomethanes with Aldimines. *J. Am. Chem. Soc.* **2015**, *137*, 13768–13771. [CrossRef]
56. Kizu, T.; Uraguchi, D.; Ooi, T. Independence from the Sequence of Single-Electron Transfer of Photoredox Process in Redox-Neutral Asymmetric Bond-Forming Reaction. *J. Org. Chem.* **2016**, *81*, 6953–6958. [CrossRef]
57. Lin, L.; Bai, X.; Ye, X.; Zhao, X.; Tan, C.-H.; Jiang, Z. Organocatalytic Enantioselective Protonation for Photoreduction of Activated Ketones and Ketimines Induced by Visible Light. *Angew. Chem., Int. Ed.* **2017**, *56*, 13842–13846. [CrossRef]
58. Cao, K.; Tan, S.M.; Lee, R.; Yang, S.; Jia, H.; Zhao, X.; Qiao, B.; Jiang, Z. Catalytic Enantioselective Addition of Prochiral Radicals to Vinylpyridines. *J. Am. Chem. Soc.* **2019**, *141*, 5437–5443. [CrossRef]

59. Guo, X.; Okamoto, Y.; Schreier, M.R.; Ward, T.R.; Wenger, O.S. Enantioselective synthesis of amines by combining photoredox and enzymatic catalysis in a cyclic reaction network. *Chem. Sci.* **2018**, *9*, 5052–5056. [CrossRef]

catalysts

MDPI

Review

The Combination of Lewis Acid with *N*-Heterocyclic Carbene (NHC) Catalysis

Qianfa Jia [1], Yaqiong Li [1], Yinhe Lin [1] and Qiao Ren [2,*]

1 Chongqing Key Laboratory of Inorganic Special Functional Materials, College of Chemistry and Chemical Engineering, Yangtze Normal University, Fuling, Chongqing 408100, China; qianfa_jia@163.com (Q.J.); li_ychem@yznu.cn (Y.L.); linyinhe2009@163.com (Y.L.)

2 College of Pharmaceutical Science, Southwest University, Chongqing 400715, China

* Correspondence: qren2014@swu.edu.cn; Tel.: +86-23-6825-1225

Received: 16 September 2019; Accepted: 14 October 2019; Published: 16 October 2019

Abstract: In the last ten years, the combination of Lewis acid with *N*-heterocyclic carbene (NHC) catalysis has emerged as a powerful strategy in a variety of important asymmetric synthesis, due to the ready availability of starting materials, operational simplicity and mild reaction conditions. Recent findings illustrate that Lewis acid could largely enhance the efficiency and enantioselectivity, reverse the diastereoselectivity, and even influence the pathway of the same reaction partners. Herein, this review aims to reveal the recent advances in NHC-Lewis acid synergistically promoted enantioselective reactions for the expeditious assembly of versatile biologically important chiral pharmaceuticals and natural products.

Keywords: *N*-heterocyclic carbenes (NHC); Lewis acid; cooperative catalysis; asymmetric synthesis; umpolung

1. Introduction

N-heterocyclic carbenes (NHCs) are roughly categorized into three sections in accordance with their properties and applications: (i) excellent ligands for transition metals; (ii) coordination to *p*-block elements and (iii) organocatalysts [1–7]. The first *N*-heterocyclic carbene stabilized by two bulky adamantyl substituents was isolated and characterized by the Arduengo group in 1991 [8], which opened up an intriguing class of organic compounds for investigation. So far, a variety of novel carbenes have been revealed and synthesized, for instance, the dominant thiazolium-, imidazolium- and triazolium-based carbenes [9–12]. As powerful and efficient organocatalysts, they have been employed successfully for the synthesis of highly complex molecular architectures [13–18]. However, the stereoselectivities and/or regioselectivities of assembled products were limited for the single mode of activation during NHC catalyzed processes. Inspired by the importance and advantages of the cooperative catalysis strategy [19–25], which could activate the starting materials simultaneously with satisfying enantio- and stereoselectivities, *N*-heterocyclic carbenes as an important class of Lewis bases can cooperate with various Lewis acids to enhance yield and reverse selectivity or regioselectivity [26]. This strategy has emerged as a powerful approach for the direct access to various carbocyclic and heterocyclic compounds [27–32].

The recent developments in NHC/Lewis acid cooperative catalysis for the synthesis of some important enantioenriched molecules will be discussed in this review. They are categorized into several sections according to the species of the Lewis acid, including LiCl, $Mg(Ot\text{-}Bu)_2$, $Sc(OTf)_3$, $La(OTf)_3$, $Ti(Oi\text{-}Pr)_4$, etc. Due to the unique umpolung capacity, NHCs are widely applied in a variety of asymmetric transformations, resulting in the construction of versatile active acyl anions, enolates, homoenolates and α,β-unsaturated acylazolium equivalents from the corresponding carbonyl compounds (Scheme 1) [33–41], while Lewis acids as co-catalysts improve the reactivities or activate inactive electrophiles. The related exciting discoveries involving NHCs/Lewis acid cooperative catalysis are presented herein.

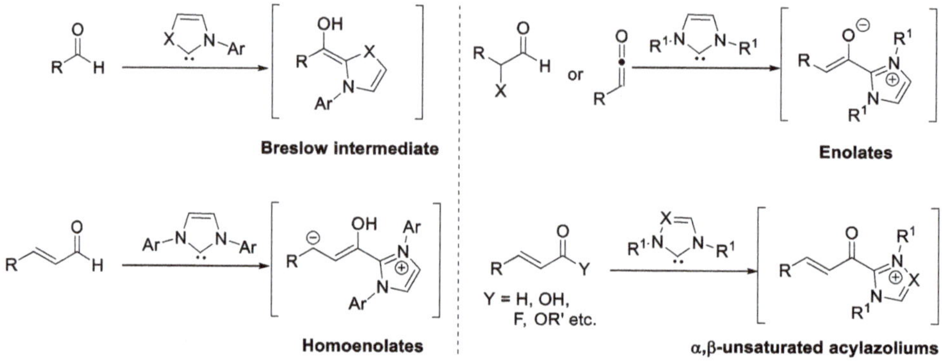

Scheme 1. Carbonyl compounds and corresponding *N*-heterocyclic carbene (NHC)-bound intermediates.

2. Cooperative NHC/Mg Catalysis

Though NHCs have been proved to be good ligands for many transition metals on account of their strong donor properties, the Scheidt group reported a pioneering investigation and made an important breakthrough, which defies conventional wisdom with respect to the potential incompatibility of Lewis acids and bases. γ-lactam derivatives 2 with high enantioselectivities and diastereoselectivities were obtained by a formal [3 + 2] cycloaddition of *N-acyl* hydrazones 1 and cinnamaldehyde (Scheme 2). The results indicated that the employment of $Mg(Ot\text{-}Bu)_2$ could increase the yield of γ-lactam distinctly from 31% to 78% and enhance the enantioselectivity slightly in the presence of 5 mol% of NHC catalyst. Subsequently, the substrates scope was surveyed and a broad range of functionalities turned out to be well accommodated. In these processes, carbene precursor A was deprotonated by the base 1,5,7-triazabicyclo[4.4.0]dec-5-ene (TBD) followed by addition to α,β-unsaturated aldehyde to produce homoenolate equivalent 4 by raising the HOMO energy. Simultaneously, magnesium (II) di-*tert*-butoxide was selected as the optimal Lewis acid to activate *N-acyl* hydrazones by lowering the LUMO energy. Further kinetic studies indicated that the reaction initial-rate emerged an inverse first-order relation for Mg (II) concentration [42].

The Zhao group reported the pioneering work on the kinetic resolution of tertiary alcohols 6 by NHC-catalyzed esterification under oxidative conditions (Scheme 3). The presence of $Mg(OTf)_2$ and $NaBF_4$ in this catalytic system turned out to be efficient to improve both selectivity and reaction rate. A broad range of substrates was investigated and displayed the practicability of this protocol. In most cases, the unreacted starting tertiary alcohols 8 were recovered in high to excellent enantioselectivities. The proposed mechanism revealed that the Lewis acid $Mg(OTf)_2$ may activate the substrate in a cooperative way to benefit the presumable attack of tertiary alcohol on acyl azolium intermediate from the opposite side of the catalyst's chiral backbone. However, the oxindole structure was proved to be critical for this system to work smoothly [43].

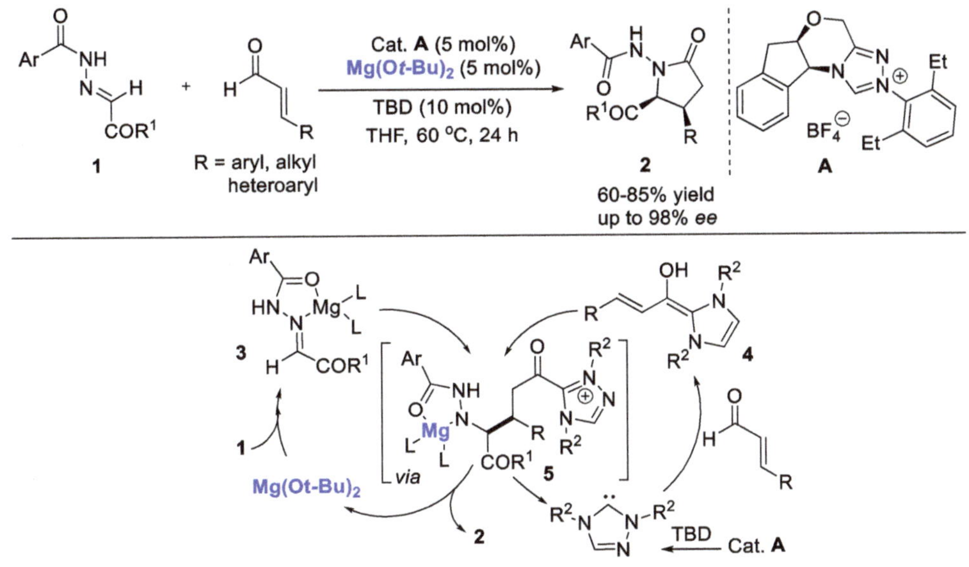

Scheme 2. NHC/Mg(O*t*-Bu)$_2$ strategy for the stereoselective synthesis of γ-lactams.

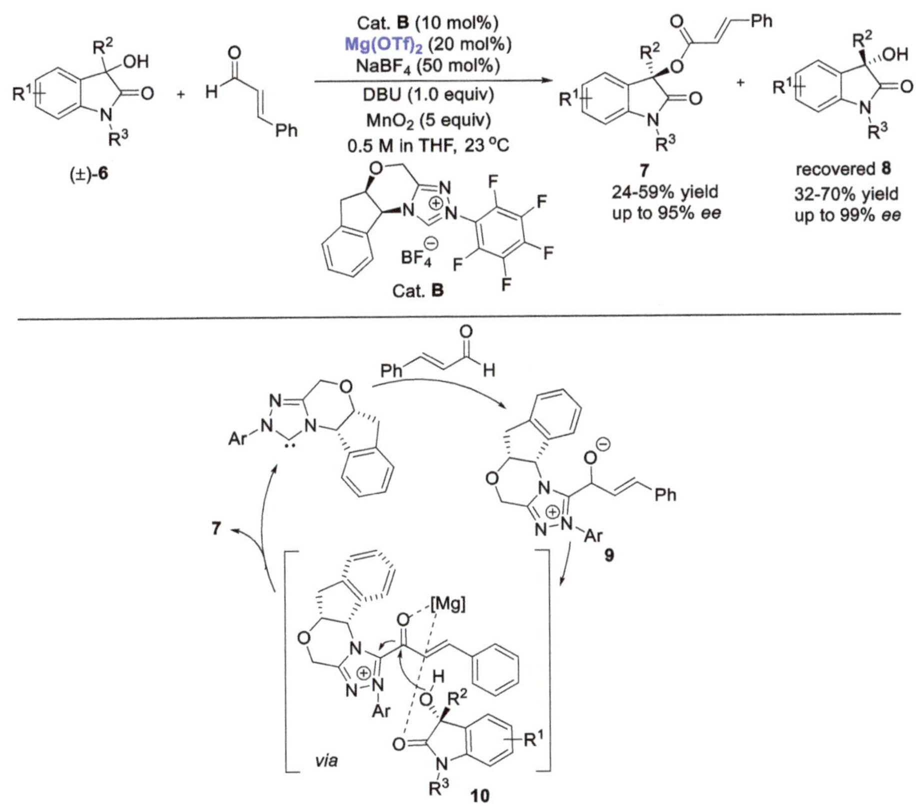

Scheme 3. Kinetic resolution of tertiary alcohols by NHC and Lewis acid.

In 2018, Wang and co-workers disclosed a [3 + 3] atroposelective annulation of alkynyl acyl azoliums 15 with 1,3-dicarbonyls 12 (Scheme 4). 3,3′,5,5′-tetratert-butyldiphenoquinone (DQ) as an external oxidant, acted as a two-electron acceptor to deliver the Breslow intermediate 14 to alkynyl acyl azolium 15. This approach accessed an array of axially chiral pyranones 13 in a good to excellent level of enantioselectivities. The Lewis acid $Mg(OTf)_2$ was proved to be critical to promote the ketoenolate's 'C' attack to alkynyl acyl azolium instead of the direct 'O' attack. Further transformation of the axially chiral pyranone-aryls successfully afforded the commonly used axial biaryls via Diels-Alder reaction. It is worth mentioning that the key intermediate alkynyl acyl azoliums 15 derived from ynals were discovered for the first time in NHC catalysis [44]. Encouraged by the initial exploration of the alkynyl acylazolium intermediates, they reported another formal [3 + 3] annulation of ynals with amidines to afford a series of functionalized 1,2,6-trisubstituted pyrimidin-4-ones under mild reaction conditions. Notably, the chemical yields of pyrimidin-4-ones were increased significantly with the addition of $Mg(OTf)_2$, which activated the *N*-substituted amidines and alkynyl acyl azolium intermediates simultaneously [45].

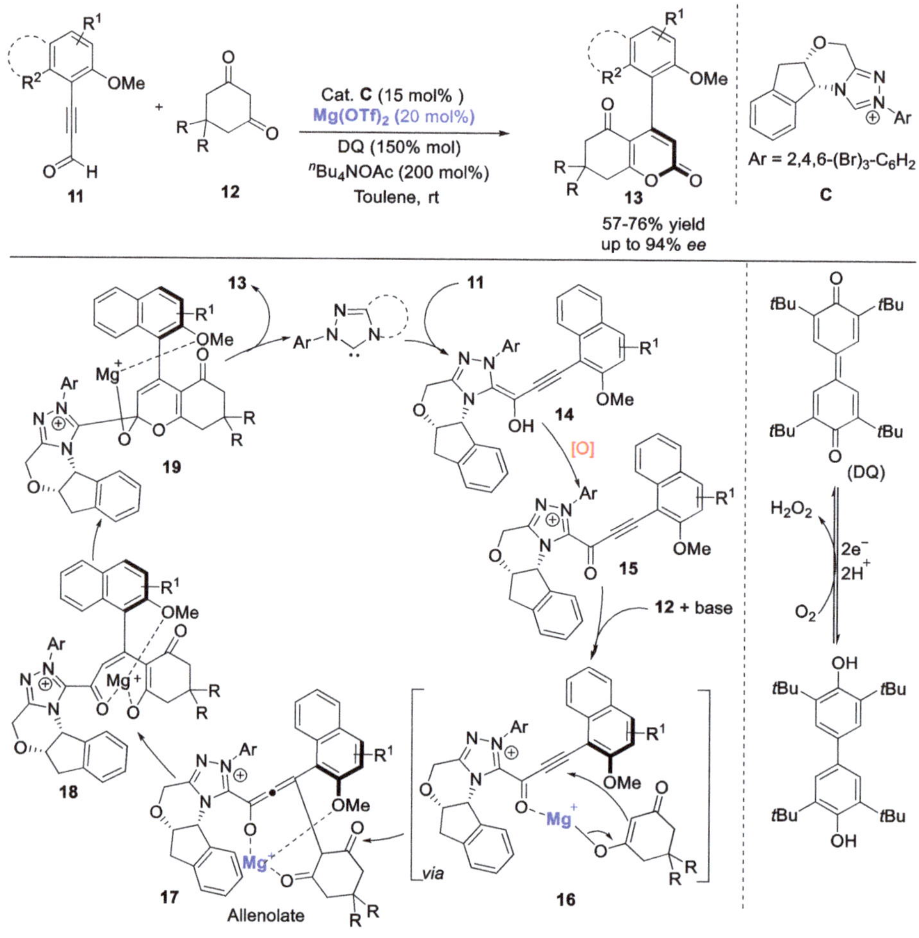

Scheme 4. Enantioselective [3 + 3] atroposelective annulation catalyzed by *N*-heterocyclic carbenes.

3. Cooperative NHC/Ti Catalysis

The Scheidt group found that an appropriate Lewis acid could reverse the diastereoselectivity in NHC-catalyzed enals with chalcones 20 through preorganization of the reactants (Scheme 5) [46]. Though the Bode group previously reported that NHC-catalyzed reactions of chalcone derivatives and enals provided *trans*-cyclopentene products, *cis*-cyclopentenes 21 were also obtained only using *(E)*-ethyl 4-oxo-2-butenoate as the partner of substrates. Optically active *cis* 1,3,4-trisubstituted cyclopentenes were observed when employing titanium tetraisopropoxide as the Lewis acid and involving homoenolates 22 generated by NHC catalysis. While other Lewis acids such as magnesium, zinc or scandium triflate completely inhibited homoenolate addition, the usage of $Mg(Ot\text{-}Bu)_2$ afforded predominately the *trans* cyclopentenes in moderate yield. It is noteworthy that the catalytic amounts of isopropyl alcohol could improve both the yield and the rate by separating the titanium catalyst from the intermediate easily. In the exploration of mechanism, Scheidt proposed that the titanium Lewis acid promotes the generation of homoenolate equivalents 22, and then activates the chalcone to coordinate with the homoenolate-titanium intermediate 23. Subsequently, two successive C–C bond formation, acylation and decarboxylation provides the cyclopentene. Density functional theory (DFT) studies by Domingo illustrated that the titanium complex could not only accelerate the annulation reaction by lowing the Gibbs free energy, but also favor pre-organizing the spatial alignment of homoenolate and chalcone, in line with the proposed reaction pathway [47].

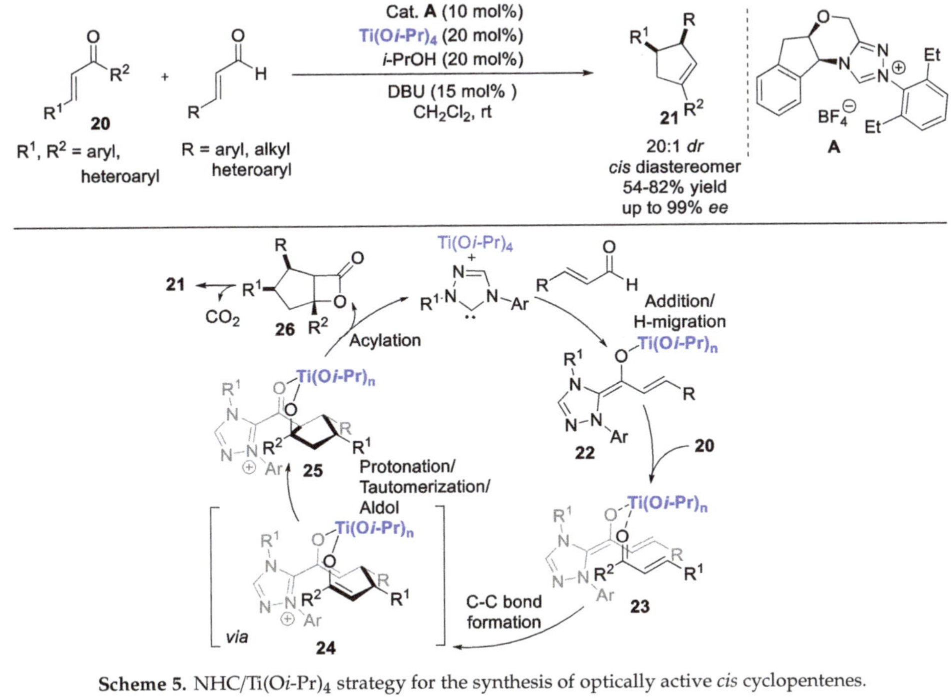

Scheme 5. NHC/Ti$(Oi\text{-}Pr)_4$ strategy for the synthesis of optically active *cis* cyclopentenes.

To extend the scope of substrates, Scheidt et al. developed the enantioselective dimerization of enals using cooperative catalysis NHC/Ti$(Oi\text{-}Pr)_4$ (Scheme 6). The NHC/titanium-mediated 1,4-addition of homoenolate equivalents to enals was favored over the traditional 1,2-addition, resulting in the formation of γ-lactones [48].

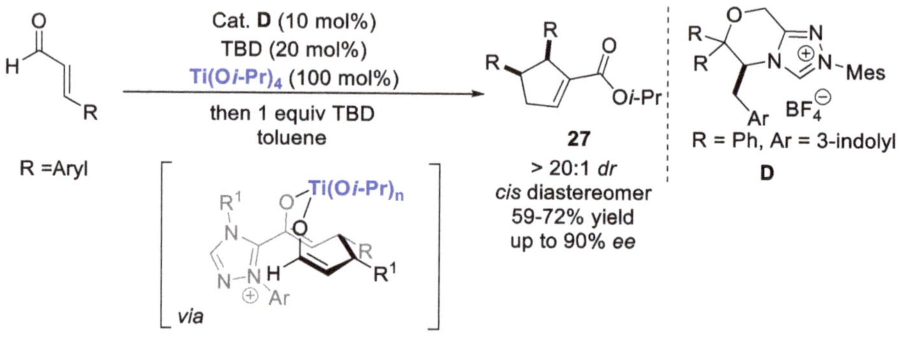

Scheme 6. NHC-catalyzed enantioselective enal dimerization.

Meanwhile, the Scheidt group uncovered the β,γ-unsaturated α-ketoesters 28 as a suitable class of homoenolate acceptors with enals to give the highly functionalized cyclopentanes 29 (Scheme 7) [49]. Importantly, the Lewis acid $Ti(Oi\text{-}Pr)_4$ played a crucial role in the initial coordination to the enals, which promoted the formation of extended Breslow intermediate 30 and subsequent coordination to another enal or the β,γ-unsaturated α-ketoester. The high enantiopure products were received through the C–C bond formation, protonation/tautomerization, intramolecular aldol reaction and then acylation/elimination or transesterification. In addition, no reaction happened drastically without the titanium (IV) *iso*-propoxide, which confirmed the significance of the Lewis acid during the process.

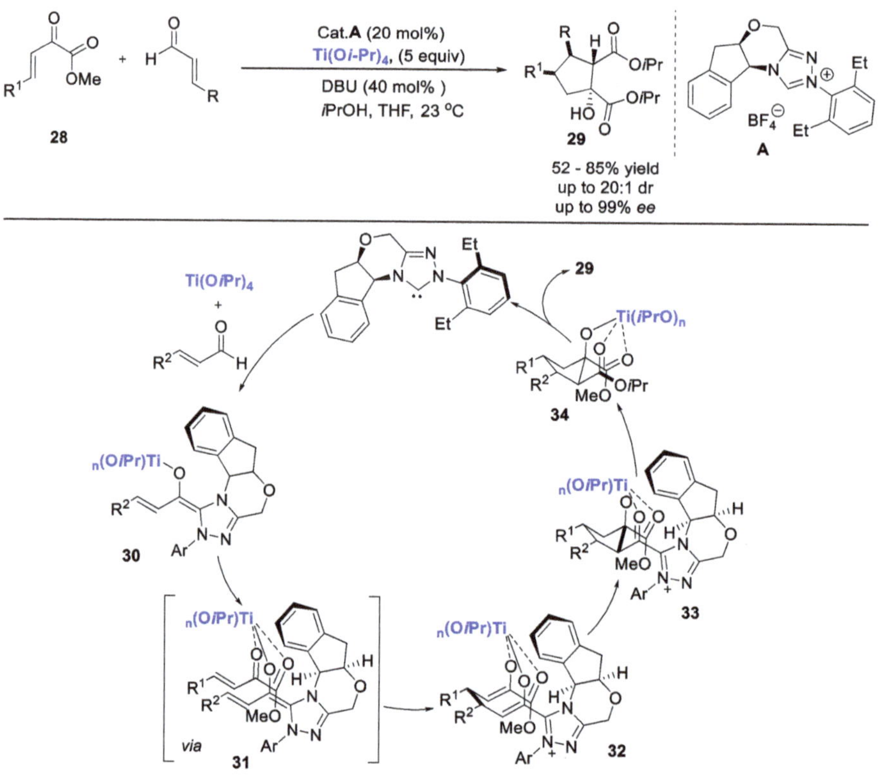

Scheme 7. NHC-catalyzed addition of homoenolates to β,γ-unsaturated α-ketoesters.

In 2013, Snyder group disclosed a one-step cooperative NHC/Lewis acid catalysis to forge the entire tricyclic butenolide cores, which belong to securinega alkaloids and exhibit various intriguing biological activities (Scheme 8). A linear enynal 35 was selected as the appropriate precursor for the in-situ generated homoenolate equivalent, which underwent an intramolecular addition to ketone and followed by lactonization to afford the desired product. The reaction conditions were optimized by investigating different NHC catalyst structures, species of Lewis acid or the concentration of solutions. Eventually, the tricyclic butenolide 36 product was isolated by slowly adding starting material into a suspension of catalyst and Ti(O*i*-Pr)$_4$ in toluene with the concentration of 0.03 M. Surprisingly, an exogenous base was not necessary, probably due to the Ti(O*i*-Pr)$_4$ was basic enough to facilitate the generation of the active carbene. In this dual catalysis strategy, ynals were used as nucleophilic homoenolate precursors to synthesize securinega alkaloids in only nine steps from commercial materials, which could provide opportunities for the further development of NHC-catalyzed reactions in total synthesis of fused bicyclic butenolide domains [50].

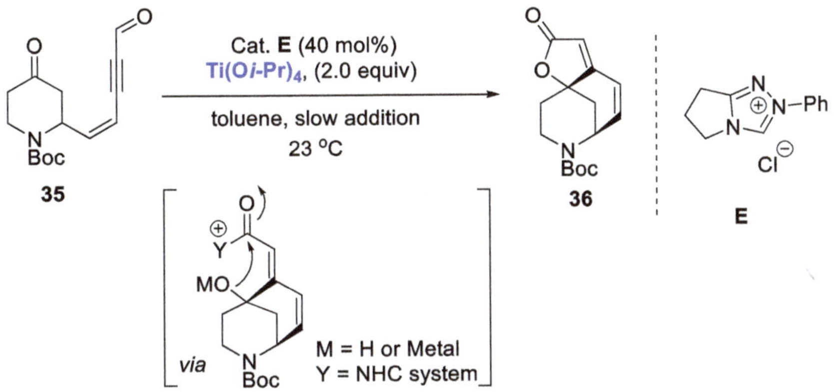

Scheme 8. NHC/Ti(O*i*-Pr)$_4$ strategy for the synthesis of securinega alkaloids.

Later on, Cheng group reported the NHC/Ti(O*i*-Pr)$_4$ co-catalyzed dimerization of 2-formylcinnamates 37 for the access to isochromenone derivatives 38 via an unexpected pathway. However, a mixture of two diastereoisomers with *cis*- and *trans*-isochromeno[4,3-c]isochromene-6,12-diacetates were isolated in the absence of Ti(O*i*-Pr)$_4$ (Scheme 9). The combination of NHC and Ti(O*i*-Pr)$_4$ changed the reaction pathway of the dimerization completely [51]. The proposed mechanism revealed that the Breslow intermediate 39 was firstly generated via the deprotonation, nucleophilic addition and proton transfer. It then underwent the second nucleophilic addition to 2-formylcinnamates to generate α-hydroxyl ketones 40, which was activated by titanium(IV) and underwent the aerobic oxidation to afford the 1,2-diketone 42. Intriguingly, the nucleophilic addition of NHC to the carbonyl of 1,2-diketone, the following intramolecular rearrangement and copy rearrangement produced the peroxide intermediates 45. 1,8-Diazabicyclo[5.4.0]undec-7-ene (DBU) as a powerful nucleophile attacked the peroxide intermediate to facilitate the cleavage of O–O bond and elimination of the acetate moiety resulting in the formation of the enolates 46. At last, the intramolecular lactonization and release of the NHC delivered the isochromenone derivatives 38. In this case, the Lewis acid changed the reaction pathway and provided opportunities for the application of NHC/Lewis acid cooperative catalysis.

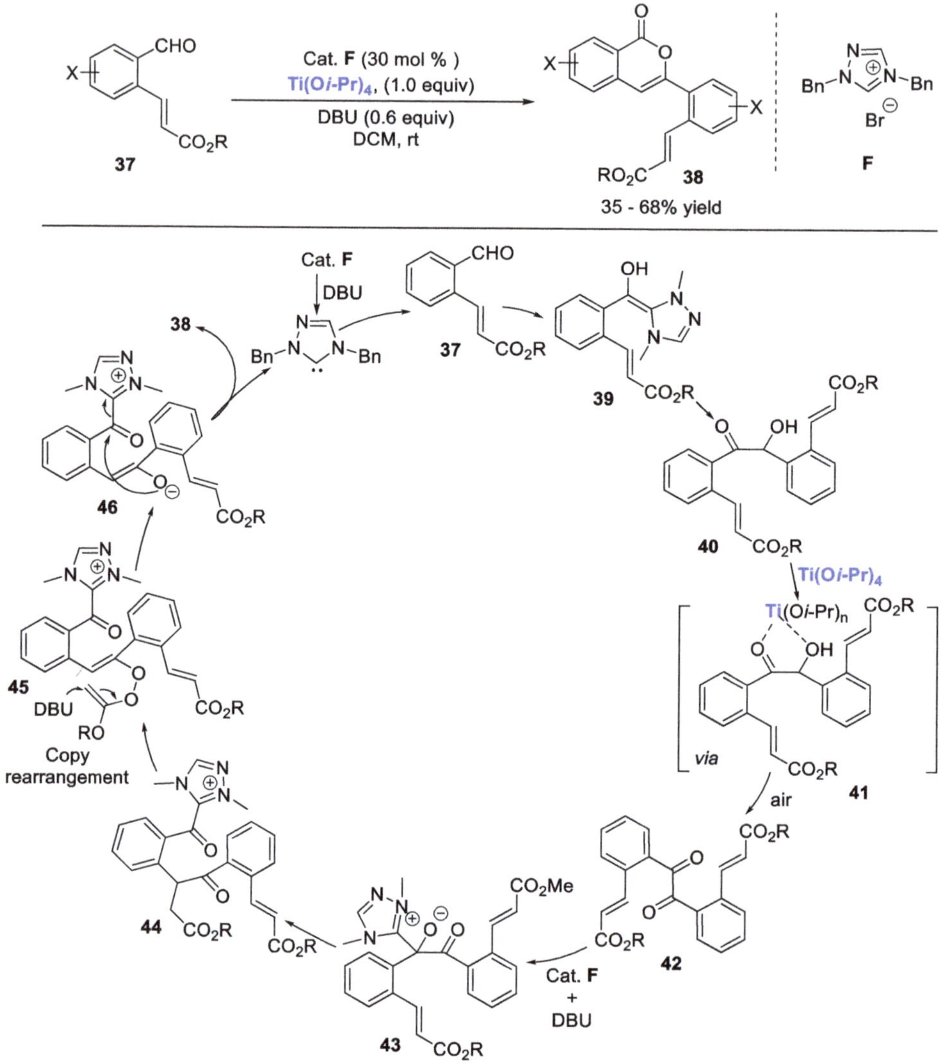

Scheme 9. Dimerization of 2-formylcinnamates by NHC/Lewis acid cooperative catalysis.

Moreover, the Cheng group has systematically studied the chiral *N*-heterocyclic carbene/Lewis acid co-catalyzed intermolecular dimerization of 2-aroylvinylcinnamaldehydes, 2-aroylvinylcinnamaldehydes 47 with aromatic aldehydes 48 and 2-(aroylvinyl)benzaldehydes 51 with enals (Scheme 10). A variety of synthetically unavailable functionalized chiral indeno[1,2-c]furan-1-ones 49, tetrahydroindeno[1,2-c]furan-1-ones 50, 4,5-dihydro-1,4-methanobenzo[c]oxepin-3-ones 52 and 2,8-dihydrocyclopenta[*a*]indenes 53 were synthesized with good yields, excellent enantioselectivities and high diastereoselectivities. The Lewis acid Ti(O*i*-Pr)$_4$ as a co-catalyst has been exemplified by activating the reaction partners simultaneously and then inducing the stereoselectivities preferentially [52–54].

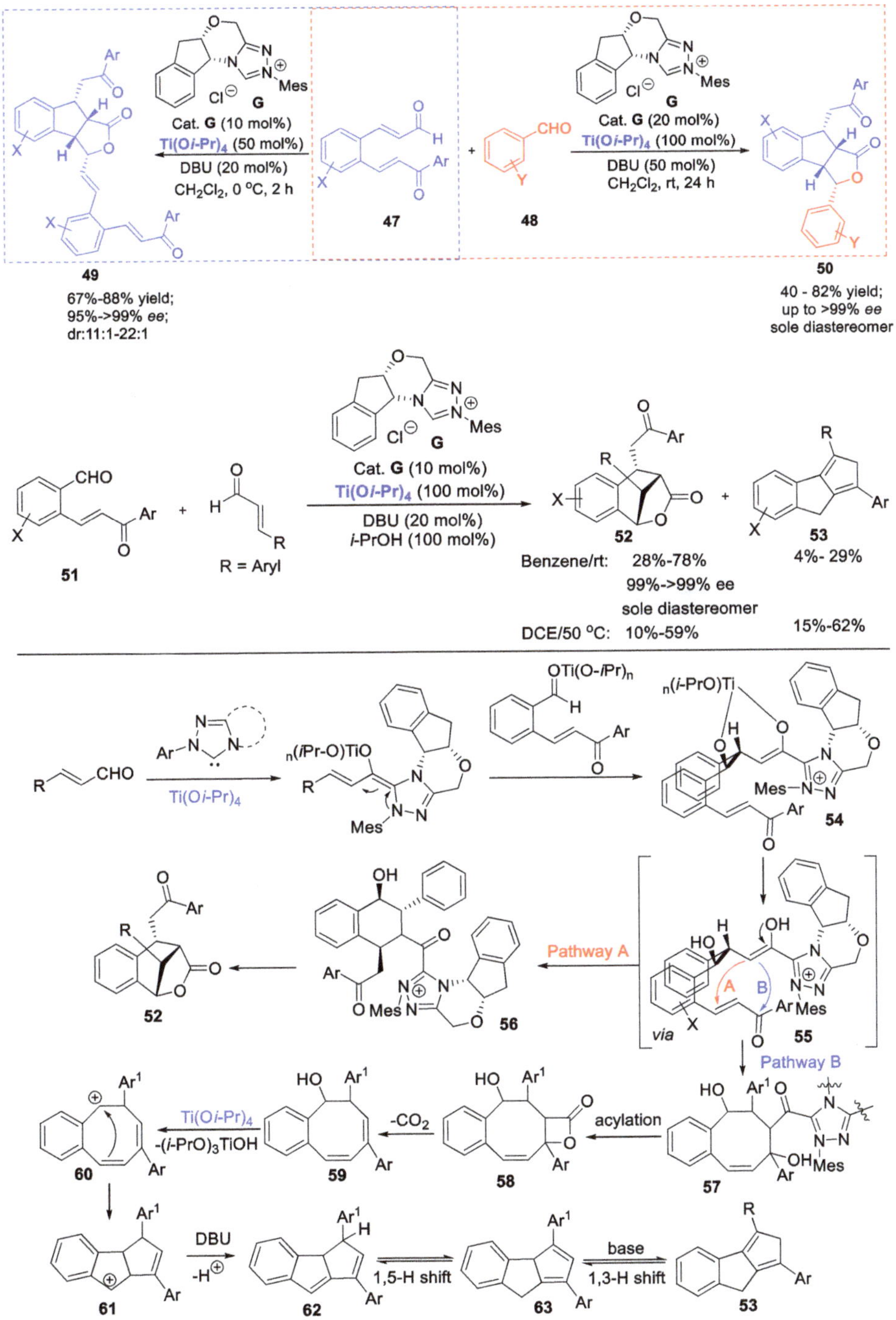

Scheme 10. The reactions of 2-aroylvinylcinnamaldehydes or 2-(aroylvinyl)benzaldehydes involved in NHC/Ti(O*i*-Pr)$_4$ cooperate catalysis.

Zhao and co-workers developed a divergent annulation reaction of heterocyclic enones 64 with enals to synthesis ε-lactones 65 or spiro-heterocycles (Scheme 11). Ti(O*i*-Pr)$_4$ as the optimal Lewis acid was surveyed to enhance the conversion of the catalytic system and deliver the [3 + 4] annulation product ε-lactones 65 in good yield and excellent chemo- and stereo-selectivities [55]. Surprisingly, the indole-based enones afforded the indole-fused ε-lactones as a single diastereomer, potentially due to the control of chiral catalyst. This work displayed the catalyst-controlled chemoselective process in NHC catalysis and extended the capability of NHC/Lewis acid cooperative catalysis.

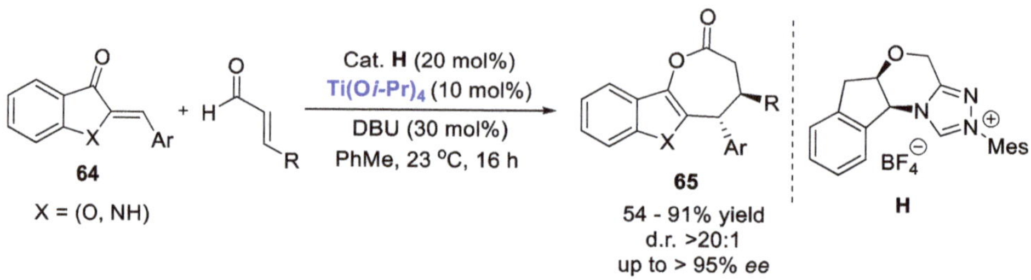

Scheme 11. Stereoselective synthesis of ε-lactones by NHC-catalyzed annulation.

4. Cooperative NHC/Li Catalysis

Since the combination of NHC catalysis with Lewis acids demonstrated high efficiency to facilitate synthetic transformations, the Scheidt group has made a number of pioneer works and reported the first cooperative catalytic system consisting of achiral NHC and LiCl to promote an intermolecular conjugate addition of primary and secondary alcohols with activated alkenes (Scheme 12). The impact of lithium cation was probed that the addition of 1.0 equivalent of 12-crown-4 leaded a decreased yield. Therefore, LiCl was added to generate the desired β-alkoxy ketone 68 in 95% yield. Unfortunately, the enantioselective versions with the chiral NHC/LiCl co-catalysis were carried out resulting in racemic products or low *ee* values even if high yields. The mechanism investigation illustrated that the free *N*-heterocyclic carbene derived from IMes acted as a Brønsted base and accessed the NHC-alcohol complex 69 as a crucial intermediate. Remarkably, no oligomerization products were observed presumably because the lithium chloride as a Lewis acid activated the enones toward the 1,4-additon of the alcohols. Ultimately, the overall yield was improved [56].

Subsequently, Scheidt and coworkers reported an enantioselective annulation of isatins 71 with enals for accessing spirooxindole lactones 72 in good yields and high enantioselectivities (Scheme 13) [57]. Soon later, Sunoj and coworkers focused on the mechanism and origin of stereoselectivity in the chiral NHC/Lewis acid co-catalyzed synthesis of spirooxindole lactones. The addition of chiral NHC to α, β-unsaturated aldehydes generates the homoenolate equivalent as nucleophilic species and then inducts the enantioselectivity of the annulation process. On the other hand, the lithium counterion interacts with both the 1,2-dicarbonyl of the isatin and the NHC-bound homoenolate, resulting in an enantioselective addition of the *re* face with the enhancement on the level of enantioselectivity, in line with the experimental observations [58]. At last, this approach was successfully applied into the concise synthesis of maremycin B, which contains a 3-hydroxy indole structure scaffold and exhibits the excellent anticancer activity [57].

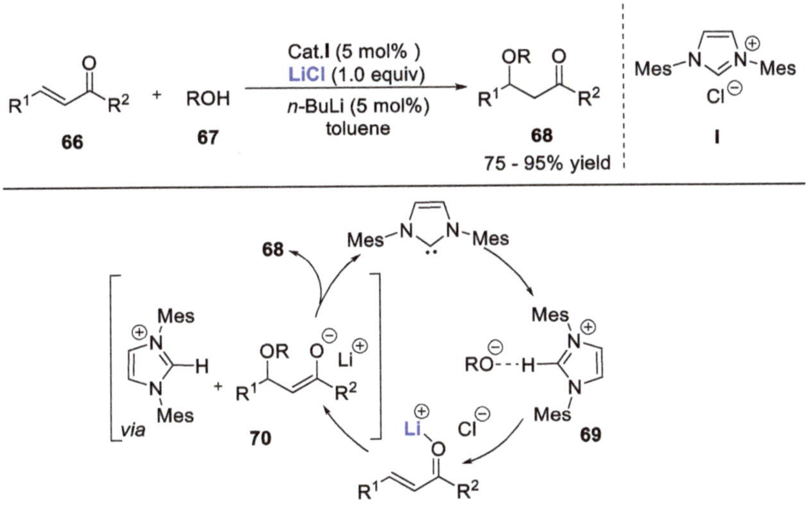

Scheme 12. NHC-catalyzed conjugate additions of alcohols.

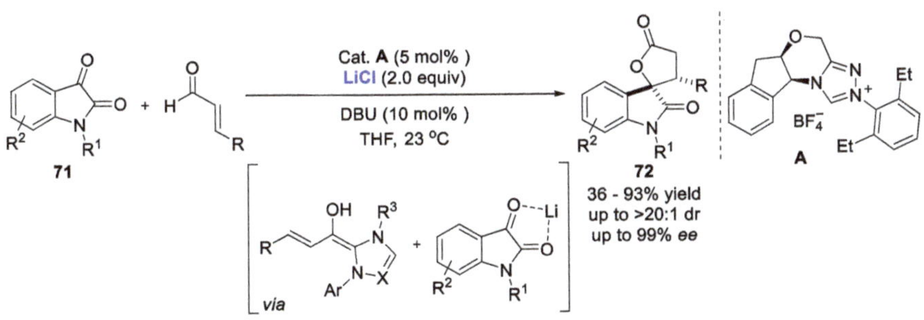

Scheme 13. NHC/LiCl strategy for the stereoselective synthesis of spirooxindole lactones.

The appropriate Lewis acid has a significant influence on the NHC-catalyzed umpolung reactions. In 2013, the She group reported an elegant work that the NHC/Lewis acid catalytic system mediated the [3 + 2] annulation of alkynyl aldehydes 73 with β,γ-unsaturated α-ketoester 74 (Scheme 14). No desired product was observed in initial studies employing only NHC-catalyst in the absence of the Lewis acid. Notably, the yields of butenolides were enhanced and the starting materials were consumed completely in a short time in the presence of LiCl. In addition, the enantioselective studies of this methodology were carried out by screening several available chiral carbenes to realize the asymmetric version of this formal [3 + 2] cyclization. Although a moderate enantioselectivity was observed, this result prompted further exploration on the NHC/Lewis acid mediated enantioselective reactions [59]. Shortly thereafter, Scheidt and co-workers further explored a chiral NHC-catalyzed cascade reaction of α,β-alkynals with α-ketoesters by using the same cooperative catalysis strategy. The enantioselectivity of this formal [3 + 2] annulation reaction was induced by a saturated imidazolium J (SImes·Cl) and chiral phosphoric acid. Remarkably, the introduction of lithium cation organized the transition state by means of activating the phosphate and α-ketoesters simultaneously [60]. Du and Lu group reported another formal [3 + 2] annulations of alkynyl aldehydes with isatins. A variety of spirooxindole butenolides and spirooxindole furan-3(2H)-ones were formed by the NHC/LiCl co-catalyzed transformation via the a^3-d^3 umpolung of alkynyl aldehydes and a^1-d^1 umpolung process respectively. The asymmetric version of this formal [3 + 2] cyclization reaction has also been realized using the chiral carbene precursor I as catalyst, and the spirooxindole butenolide as a single regioisomer was afforded in 78% yield and 73% *ee* value [61]. In short, the common nature of these elegant studies is the umpolung of the β-position of alkynyl aldehydes by *N*-heterocyclic carbenes to afford a unique "allenolate" nucleophile, whilst LiCl as the optimal Lewis acid activates the carbonyl of various electrophilic reagents.

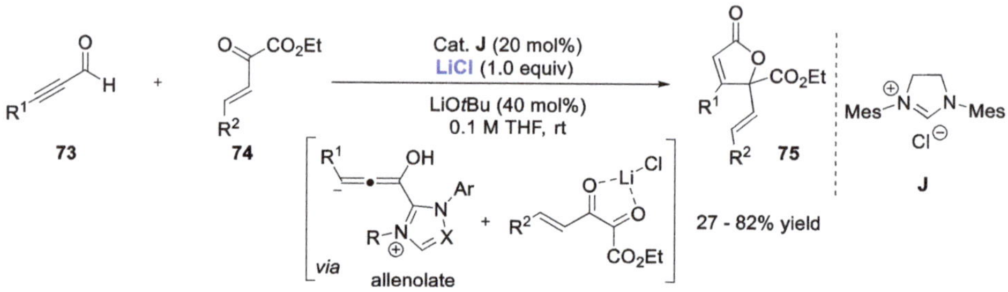

Scheme 14. NHC-catalyzed/Lewis acid mediated conjugate umpolung of alkynyl aldehydes.

Studer et al. reported the cooperative NHC/LiCl catalyzed the conjugate additions of tertiary prochiral C-nucleophiles to α,β-unsaturated acyl azolium in situ generated from the oxidation of the Breslow intermediate (Scheme 15). β-diketones, β-ketoesters, and malonates 76 bearing a β-oxyalkyl substituent at the α position reacted smoothly with enals to afford highly substituted cyclopentanes 77 in high yields and excellent diastereo- and enantioselectivities. The proposed mechanism revealed that this organic cascade process consisted of the deprotonation of NHC precursor K, Michael addition and subsequent asynchronous formal [2 + 2] aldol lactonization with the regeneration of the NHC catalyst to give the desired highly substituted β-lactones [62].

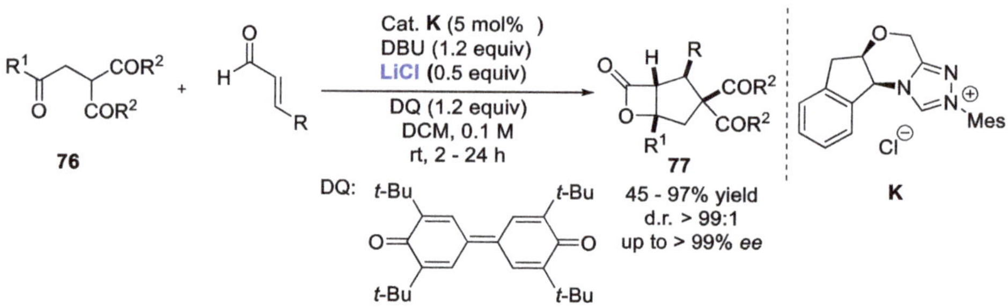

Scheme 15. Asymmetric synthesis of highly substituted β-lactones through oxidative carbene catalysis with LiCl as cooperative Lewis acid.

In 2015, Studer and Ye reported the similar results that the cooperative oxidative NHC/Lewis acid enantioselective catalysis gave highly substituted δ-lactones 79 through the reactions of enals with ε-oxo-γ,δ-malonates 78 containing two Michael acceptors (Scheme 16). The suggested mechanism of these two cascade reaction demonstrated that the oxidation of vinyl Breslow intermediate afforded the α,β-unsaturated acyl azolium intermediate, which was then attacked by the deprotonated malonates to give the enolate intermediate. Subsequently, the second intramolecular Michael-type cyclization and lactonization afforded the cyclopentane- and cyclohexane-fused δ-lactones with the release of the catalyst. Meanwhile, the LiCl was likely to coordinate with the enolate of malonates and the oxygen atom of the α,β-unsaturated acylazolium intermediate by lowering the LUMO energy. Therefore, LiCl turned out to be essential for facilitating the formation of new C–C bond and the outcome of high yields and excellent stereoselectivities [63,64].

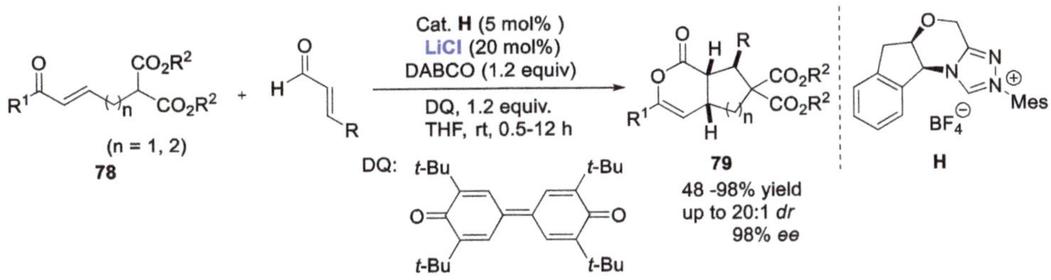

Scheme 16. Enantioselective synthesis of bicyclic δ-lactones via NHC-catalyzed cascade reaction.

In 2016, Zhong reported an enantioselective annulation of α,β-unsaturated aldehydes with 1,3-dicarbonyl compounds 80 by cooperative *N*-heterocyclic carbene/Lewis acid catalysis strategy (Scheme 17). LiCl and 4Å molecular sieves were found to be the best in the optimization studies. The *ee* value of the desired dihydropyranone 81 was improved distinctly from 60% to 87% in the presence of LiCl. It was noteworthy that the ambient air acted as the sole oxidant in this asymmetric annulation reaction. Some control experiments were carried out to illustrate that the molecular O_2 indeed could promote the oxidation of homoenolate equivalent to the α,β-unsaturated acyl triazolium. In this regards, the aerobic oxidative NHC/Lewis acid catalyzed enantioselective annulations provided an efficient, concise and green version in asymmetric synthesis [65]. Then, Du and Zheng independently reported the formal NHC/Lewis acid catalyzed [3 + 3] annulation of 1,3-dicarbonyl compounds with isatin-derived 2-bromoenals and β-cyano-substituted α,β-unsaturated aldehydes, respectively [66]. Compared with the isatin-derived enals, which was unstable under air and always difficult to separate from the Z/E mixtures, the isatin-derived 2-bromoenals were proved to be more stable under air and reacted well with 1,3-dicarbonyl compounds under NHC/base conditions. However, it was still ambiguous that whether the β-cyano-substituted α,β-unsaturated aldehydes could be attacked by the NHC catalyst or not. The enantioselective annulations of β-cyano-substituted α,β-unsaturated aldehydes with malonates were investigated under NHC-catalyzed oxidative conditions. In these two reactions, the addition of LiCl enhanced the reaction yields and stereoselectivities significantly to give the desired spirooxindole δ-lactones and dihydropyran-4-carbonitrile compounds [67]. Then, Dong, Du and colleagues reported the first application of esters as alkynyl acyl azolium precursors that have been utilized to undergo a formal [2 + 3] annulation with amidomalonates through dimethylaminopyridine (DMAP)/LiCl and NHC/LiCl cooperative catalysis. A wide range of (Z)-5-amino-3-furanones was synthesized with moderate to high yields (41%–99% yield) and high regioselectivities [68].

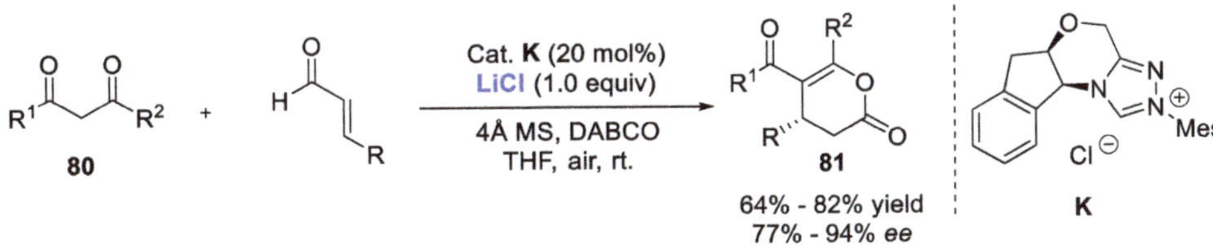

Scheme 17. NHC/Lewis acid catalyzed enantioselective aerobic annulation of α,β-unsaturated aldehydes with 1,3-dicarbonyl compounds.

Compared with enals, the carboxylic acids are more readily available and stable. They could be easily activated and in situ converted to the key NHC-bound intermediates with the assist of an array of coupling reagents, such as carbonyldiimidazole (CDI), 2-(7-aza-1H-benhexafluorophoszotriazole-1-yl)-1,1,3,3-tetramethyluroniumphate (HATU), and pivaloyl chloride. Yao and co-workers reported that α,β-unsaturated carboxylic acids could be transformed into α,β-unsaturated acyl azolium in the presence of HATU via the in situ activation strategy. Moreover, the introduction of LiCl could improve the enantioselectivities of dihydropyranone products [69]. In 2017, the Biju group demonstrated an intramolecular NHC-catalyzed aldol-lactonization of ketoacids 82 using the dynamic kinetic resolution (DKR) strategy (Scheme 18). The kinetics study indicated that the reaction proceeded via DKR process because more than 50% β-lactones were obtained in 6 h. Further transformation of the cyclopentane-fused β-lactone products 83 with primary amines resulted in succinimide derivatives containing four contiguous stereocenters in excellent yields and diastereoselectivities and good enantiopurities [70].

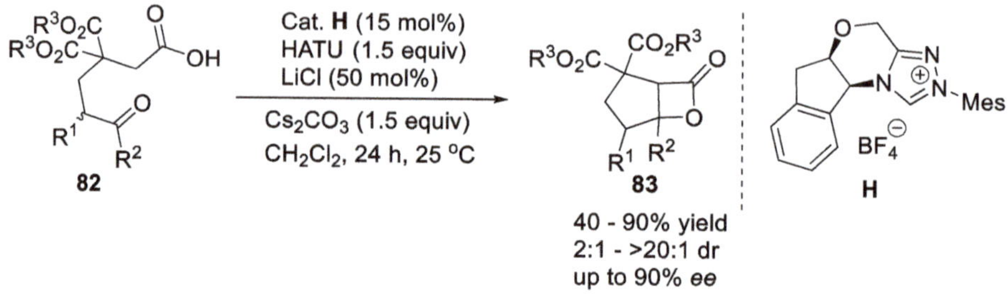

Scheme 18. NHC-catalyzed aldol-lactonization of ketoacids via dynamic kinetic resolution.

In 2017, Huang and Fu Group reported an oxidative amidation of aldimine 84 by NHC catalysis with LiCl as cooperative Lewis acid under ambient air (Scheme 19). The proposed reaction pathway indicated that the NHC-bounded aldimine intermediate was produced firstly by the umpolung of aldimine under NHC catalysis with the assistance of the LiCl, and the structure of the intermediate was confirmed by X-ray diffraction analysis. Then the intermediate underwent dearomatization and deprotonation process to form an imine-derived Breslow intermediate, which then added to dioxygen from the air and cleaved the O–O bond under basic condition to afford amides 85 with the expulsion of the free carbene. Overall, an economical and efficient methodology was developed for the synthesis of some biological molecules by the cooperative catalysis with ambient air or O_2 as the sole oxidant [71].

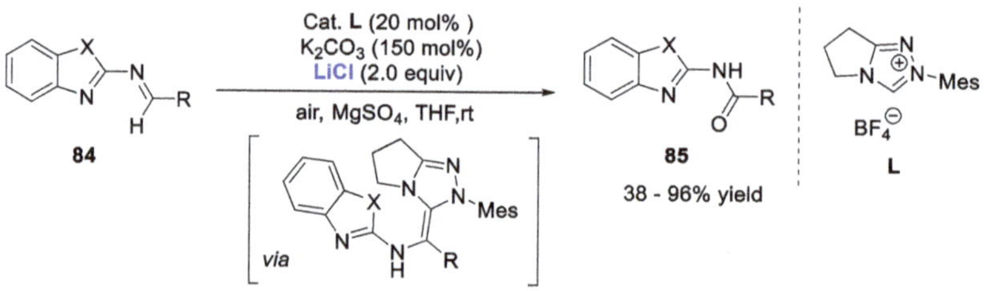

Scheme 19. Access to amide from aldimine via aerobic oxidative carbene catalysis and LiCl as cooperative Lewis acid.

Later, the same authors employed potassium 2-oxo-3-enoates 86 as outstanding and practical surrogates for α,β-unsaturated aldehydes in NHC-catalyzed asymmetric reactions. These salts could be prepared at scale and purified to undergo NHC-catalyzed reactions with enones 87, isatins 89, and 1,3-dicarbonyl compounds 91 respectively, affording various corresponding cyclopentenes 88, spirooxindole lactones 90 and lactones 92 with broad substrate scopes and good to excellent enantioselectivities [72]. In 2019, this group further developed a novel NHC-catalyzed [3 + 3] annulation of potassium 2-oxo-3-enoates 86 with 2-ethylidene 1,3-indandione 93 to give 2,2-diacyl spirocyclohexanones 94 in good to excellent yields. Lewis acid LiCl was added in these reactions to activate the potassium 2-oxo-3-enoates via the collaborative strategy (Scheme 20) [73].

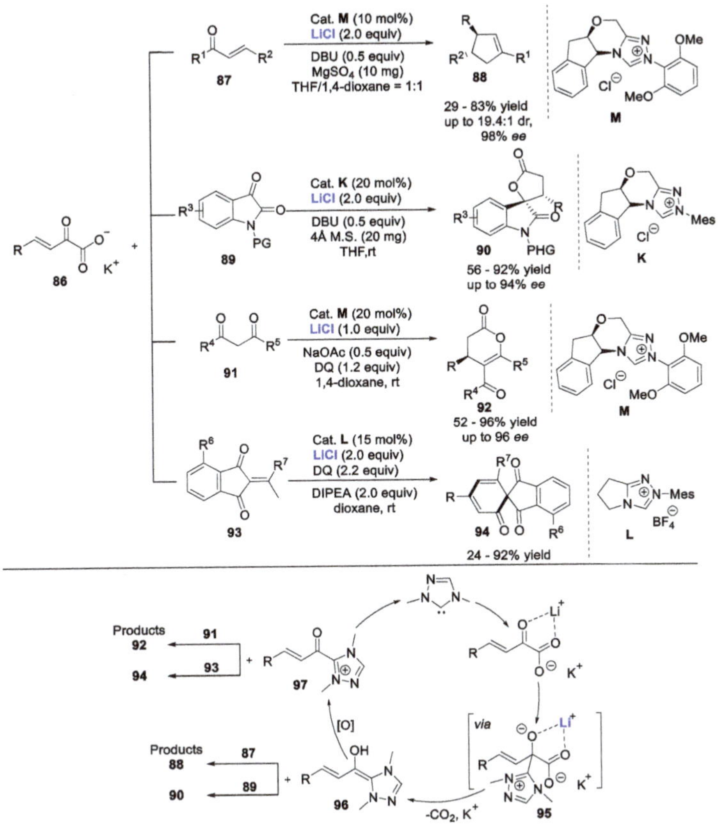

Scheme 20. Potassium 2-oxo-3-enoates as effective and versatile surrogates for α,β-unsaturated aldehydes in NHC-catalyzed asymmetric reactions.

In 2018, the Enders group reported a new NHC-catalyzed domino process of enals with reactive malonates 98 (Scheme 21). The huge challenge of this cascade reaction was how to control the reactivities of multifold nucleophilic and electrophilic of the substrates [74]. Hence the malonates bearing an ortho-hydroxy phenyl group and enals were selected as starting materials to validate the feasibility of the domino processes. Gratifyingly, the desired cyclopenta[c]-fused chromenones 99 were isolated in an acceptable yields and high enantioselectivities with the assistance of LiCl as a cooperative Lewis acid. Since there are two possible mechanisms to illuminate the reaction pathway, DFT calculations and control experiments were carried out to verify that the reaction has been subjected to the domino Michael/aldol/lactonization/dehydration process. Notably, low chemical yields and *ee* values were obtained in the absence of LiCl. This domino reaction clearly showed the power of NHC/Lewis acid cooperative catalysis involving reactive reagents.

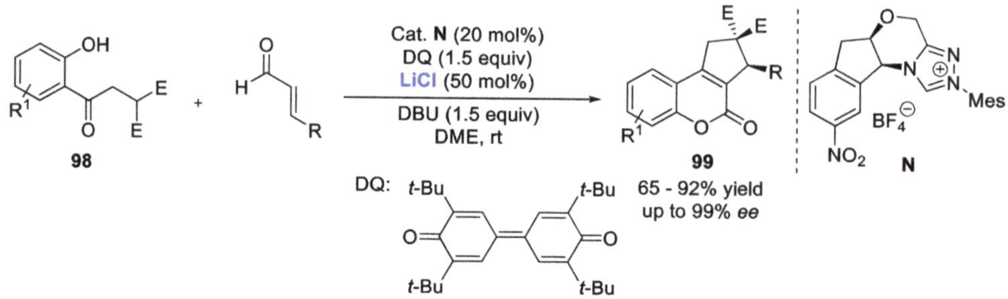

Scheme 21. NHC-catalyzed quadruple domino reactions: asymmetric synthesis of cyclopenta[c]chromenones.

Recently, Naumann, Buchmeiser and co-workers established an NHC/LiCl cooperative catalysis for the synthesis of linear poly(oxazolidin-2-one)s (POxa) 102 (Scheme 22). Diepoxides 101, aromatic as well as aliphatic diisocyanates 100, and NHC-CO_2 adducts were employed in the polymerization reaction. More importantly, the Lewis acid LiCl was selected as cocatalyst to secure high-molecular-weight POxa and control the polymerization in a reasonable degree [75].

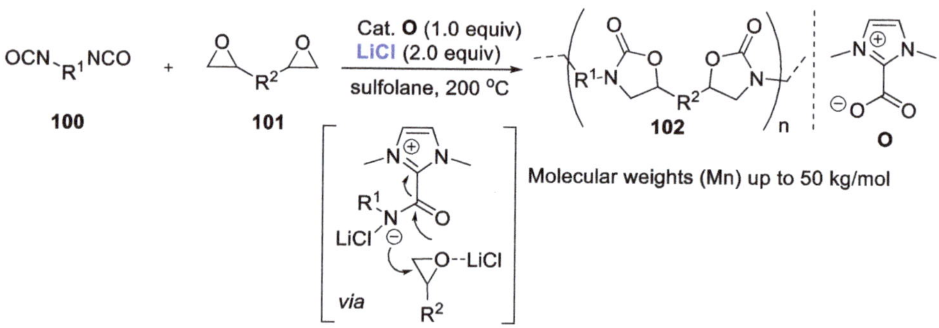

Scheme 22. Synthesis of linear poly(oxazolidin-2-one)s by cooperative catalysis based on NHC/LiCl.

5. Cooperative NHC/Sc Catalysis

In 2012, the scandium-based Lewis acid was first applied in the cooperative NHC catalysis by the Chi group [76]. The authors successfully circumvented the difficulties in improving the reactivity and enantioselectivity of the remote γ-carbon of enals (Scheme 23). β-phenyl substituted butenal was chosen as the starting material to avoid the competing pathway of the NHC-mediated enal reactions. The combination of $Sc(OTf)_3$ and $Mg(OTf)_2$ offered a small but consistent additional *ee* enhancement. Remarkably, only 5%–23% *ee* were observed in all cases when the reactions were conducted without $Sc(OTf)_3$, which demonstrated that the potential coordination of $Sc(OTf)_3$ played a critical role in the chiral induction.

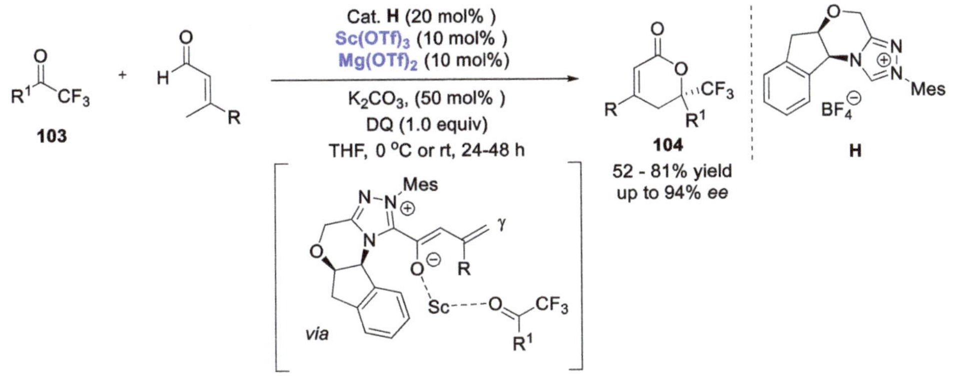

Scheme 23. Oxidative γ-addition of enals to trifluoromethyl ketones by NHC/$Sc(OTf)_3$ cooperative catalysis.

Inspired by these results, Wang and co-workers reported an enantioselective intermolecular dynamic kinetic resolution (DKR) catalyzed by an *N*-heterocyclic carbene and a Lewis acid cooperatively (Scheme 24) [77]. The enantiomerically pure DKR products σ-lactones 106 were isolated from the reaction of β-phenyl substituted butenal and β-halo-α-ketoesters 105 under oxidative conditions with excellent enantio- and diastereocontrol. The postulated reaction pathway was illustrated that the key intermediate vinyl enolate 109 arose from the γ-deprotonation of the oxidatively generated unsaturated acyl azolium intermediate 108. Then, this intermediate underwent the nucleophilic addition to activate ketones, resulting in the regioselective γ-addition and construction of the cyclization products. The Lewis acid $Sc(OTf)_3$ or $Mg(OTf)_2$ were known to exhibit good affinities for carbonyl oxygen and carboxylates, and potentially involved in the multisite coordination to bring the ketone electrophile into close proximity with vinyl enolate intermediate. In general, the effect of this coordination may amplify the otherwise weak chiral induction by the chiral NHC catalyst.

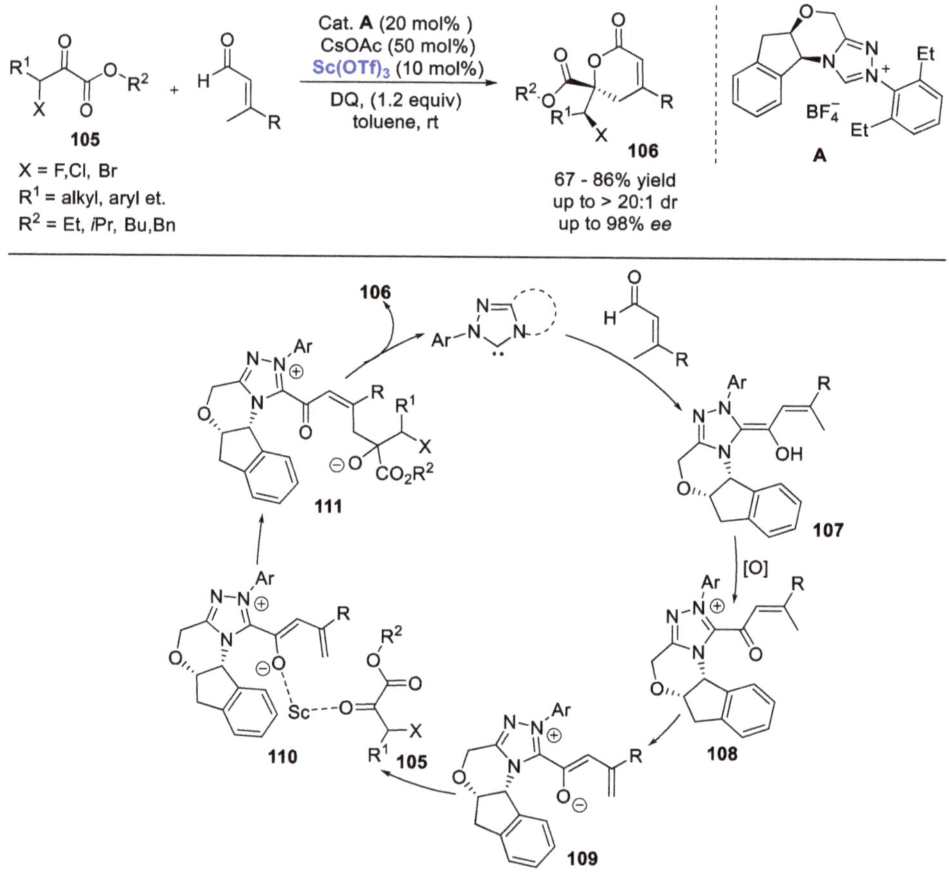

Scheme 24. Intermolecular dynamic kinetic resolution cooperatively catalyzed by an N-heterocyclic carbene and a Lewis acid.

6. Miscellaneous

In 2014, Yao and coworkers reported a novel Lewis acid La(OTf)3 as a co-catalyst in the NHC-catalyzed [4 + 2] annulation of 2-bromo-2-enals with isatin derivatives 112 (Scheme 25). This dual catalysis process was initiated by the addition of the NHC to 2-bromo-2-enals to give the Breslow intermediate, which then underwent a^3 to d^3 umpolung and debromination to generate α,β-unsaturated acyl azolium intermediate without the addition of external oxidant. The α,β-unsaturated acyl azolium was then deprotonated at the γ-position to provide the vinyl enolate intermediate. Subsequently, the nucleophilic addition and intramolecular lactonization occurred between the vinyl enolate and isatin 112. Spirocyclic oxindole-dihydropyranones 113 were prepared in good yields and with excellent enantioselectivities. Similar to Chi's and Wang's report (vide supra), only 19–41% *ee* were observed in all cases in the absence of Lewis acid La(OTf)3 as a co-catalyst [78].

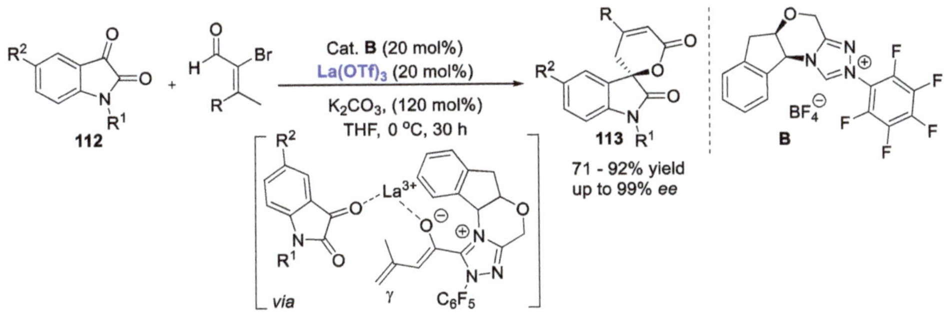

Scheme 25. NHC/La(OTf)$_3$ strategy for the stereoselective synthesis of spirocyclic oxindole-dihydropyranones.

Ye and colleagues investigated the reactions of α,β-unsaturated carboxylic acids with isatin-derived ketamines under NHC catalysis. Initially, the desired spirocyclic oxindolodihydropyridinone product was isolated in low yield with only 7% *ee* in the absence of Lewis acid. Notably, the Lewis acid La(OTf)$_3$ performed well to improve the yield and enantioselectivity [79]. In 2017, Huang and co-workers presented an enantioselective β-protonation of enals via a shuttling strategy (Scheme 26). A variety of Lewis acids have been screened and displayed a strong impact on the enantioselectivity. Finally, Cu(OTf)$_2$ resulted in the highest yield and *ee* value. The Lewis acid could coordinate with enals and mercaptans 114 through not only stabilizing a homoenolate intermediate but also facilitating protonation by increasing the acidity of the thiol. In a word, the combination of a chiral *N*-heterocyclic carbene (NHC) catalyst and a strong Brønsted/Lewis acid cocatalyst solved the longstanding challenge of enantioselective remote β-protonation of homoenolates with excellent reactivity and enantioselectivity [80].

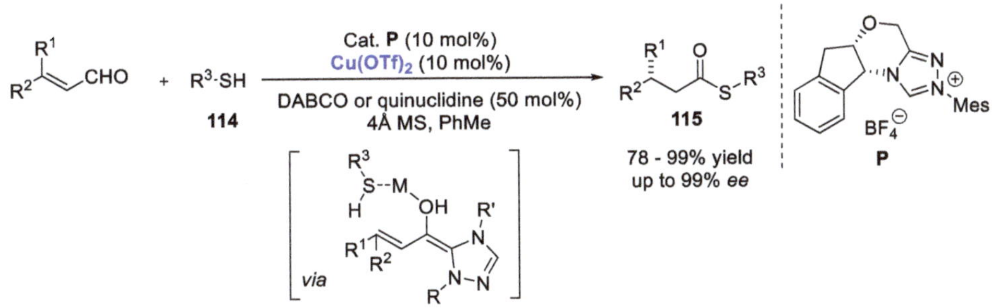

Scheme 26. Enantioselective β-protonation of enals via a shuttling strategy.

7. Conclusions

In this review, a number of catalytic applications have demonstrated that the combination of NHC catalysis with Lewis acid is a unique and efficient strategy for access to a wide range of highly functionalized complex and enantiomerically enriched structural motifs. The use of a Lewis acid in combination with a NHC catalyst enable us to (1) increase the yield and enantioselectivity, (2) reverse diastereo- and regioselectivity, (3) change the reaction pathway and (4) activate a previous inactive electrophile in NHC-generated processes. Overall, the dual catalytic approaches would expand the utility of NHC/Lewis acid methodology and construct other synthetically useful products with diverse particularly important skeletons. Further development of the cooperative catalysis in the total synthesis of natural products and pharmaceuticals will be the target of future studies.

Author Contributions: Q.J. and Q.R. wrote the manuscript. Y.L. (Yaqiong Li) and Y.L. (Yinhe Lin) contributed to discussions and edited the manuscript.

Funding: The project is funded by the National Natural Science Foundation of China (No. 21602179) and Science Foundation of Yangtze Normal University (No. 2017KYQD123, 2018QNRC17).

Conflicts of Interest: The authors declare no conflict of interest.

References

1. Doddi, A.; Peters, M.; Tamm, M. *N*-Heterocyclic Carbene Adducts of Main Group Elements and Their Use as Ligands in Transition Metal Chemistry. *Chem. Rev.* **2019**, *119*, 6994–7112. [CrossRef] [PubMed]
2. Hopkinson, M.N.; Richter, C.; Schedler, M.; Glorius, F. An overview of *N*-heterocyclic carbenes. *Nature* **2014**, *510*, 485–496. [CrossRef] [PubMed]
3. Díez-González, S.; Marion, N.; Nolan, S.P. *N*-Heterocyclic Carbenes in Late Transition Metal Catalysis. *Chem. Rev.* **2009**, *109*, 3612–3676. [CrossRef] [PubMed]
4. Herrmann, W.A.; Elison, M.; Fischer, J.; Ko¨cher, C.; Artus, G.R.J. Metal Complexes of *N*-Heterocyclic Carbenes-a New Structural Principle for Catalysts in Homogeneous Catalysis. *Angew. Chem. Int. Ed. Engl.* **1995**, *34*, 2371–2374. [CrossRef]
5. Hahn, F.E.; Jahnke, M.C. Heterocyclic Carbenes: Synthesis and Coordination Chemistry. *Angew. Chem. Int. Ed.* **2008**, *47*, 3122–3172. [CrossRef]
6. Enders, D.; Niemeier, O.; Henseler, A. Organocatalysis by *N*-Heterocyclic Carbenes. *Chem. Rev.* **2007**, *107*, 5606–5655. [CrossRef]
7. Marion, N.; Nolan, S.P.; Diez-Gonzales, S.; Diez-Gonzales, S. *N*-Heterocyclic Carbenes as Organocatalysts. *Angew. Chem. Int. Ed.* **2007**, *38*, 2988–3000. [CrossRef]
8. Arduengo, A.J., III; Harlow, R.L.; Kline, M. A Stable Crystalline Carbene. *J. Am. Chem. Soc.* **1991**, *113*, 361–363. [CrossRef]
9. Lebeuf, R.; Hirano, K.; Glorius, F. Palladium-Catalyzed C-Allylation of Benzoins and an NHC-Catalyzed Three Component Coupling Derived Thereof: Compatibility of NHC- and Pd-Catalysts. *Org. Lett.* **2008**, *10*, 4243–4246. [CrossRef]
10. Kamijo, S.; Yamamoto, Y. Recent Progress in the Catalytic Synthesis of Imidazoles. *Chem. Asian J.* **2007**, *38*, 568–578. [CrossRef]
11. Kerr, M.S.; De Alaniz, J.R.; Rovis, T. An Efficient Synthesis of Achiral and Chiral 1,2,4-Triazolium Salts: Bench Stable Precursors for *N*-Heterocyclic Carbenes. *J. Org. Chem.* **2005**, *70*, 5725–5728. [CrossRef] [PubMed]
12. Vora, H.U.; Lathrop, S.P.; Reynolds, N.T.; Kerr, M.S.; De Alaniz, J.R.; Rovis, T. Preparation of Chiral and Achiral Triazolium Salts: Carbene Precursors with Demonstrated Synthetic Utility. *Org. Synth.* **2010**, *87*, 350–361.
13. Hovey, M.T.; Jaworski, A.A.; Scheidt, K.A. *N*-Heterocyclic Carbene Catalysis in Natural Product and Complex Target Synthesis. *N-Heterocycl. Carbenes Organocatal.* **2019**, 345–404.
14. Izquierdo, J.; Hutson, G.E.; Cohen, D.T.; Scheidt, K.A. A continuum of progress: Applications of N-hetereocyclic carbene catalysis in total synthesis. *Angew. Chem. Int. Ed.* **2012**, *51*, 11686–11698. [CrossRef] [PubMed]
15. Bugaut, X.; Glorius, F. Organocatalytic umpolung: *N*-heterocyclic carbenes and beyond. *Chem. Soc. Rev.* **2012**, *41*, 3511. [CrossRef]
16. Flanigan, D.M.; Romanov-Michailidis, F.; White, N.A.; Rovis, T. Organocatalytic Reactions Enabled by *N*-Heterocyclic Carbenes. *Chem. Rev.* **2015**, *115*, 9307–9387. [CrossRef]
17. Zhao, M.; Zhang, Y.T.; Chen, J.; Zhou, L. Enantioselective Reactions Catalyzed by *N*-Heterocyclic Carbenes. *Asian J. Org. Chem.* **2018**, *7*, 54–69. [CrossRef]
18. Zhang, J.; Shi, M. *N*-Heterocyclic Carbenes Catalyzed Cycloadditions. *Organocatalyt. Cycloaddit. Synth. Carbo-Heterocycles* **2018**, 237–307. [CrossRef]
19. Zhao, X.; DiRocco, D.A.; Rovis, T. *N*-Heterocyclic carbene and Brønsted acid cooperative catalysis: Asymmetric synthesis of trans-γ-lactams. *J. Am. Chem. Soc.* **2011**, *133*, 12466–12469.
20. Guo, C.; Fleige, M.; Janssen-Müller, D.; Daniliuc, C.G.; Glorius, F. Cooperative *N*-Heterocyclic Carbene/Palladium-Catalyzed Enantioselective Umpolung Annulations. *J. Am. Chem. Soc.* **2016**, *138*, 7840–7843. [CrossRef]

21. Liu, G.; Wilkerson, P.D.; Toth, C.A.; Xu, H. ChemInform Abstract: Highly Enantioselective Cyclizations of Conjugated Trienes with Low Catalyst Loadings: A Robust Chiral *N*-Heterocyclic Carbene Enabled by Acetic Acid Cocatalyst. *Org. Lett.* **2012**, *14*, 858–861. [CrossRef] [PubMed]
22. DiRocco, D.A.; Rovis, T. Catalytic asymmetric intermolecular Stetter reaction of enals with nitroalkenes: Enhancement of catalytic efficiency through bifunctional additives. *J. Am. Chem. Soc.* **2011**, *133*, 10402–10405. [CrossRef] [PubMed]
23. Zhao, J.; Mück-Lichtenfeld, C.; Studer, A.; Mück-Lichtenfeld, C. Cooperative *N*-Heterocyclic Carbene (NHC) and Ruthenium Redox Catalysis: Oxidative Esterification of Aldehydes with Air as the Terminal Oxidant. *Adv. Synth. Catal.* **2013**, *355*, 1098–1106. [CrossRef]
24. Jin, Z.; Xu, J.; Yang, S.; Song, B.-A.; Chi, Y.R. Enantioselective Sulfonation of Enones with Sulfonyl Imines by Cooperative *N*-Heterocyclic-Carbene/Thiourea/Tertiary-Amine Multicatalysis. *Angew. Chem.* **2013**, *125*, 12580–12584. [CrossRef]
25. Fu, Z.; Sun, H.; Chen, S.; Tiwari, B.; Li, G.; Chi, Y.R. Controlled β-protonation and [4 + 2] Cycloaddition of Enals and Chalcones *via* *N*-heterocyclic Carbene/acid Catalysis: Toward Substrate Independent Reaction Control. *Chem. Commun.* **2013**, *49*, 261–263. [CrossRef]
26. Cohen, D.T.; Scheidt K, A. Cooperative Lewis acid/*N*-heterocyclic carbene catalysis. *Chem. Sci.* **2012**, *3*, 53–57. [CrossRef]
27. Chen, X.-Y.; Li, S.; Vetica, F.; Kumar, M.; Enders, D. *N*-Heterocyclic-Carbene-Catalyzed Domino Reactions via Two or More Activation Modes. *IScience* **2018**, *2*, 1–26. [CrossRef]
28. Hirano, K.; Piel, I.; Glorius, F. Dual Activation in *N*-Heterocyclic Carbene-organocatalysis. *Chem. Lett.* **2011**, *40*, 786–791. [CrossRef]
29. Grossmann, A.; Enders, D. *N*-Heterocyclic Carbene Catalyzed Domino Reactions. Angew. *Chem. Int. Ed.* **2012**, *51*, 314–325. [CrossRef]
30. Wang, M.H.; Scheidt, K.A. Cooperative Catalysis and Activation with *N*-Heterocyclic Carbenes. *Angew. Chem. Int. Ed.* **2016**, *55*, 14912–14922. [CrossRef]
31. Ding, Q.; Fan, X.; Wu, J. Recent Developments for the Synthesis of Heterocycles Catalyzed by *N*-Heterocyclic Carbene. *Curr. Org. Chem.* **2014**, *18*, 700–718. [CrossRef]
32. Ren, Q.; Li, M.; Yuan, L.; Wang, J. Recent advances in *N*-heterocyclic carbene catalyzed achiral synthesis. *Org. Biomol. Chem.* **2017**, *15*, 4731–4749. [CrossRef] [PubMed]
33. Ryan, S.J.; Candish, L.; Lupton, D.W. Acyl anion free *N*-heterocyclic carbene organocatalysis. *Chem. Soc. Rev.* **2013**, *42*, 4906. [CrossRef] [PubMed]
34. Vora, H.U.; Rovis, T. Asymmetric *N*-Heterocyclic Carbene (NHC) Catalyzed Acyl Anion Reactivity. *Aldrichim. Acta* **2011**, *44*, 3–11.
35. Johnson J, S. Catalyzed Reactions of Acyl Anion Equivalents. Angew. *Chem. Int. Ed.* **2004**, *43*, 1326–1328. [CrossRef]
36. Ryan, S.J.; Candish, L.; Lupton, D.W. *N*-Heterocyclic Carbene-Catalyzed Generation of α,β-Unsaturated Acyl Imidazoliums: Synthesis of Dihydropyranones by their Reaction with Enolates. *J. Am. Chem. Soc.* **2009**, *131*, 14176–14177. [CrossRef]
37. Menon, R.S.; Biju, A.T.; Nair, V. Recent Advances in Employing Homoenolates Generated by *N*-Heterocyclic Carbene (NHC) Catalysis in Carbon-carbon Bond-forming Rreactions. *Chem. Soc. Rev.* **2015**, *44*, 5040–5052. [CrossRef]
38. Zhang, C.; Hooper, J.F.; Lupton, D.W. *N*-Heterocyclic Carbene Catalysis via the α,β-Unsaturated Acyl Azolium. *ACS Catal.* **2017**, *7*, 2583–2596. [CrossRef]
39. Mondal, S.; Yetra, S.R.; Mukherjee, S.; Biju, A.T. NHC-Catalyzed Generation of α,β-Unsaturated Acylazoliums for the Enantioselective Synthesis of Heterocycles and Carbocycles. *Acc. Chem. Res.* **2019**, *52*, 425–436. [CrossRef]
40. Dzieszkowski, K.; Rafiński, Z. *N*-Heterocyclic Carbene Catalysis under Oxidizing Conditions. *Catalysts* **2018**, *8*, 549. [CrossRef]
41. Albanese, D.; Gaggero, N. An Overview on the *N*-Heterocyclic Carbene-Catalyzed Aza-Benzoin Condensation Reaction. *Catalysts* **2018**, *8*, 181. [CrossRef]
42. Raup, D.E.; Cardinal-David, B.; Holte, D.; Scheidt, K.A. Cooperative Catalysis by Carbenes and Lewis Acids in a Highly Stereoselective Route to γ-Lactams. *Nat. Chem.* **2010**, *2*, 766. [CrossRef] [PubMed]
43. Lu, S.; Poh, S.B.; Siau, W.-Y.; Zhao, Y. Kinetic Resolution of Tertiary Alcohols: Highly Enantioselective Access to 3-Hydroxy-3-Substituted Oxindoles. *Angew. Chem.* **2013**, *125*, 1775–1778. [CrossRef]

44. Zhao, C.; Guo, D.; Munkerup, K.; Huang, K.W.; Li, F.; Wang, J. Enantioselective [3 + 3] atroposelective annulation catalyzed by *N*-heterocyclic carbenes. *Nat. Commun.* **2018**, *9*, 611. [CrossRef] [PubMed]
45. Xie, Y.; Wang, J. *N*-Heterocyclic carbene-catalyzed annulation of ynals with amidines: Access to 1,2,6-trisubstituted pyrimidin-4-ones. *Chem. Commun.* **2018**, *54*, 4597–4600. [CrossRef] [PubMed]
46. Cardinal-David, B.; Raup, D.E.A.; Scheidt, K.A. Cooperative *N*-Heterocyclic Carbene/Lewis Acid Catalysis for Highly Stereoselective Annulation Reactions with Homoenolates. *J. Am. Chem. Soc.* **2010**, *132*, 5345–5347. [CrossRef]
47. Domingo, L.R.; Zaragozá, R.J.; Arnó, M. Understanding the Cooperative NHC/LA Catalysis for Stereoselective Annulation Reactions with Homoenolates. A DFT Study. *Org. Biomol. Chem.* **2011**, *9*, 6616–6622. [CrossRef]
48. Cohen, D.T.; Cardinal-David, B.; Roberts, J.M.; Sarjeant, A.A.; Scheidt, K.A. NHC-Catalyzed/Titanium(IV)–Mediated Highly Diastereo- and Enantioselective Dimerization of Enals. *Org. Lett.* **2011**, *13*, 1068–1071. [CrossRef]
49. Cohen, D.T.; Cardinal-David, B.; Scheidt, K.A. Lewis acid activated synthesis of highly substituted cyclopentanes by the *N*-heterocyclic carbene catalyzed addition of homoenolate equivalents to unsaturated ketoesters. *Angew. Chem. Int. Ed.* **2011**, *50*, 1678–1682. [CrossRef]
50. ElSohly, A.M.; Wespe, D.A.; Poore, T.J.; Snyder, S.A. An Efficient Approach to the Securinega Alkaloids Empowered by Cooperative *N*-Heterocyclic Carbene/Lewis Acid Catalysis. *Angew. Chem.* **2013**, *125*, 5901–5906. [CrossRef]
51. Dang, H.-Y.; Wang, Z.-T.; Cheng, Y. Changing Reaction Pathways of the Dimerization of 2-Formylcinnamates by *N*-Heterocyclic Carbene/Lewis Acid Cooperative Catalysis: An Unusual Cleavage of the Carbon–Carbon Bond. *Org. Lett.* **2014**, *16*, 5520–5523. [CrossRef] [PubMed]
52. Wang, Z.-Y.; Ding, Y.-L.; Wang, G.; Cheng, Y. Chiral *N*-heterocyclic carbene/Lewis acid cooperative catalysis of the reaction of 2-aroylvinylcinnamaldehydes: A switch of the reaction pathway by Lewis acid activation. *Chem. Commun.* **2016**, *52*, 788–791. [CrossRef] [PubMed]
53. Wang, G.; Wang, Z.-Y.; Niu, S.-S.; Rao, Y.; Cheng, Y. The Reaction of 2-Aroylvinylcinnamaldehydes with Aromatic Aldehydes by Dual Catalysis with a Chiral *N*-Heterocyclic Carbene and a Lewis Acid: Enantioselective Construction of Tetrahydroindeno[1,2-c]furan-1-ones. *J. Org. Chem.* **2016**, *81*, 8276–8286. [CrossRef] [PubMed]
54. Wang, Z.-Y.; Ding, Y.-L.; Li, S.-N.; Cheng, Y. *N*-Heterocyclic Carbene/Lewis Acid Dual Catalysis for the Divergent Construction of Enantiopure Bridged Lactones and Fused Indenes. *J. Org. Chem.* **2016**, *81*, 11871–11881. [CrossRef] [PubMed]
55. Wang, M.; Rong, Z.-Q.; Zhao, Y. Stereoselective synthesis of ε-lactones or spiro-heterocycles through NHC-catalyzed annulation: Divergent reactivity by catalyst control. *Chem. Commun.* **2014**, *50*, 15309–15312. [CrossRef]
56. Phillips, E.M.; Riedrich, M.; Scheidt, K.A. *N*-Heterocyclic Carbene-Catalyzed Conjugate Additions of Alcohols. *J. Am. Chem. Soc.* **2010**, *132*, 13179–13181. [CrossRef]
57. Dugal-Tessier, J.; O'Bryan, E.A.; Schroeder, T.B.H.; Cohen, D.T.; Scheidt, K.A. An *N*-heterocyclic carbene/Lewis acid strategy for the stereoselective synthesis of spirooxindole lactones. *Angew. Chem. Int. Ed.* **2012**, *51*, 4963–4967. [CrossRef]
58. Reddi, Y.; Sunoj, R.B. Origin of Stereoselectivity in Cooperative Asymmetric Catalysis Involving *N*-Heterocyclic Carbenes and Lewis Acids toward the Synthesis of Spirooxindole Lactone. *ACS Catal.* **2016**, *7*, 530–537. [CrossRef]
59. Qi, J.; Xie, X.; Han, R.; Ma, D.; Yang, J.; She, X. *N*-Heterocyclic Carbene (NHC)-Catalyzed/Lewis Acid Mediated Conjugate Umpolung of Alkynyl Aldehydes for the Synthesis of Butenolides: A Formal [3 + 2] Annulation. *Chem. Eur. J.* **2013**, *19*, 4146–4150. [CrossRef]
60. Lee, A.; Scheidt, K.A. A Cooperative *N*-Heterocyclic Carbene/Chiral Phosphate Catalysis System for Allenolate Annulations. *Angew. Chem.* **2014**, *126*, 7724–7728. [CrossRef]
61. Zhang, Y.; Lu, Y.; Tang, W.; Lu, T.; Du, D. Cooperative *N*-Heterocyclic Carbene (NHC)-Lewis Acid-mediated Regioselective Umpolung Formal [3 + 2] Annulations of Alkynyl Aldehydes with Isatins. *Org. Biomol. Chem.* **2014**, *12*, 3009–3015. [CrossRef] [PubMed]
62. Bera, S.; Samanta, R.C.; Daniliuc, C.G.; Studer, A. Berichtigung: Asymmetric Synthesis of Highly Substituted β-Lactones through Oxidative Carbene Catalysis with LiCl as Cooperative Lewis Acid. *Angew. Chem.* **2014**, *126*, 12199. [CrossRef]

63. Bera, S.; Daniliuc, C.G.; Studer, A. Enantioselective Synthesis of Substituted δ-Lactones by Cooperative Oxidative *N*-Heterocyclic Carbene and Lewis Aacid Catalysis. *Org. Lett.* **2015**, *17*, 4940–4943. [CrossRef] [PubMed]
64. Liang, Z.-Q.; Wang, D.-L.; Zhang, H.-M.; Ye, S. Enantioselective Synthesis of Bicyclic δ-Lactones via *N*-Heterocyclic Carbene-Catalyzed Cascade Reaction. *Org. Lett.* **2015**, *17*, 5140–5143. [CrossRef] [PubMed]
65. Xie, D.; Shen, D.; Chen, Q.; Zhou, J.; Zeng, X.; Zhong, G. *N*-Heterocyclic Carbene/Lewis Acid Catalyzed Enantioselective Aerobic Annulation of α,β-Unsaturated Aldehydes with 1,3-Dicarbonyl Compounds. *J. Org. Chem.* **2016**, *81*, 6136–6141. [CrossRef] [PubMed]
66. Xu, J.; Zhang, W.; Liu, Y.; Zhu, S.; Liu, M.; Hua, X.; Chen, S.; Lu, T.; Du, D. Formal [3 + 3] Annulation of Isatin-derived 2-Bromoenals with 1,3-Dicarbonyl Compounds Enabled by Lewis Acid/*N*-Heterocyclic Carbene Cooperative Catalysis. *RSC Adv.* **2016**, *6*, 18601–18606. [CrossRef]
67. Wu, Q.; Li, C.; Wang, W.; Wang, H.; Pan, D.; Zheng, P. NHC-catalyzed enantioselective synthesis of dihydropyran-4-carbonitriles bearing all-carbon quaternary centers. *Org. Chem. Front.* **2017**, *4*, 2323–2326. [CrossRef]
68. Cao, J.; Sun, K.; Dong, S.; Lu, T.; Dong, Y.; Du, D. Esters as Alkynyl Acyl Ammonium and Azolium Precursors: A Formal [2 + 3] Annulation with Amidomalonates via Lewis Base/Lewis Acid Cooperative Catalysis. *Org. Lett.* **2017**, *19*, 6724–6727. [CrossRef]
69. Que, Y.; Lu, Y.; Wang, W.; Wang, Y.; Wang, H.; Yu, C.; Li, T.; Wang, X.-S.; Shen, S.; Yao, C. An Enantioselective Assembly of Dihydropyranones through an NHC/LiCl-Mediated in situ Activation of α,β-Unsaturated Carboxylic Acids. *Chem. Asian J.* **2016**, *11*, 678–681. [CrossRef]
70. Mondal, S.; Mukherjee, S.; Das, T.K.; Gonnade, R.; Biju, A.T. *N*-Heterocyclic Carbene-Catalyzed Aldol-Lactonization of Ketoacids via Dynamic Kinetic Resolution. *ACS Catal.* **2017**, *7*, 3995–3999. [CrossRef]
71. Wang, G.; Fu, Z.; Huang, W. Access to Amide from Aldimine via Aerobic Oxidative Carbene Catalysis and LiCl as Cooperative Lewis Acid. *Org. Lett.* **2017**, *19*, 3362–3365. [CrossRef] [PubMed]
72. Gao, Y.; Ma, Y.; Xu, C.; Li, L.; Yang, T.; Sima, G.; Fu, Z.Q.; Huang, W. Potassium 2-oxo-3-enoates as Effective and Versatile Surrogates for α,β-Unsaturated Aldehydes in NHC-Catalyzed Asymmetric Reactions. *Adv. Synth. Catal.* **2018**, *360*, 479–484. [CrossRef]
73. Gao, Y.; Liu, D.; Fu, Z.; Huang, W. Facile Synthesis of 2, 2-Diacyl Spirocyclohexanones via an *N*-Heterocyclic Carbene-Catalyzed Formal [3C + 3C] Annulation. *Org. Lett.* **2019**, *21*, 926–930. [CrossRef] [PubMed]
74. Liu, Q.; Chen, X.-Y.; Puttreddy, R.; Rissanen, K.; Enders, D. *N*-Heterocyclic Carbene Catalyzed Quadruple Domino Reactions: Asymmetric Synthesis of Cyclopenta[c]chromenones. *Angew. Chem. Int. Ed.* **2018**, *57*, 17100–17103. [CrossRef]
75. Altmann, H.J.; Clauss, M.; König, S.; Frick-Delaittre, E.; Koopmans, C.; Wolf, A.; Guertler, C.; Naumann, S.; Buchmeiser, M.R. Synthesis of Linear Poly(oxazolidin-2-one)s by Cooperative Catalysis Based on *N*-Heterocyclic Carbenes and Simple Lewis Acids. *Macromolecules* **2019**, *52*, 487–494. [CrossRef]
76. Mo, J.; Chen, X.; Chi, Y.R. Oxidative γ-Addition of Enals to Trifluoromethyl Ketones: Enantioselectivity Control via Lewis Acid/*N*-Heterocyclic Carbene Cooperative Catalysis. *J. Am. Chem. Soc.* **2012**, *134*, 8810–8813. [CrossRef]
77. Wu, Z.; Li, F.; Wang, J. ChemInform Abstract: Intermolecular Dynamic Kinetic Resolution Cooperatively Catalyzed by an *N*-Heterocyclic Carbene and a Lewis Acid. *Angew. Chem. Int. Ed.* **2015**, *54*, 1629–1633. [CrossRef]
78. Xiao, Z.; Yu, C.; Li, T.; Wang, X.-S.; Yao, C. *N*-Heterocyclic Carbene/Lewis Acid Strategy for the Stereoselective Synthesis of Spirocyclic Oxindole-Dihydropyranones. *Org. Lett.* **2014**, *16*, 3632–3635. [CrossRef]
79. Jia, W.Q.; Zhang, H.M.; Zhang, C.L.; Gao, Z.H.; Ye, S. *N*-Heterocyclic carbene-catalyzed [4 + 2] annulation of α,β-unsaturated carboxylic acids: Enantioselective synthesis of dihydropyridinones and spirocyclic oxindolodihydropyridinones. *Org. Chem. Front.* **2016**, *3*, 77–81. [CrossRef]
80. Chen, J.; Yuan, P.; Wang, L.; Huang, Y. Enantioselective β-Protonation of Enals via a Shuttling Strategy. *J. Am. Chem. Soc.* **2017**, *139*, 7045–7051. [CrossRef]

MDPI
St. Alban-Anlage 66
4052 Basel
Switzerland
Tel. +41 61 683 77 34
Fax +41 61 302 89 18
www.mdpi.com

Catalysts Editorial Office
E-mail: catalysts@mdpi.com
www.mdpi.com/journal/catalysts

www.ingramcontent.com/pod-product-compliance
Lightning Source LLC
Chambersburg PA
CBHW061621210926
51850CB00009B/2397